ALLE · ZEIT · WACH · 1842

P. Hughes · E. Fanning · M. Kohiyama (Eds.)

DNA Replication: The Regulatory Mechanisms

With 137 Figures

Springer-Verlag
Berlin Heidelberg New York
London Paris Tokyo
Hong Kong Barcelona
Budapest

Dr. Patrick Hughes
Institut Jacques Monod, CNRS
Université Paris 7
2 place Jussieu
75251 Paris Cedex 05
France

Dr. Ellen Fanning
Institut für Biochemie
Karlstraße 23
8000 München 2
FRG

Dr. Masamichi Kohiyama
Institut Jacques Monod, CNRS
Université Paris 7
2 place Jussieu
75251 Paris Cedex 05
France

ISBN-13: 978-3-642-76990-0 e-ISBN-13: 978-3-642-76988-7
DOI: 10.1007/978-3-642-76988-7

Softcover reprint of the hardcover 1st edition 1992

31/3145-5 4 3 2 1 0 – Printed on acid-free paper

Preface

The discovery of DNA polymerase I by A. Kornberg marked the starting point of a molecular approach to the study of DNA replication, a key event in the cell cycle. Since then, many enzymes and a multiplicity of biochemical mechanisms, apparently involved in some aspect of the initiation or elongation of DNA chains, have been discovered. Today, we thus find ourselves with a considerable variety of mechanisms and molecules that control DNA replication. Although our knowledge is far from complete and many elusive regulatory molecules remain beyond our grasp, it was one of the hopes of this meeting that researchers might reflect together on the relevance of accumulated data on regulation of replication and discover common ground. Probably, this type of approach is nowhere more necessary than in the sphere of *E.coli* DNA replication where recent discoveries have both confirmed and surpassed the original "Replicon hypothesis" of Jacob and colleagues. In yeast and higher eucaryotes, the apparent redundancy in putative origins and initiators has made an estimation of the importance of each identified element difficult to assess. In fact readers of these proceedings will appreciate that origin identification in mammalian chromosomes is still a controversial subject. On the other hand considerable advances have been made in our understanding of virus and phage DNA replication and this continues to deepen and broaden our understanding of the controls of cellular DNA replication. The unexpected role of heat shock proteins is a recent example. In assembling the contributions to this volume, we have attempted to avoid the usual dichotomy prokaryotes/eucaryotes in the hope that, as you slalom along the double helix, you may find a path intertwined and crisscrossed by trades from many systems.

PATRICK HUGHES
ELLEN FANNING
MASAMICHI KOHIYAMA
JULY 1991

Acknowledgements

We would like to thank the Human Frontier Science Program (HFSP) for financing this first meeting on the regulatory mechanisms of DNA replication. We would also like to thank FEBS for their support in supplying fellowships to young scientists to attend the conference. Also, a number of people have generously contributed their time and effort to the organization of the meeting and the preparation of this volume. These include the organizers Drs Eric Boye, Jun-ichi Tomizawa and Judith Campbell. We also appreciate the combined efforts of Irène Sorokine, Abderrahim Malki, Kamel Ben Mahrez, John Herrick and Martine Genotte in slide projection and organization.

Contents

Initiator interactions with DNA polymerases and accessory proteins

PART 2 : VIRAL AND PLASMID DNA REPLICATION

Molecular interactions during protein primed initiation

List of Contributors

Abraham, Carl. Department of Microbiology-Immunology, Albert Einstein College of Medicine, Bronx, NY 10461, USA.

Ang, D. Division of Biophysics, Department of Molecular Biology, University of Gdansk, Kladki 24, 80-822 Gdansk, POLAND.

Arthur, Avril K. Institute for Biochemistry, Karlstrasse 23, 8000 Munich 2, FEDERAL REPUBLIC OF GERMANY.

Baldacci, G. C.N.R.S.-U.P.R.272, Laboratoire de Biologie Moléculaire de la Réplication, 94800 Villejuif, FRANCE.

Bell, D. McGill Cancer Centre, McGill University, 3655 Drummond Street, Montreal,Québec H3G 1Y6, CANADA.

Bénit, P. Institut Curie-Biologie, Bâtiment 110, Centre Universitaire, 91405 Orsay, FRANCE.

Bernad, A. Centro de Biologia Molecular (CSIC-UAM), Universidad Autonoma, Canto Blanco, 28049 Madrid, SPAIN.

Bernander, Rolf. Department of Microbiology, University of Uppsala, The Biomedicale Center, Box 581, S-751 23 Uppsala, SWEDEN.

Blanco, L. Centro de Biologia Molecular (CSIC-UAM), Universidad Autonoma, Canto Blanco, 28049 Madrid, SPAIN.

Blasco, M.A. Centro de Biologia Molecular (CSIC-UAM), Universidad Autonoma, Canto Blanco, 28049 Madrid, SPAIN.

Blow, J. Julian. ICRF Cell Cycle Group, Microbiology Unit, Department of Biochemistry, University of Oxford, South Parks Road, Oxford OX1 3QU, U.K.

Bonne-Andréa, Catherine. University of Nice, INSERM U.273, Centre de Biochimie, Parc Valrose, 06034 Nice Cédex, FRANCE.

Bosher, J. Department of Biochemistry and Microbiology, University of St. Andrews, St. Andrews, Fife KY169AL. SCOTLAND.

Boulet, A. Institut Curie-Biologie, Bâtiment 110, Centre Universitaire, 91405 Orsay, FRANCE.

Bouvier, D. C.N.R.S.-U.P.R.272, Laboratoire de Biologie Moléculaire de la Réplication, 94800 Villejuif, FRANCE.

Boye, Erik. Department of Biophysics, Institute for Cancer Research, Montebello, 0310 Oslo, NORWAY.

Brewer, Bonita J. Department of Genetics, University of Washington, Seattle, WA 98195, USA.

Burgers, Peter M.J. Department of Biochemistry and Molecular Biophysics, Washington University School of Medicine, 660 S. Euclid Avenue, box 8231, St. Louis, MO 63110, USA.

Burhans, William C. Department of Cell and Developmental Biology, Roche Institute of Molecular Biology, Roche Research Center, Nutley, NJ 07110, USA.

Campbell, J.L. California Institute of Technology, Divisions of Chemistry and Biology 147-75CH, Pasadena, CA 91125, USA.

Chang, V. Section of Biochemistry, Molecular and Cell Biology, Cornell University, Ithaca, NY 14853, USA.

Christ, C. Section of Biochemistry, Molecular and Cell Biology, Cornell University, Ithaca, NY 14853, USA.

Clertant, Philippe. University of Nice, INSERM U.273, Centre de Biochimie, Parc Valrose, 06034 Nice Cédex, FRANCE.

Collins, Kathleen L. Department of Molecular Biology and Genetics, Johns Hopkins University School of Medicine, Baltimore, Maryland 21205, USA.

Damagnez, V. C.N.R.S.-U.P.R.272, Laboratoire de Biologie Moléculaire de la Réplication, 94800 Villejuif, FRANCE.

Dasgupta, Santanu. Department of Microbiology, University of Uppsala, The Biomedicale Center, Box 581, S-751 23 Uppsala, SWEDEN.

Dawson, A. Department of Biochemistry and Microbiology, University of St. Andrews, St. Andrews, Fife KY169AL. SCOTLAND.

Dehde, Silke. Institute for Biochemistry, Karlstrasse 23, 8000 Munich 2, FEDERAL REPUBLIC OF GERMANY.

DePamphilis, Melvin L. Department of Cell and Developmental Biology, Roche Institute of Molecular Biology, Roche Research Center, Nutley, NJ 07110, USA.

Derossi, D. Dipartimento di Genetica et di Biologia dei Microrganismi, Università di Milano, Via Celoria 26, 20133 Milano, ITALY.

De Recondo, Anne-Marie. C.N.R.S.-U.P.R.272, Laboratoire de Biologie Moléculaire de la Réplication, 94800 Villejuif, FRANCE.

Diaz, Ramón. Centro de Investigaciones Biologicas, Velazquez 144, 28006 Madrid, SPAIN.

Diffley, John F.X. Imperial Cancer Research Fund, Clare Hall Laboratories, Blance Lane, South Mimms, Potters Bar, Herts. EN6 3LD, U.K.

Dornreiter, Irene. Institute for Biochemistry, Karlstrasse 23, 8000 Munich 2, FEDERAL REPUBLIC OF GERMANY.

Dubey, Dharani D. Department of Zoology, Kidir Mahairdyalaya, Chakkey, Jaunpur-222 U.P., INDIA.

Elble, R. Section of Biochemistry, Molecular and Cell Biology, Cornell University, Ithaca, NY 14853, USA.

Engler, Jeffrey A. Department of Biochemistry, University of Alabama at Brimingham, Schools of Medicine and Dentistry, UAB Station, Birmingham, Alabama 35294, USA.

Erdile, Lorne F. Department of Molecular Biology and Genetics, Johns Hopkins University School of Medicine, Baltimore, Maryland 21205, USA.

Esteban, J.A. Centro de Biologia Molecular (CSIC-UAM), Universidad Autonoma, Canto Blanco, 28049 Madrid, SPAIN.

Fabre, F. Institut Curie-Biologie, Bâtiment 110, Centre Universitaire, 91405 Orsay, FRANCE.

Fanning, Ellen. Institute for Biochemistry, Karlstrasse 23, 8000 Munich 2, FEDERAL REPUBLIC OF GERMANY.

Faye, G. Institut Curie-Biologie, Bâtiment 110, Centre Universitaire, 91405 Orsay, FRANCE.

Fernández-Tresguerres, Maria Elena. Centro de Investigaciones Biologicas, Velazquez 144, 28006 Madrid, SPAIN.

Foiani, M. Dipartimento di Genetica et di Biologia dei Microrganismi, Università di Milano, Via Celoria 26, 20133 Milano, ITALY.

Francesconi, S. Dipartimento di Genetica et di Biologia dei Microrganismi, Università di Milano, Via Celoria 26, 20133 Milano, ITALY.

Fredman, Jeffrey N. Department of Biochemistry, University of Alabama at Brimingham, Schools of Medicine and Dentistry, UAB Station, Birmingham, Alabama 35294, USA.

Freire, R. Centro de Biologia Molecular (CSIC-UAM), Universidad Autonoma, Canto Blanco, 28049 Madrid, SPAIN.

Garmendia, C. Centro de Biologia Molecular (CSIC-UAM), Universidad Autonoma, Canto Blanco, 28049 Madrid, SPAIN.

Gasser, S.M. Swiss Institute for Experimental Cancer Research (ISREC), Ch. des Boveresses, CH-1066 Epalinges/Lausanne, SWITZERLAND.

Georgopoulos, C. Department of Cellular, Viral and Molecular Biology, University of Utah, School of Medicine, Salt Lake City, Utah 84132,USA.

Gibson, S. Section of Biochemistry, Molecular and Cell Biology, Cornell University, Ithaca, NY 14853, USA.

Gilbert, Ilka. Institute for Biochemistry, Karlstrasse 23, 8000 Munich 2, FEDERAL REPUBLIC OF GERMANY.

Giot, L. Institut Curie-Biologie, Bâtiment 110, Centre Universitaire, 91405 Orsay, FRANCE.

Giraldo, Rafael. Centro de Investigaciones Biologicas, Velazquez 144, 28006 Madrid, SPAIN.

Guo, Zong-Sheng. Department of Cell and Developmental Biology, Roche Institute of Molecular Biology, Roche Research Center, Nutley, NJ 07110, USA.

Gutiérrez, C. Centro de Biologia Molecular (CSIC-UAM), Universidad Autonoma, Canto Blanco, 28049 Madrid, SPAIN.

Hay, Ronald T. Department of Biochemistry and Microbiology, University of St. Andrews, St. Andrews, Fife KY169AL. SCOTLAND.

Hermoso, J.M. Centro de Biologia Molecular (CSIC-UAM), Universidad Autonoma, Canto Blanco, 28049 Madrid, SPAIN.

Herrick, John. Institut Jacques Monod, CNRS, Université Paris 7, 2 place Jussieu, 75251 Paris Cedex 05, FRANCE.

Hofmann, J.F.X. Swiss Institute for Experimental Cancer Research (ISREC), Ch. des Boveresses, CH-1066 Epalinges/Lausanne, SWITZERLAND.

Holz, Andreas. Max-Planck-Institut für Molekulare Genetik, Abt. Trautner, Ihnestrasse 63, D-1000 Berlin 33, GERMANY.

Horwitz, Marshall S. Department of Microbiology-Immunology, Albert Einstein College of Medicine, Bronx, NY 10461, USA.

Hoskins, Joel. Centre for Advanced Research in Biotechnology, National Institute of Standards and Technology, Rockville, MD 20850, USA.

Huberman, Joel A. Department of Molecular and Cellular Biology, Roswell Park Cancer Institute, Buffalo, NY 14263, USA.

Hughes, Patrick. Institut Jacques Monod, CNRS, Université Paris 7, 2 place Jussieu, 75251 Paris Cedex 05, FRANCE.

Johnson, C. Division of Biophysics, Department of Cellular, Viral and Molecular Biology, University of Utah, School of Medicine, Salt Lake City, Utah 84132,USA.

Joung, Insil. Department of Biochemistry, University of Alabama at Brimingham, Schools of Medicine and Dentistry, UAB Station, Birmingham, Alabama 35294, USA.

Karwan, Anneliese. Department of Molecular Genetics, Institute for Tumorbiology and Cancer Research, University of Vienna, Borschkegasse 8a, 1090 Wien, AUSTRIA.

Kearsey, Stephen E. Department of Zoology, University of Oxford, South Parks Road, Oxford, OX1 3PS, U.K.

Kelly, Thomas J. Department of Molecular Biology and Genetics, Johns Hopkins University School of Medicine, Baltimore, Maryland 21205, USA.

Kern, Renée. Institut Jacques Monod, CNRS, Université Paris 7, 2 place Jussieu, 75251 Paris Cedex 05, FRANCE.

Kipling, David. Department of Zoology, University of Oxford, South Parks Road, Oxford, OX1 3PS, U.K.

Knippers, R. Division of Biology, University of Konstanz, D-7750 Konstanz, FRG.

Kohiyama, Masamichi. Institut Jacques Monod, CNRS, Université Paris 7, 2 place Jussieu, 75251 Paris Cedex 05, FRANCE.

Kornberg, A. Department of Biochemistry, Beckman Center, Stanford University, Stanford, CA 94305-5307, USA.

Kunze, N. Division of Biology, University of Konstanz, D-7750 Konstanz, FRG.

Lambrecht, Ch. Division of Biology, University of Konstanz, D-7750 Konstanz, FRG.

Lázaro, J.M. Centro de Biologia Molecular (CSIC-UAM), Universidad Autonoma, Canto Blanco, 28049 Madrid, SPAIN.

Liberek, K. Division of Biophysics, Department of Molecular Biology, University of Gdansk, Kladki 24, 80-822 Gdansk, POLAND.

Linskens, Maarten H.K. Cold Spring Harbor Laboratory, Cold Spring Harbor, NY 11724, USA.

Løbner-Olesen, Anders. Department of Microbiology, The Technical University of Denmark, 2800 Lyngby, DENMARK

Locati, F. Dipartimento di Genetica et di Biologia dei Microrganismi, Università di Milano, Via Celoria 26, 20133 Milano, ITALY.

Longhese, M.P. Dipartimento di Genetica et di Biologia dei Microrganismi, Università di Milano, Via Celoria 26, 20133 Milano, ITALY.

Lucchini, G. Dipartimento di Genetica et di Biologia dei Microrganismi, Università di Milano, Via Celoria 26, 20133 Milano, ITALY.

Lyngstadass, Anita. Department of Biophysics, Institute for Cancer Research, Montebello, 0310 Oslo, NORWAY.

Maddock, A. Department of Cellular, Viral and Molecular Biology, University of Utah, School of Medicine, Salt Lake City, Utah 84132,USA.

Mah, D. McGill Cancer Centre, McGill University, 3655 Drummond Street, Montreal,Québec H3G 1Y6, CANADA.

Majumder, Sadhan. Department of Cell and Developmental Biology, Roche Institute of Molecular Biology, Roche Research Center, Nutley, NJ 07110, USA.

Malki, Abderrahim. Institut Jacques Monod, CNRS, Université Paris 7, 2 place Jussieu, 75251 Paris Cedex 05, FRANCE.

Marszalek, J. Division of Biophysics, Department of Molecular Biology, University of Gdansk, Kladki 24, 80-822 Gdansk, POLAND.

Martín, Margarita. Centro de Investigaciones Biologicas, Velazquez 144, 28006 Madrid, SPAIN.

McAlear, M. McGill Cancer Centre, McGill University, 3655 Drummond Street, Montreal,Québec H3G 1Y6, CANADA.

McKenney, Keith. Centre for Advanced Research in Biotechnology, National Institute of Standards and Technology, Rockville, MD 20850, USA.

Méndez, J. Centro de Biologia Molecular (CSIC-UAM), Universidad Autonoma, Canto Blanco, 28049 Madrid, SPAIN.

Messer, Walter. Max-Planck-Institut für Molekulare Genetik, Abt. Trautner, Ihnestrasse 63, D-1000 Berlin 33, GERMANY.

Miranda, Miriam. Department of Cell and Developmental Biology, Roche Institute of Molecular Biology, Roche Research Center, Nutley, NJ 07110, USA.

Moriya, S. Department of Genetics, Osaka University Medical School, 2-2, Yamadaoka, Suita 565, JAPAN.

Mul, Y.M. Laboratory for Physiological Chemistry, University of Utrecht, Vondellaan 24a, 3521 GG Utrecht, THE NETHERLANDS.

Muzi Falconi, M. Dipartimento di Genetica et di Biologia dei Microrganismi, Università di Milano, Via Celoria 26, 20133 Milano, ITALY.

Nallaseth, Ferez S. Department of Cell and Developmental Biology, Roche Institute of Molecular Biology, Roche Research Center, Nutley, NJ 07110, USA.

Nasheuer, Heinz-Peter. Institute for Biochemistry, Karlstrasse 23, 8000 Munich 2, FEDERAL REPUBLIC OF GERMANY.

Nawotka, Kevin A. Department of Molecular and Cellular Biology, Roswell Park Cancer Institute, Buffalo, NY 14263, USA.

Newlon, Carol S. Department of Microbiology and Molecular Genetics, UMDNJ-New Jersey Medical School, 185 South Orange Avenue, Newark, N.J. 07103, USA.

Nieto, Concepción. Centro de Investigaciones Biologicas, Velazquez 144, 28006 Madrid, SPAIN.

Nordström, Kurt. Department of Microbiology, University of Uppsala, The Biomedical Center, Box 581, S-751 23 Uppsala, SWEDEN.

Nurse, Paul. ICRF Cell Cycle Group, Microbiology Unit, Department of Biochemistry, University of Oxfor, South Parks Road, Oxford OX1 3QU, U.K.

Ogasawara, N. Department of Genetics, Osaka University Medical School, 2-2, Yamadaoka, Suita 565, JAPAN.

Opisiuk, J. Division of Biophysics, Department of Molecular Biology, University of Gdansk, Kladki 24, 80-822 Gdansk, POLAND.

Passmore, S. Section of Biochemistry, Molecular and Cell Biology, Cornell University, Ithaca, NY 14853, USA.

Pearson, C.E. McGill Cancer Centre, McGill University, 3655 Drummond Street, Montreal,Québec H3G 1Y6, CANADA.

Pettit, Steven C. Department of Biochemistry, University of Alabama at Brimingham, Schools of Medicine and Dentistry, UAB Station, Birmingham, Alabama 35294, USA.

Piatti, S. Dipartimento di Genetica et di Biologia dei Microrganismi, Università di Milano, Via Celoria 26, 20133 Milano, ITALY.

Pignède, G. C.N.R.S.-U.P.R.272, Laboratoire de Biologie Moléculaire de la Réplication, 94800 Villejuif, FRANCE.

Piseri, A. Dipartimento di Genetica et di Biologia dei Microrganismi, Università di Milano, Via Celoria 26, 20133 Milano, ITALY.

Pissagalli, A. Dipartimento di Genetica et di Biologia dei Microrganismi, Università di Milano, Via Celoria 26, 20133 Milano, ITALY.

Plevani, P. Dipartimento di Genetica et di Biologia dei Microrganismi, Università di Milano, Via Celoria 26, 20133 Milano, ITALY.

Price, G.B. McGill Cancer Centre, McGill University, 3655 Drummond Street, Montreal,Québec H3G 1Y6, CANADA.

Randall, Sandra K. Department of Molecular Biology and Genetics, Johns Hopkins University School of Medicine, Baltimore, Maryland 21205, USA.

Rhode, P.R. California Institute of Technology, Divisions of Chemistry and Biology 147-75CH, Pasadena, CA 91125, USA.

Richter, A. Division of Biology, University of Konstanz, D-7750 Konstanz, FRG.

Romig, H. Division of Biology, University of Konstanz, D-7750 Konstanz, FRG.

Russev, George. Institute of Molecular Biology, Bulgarian Academy of Sciences, Sofia 1113, BULGARIA.

Russo, Alicia A.R.. Department of Molecular Biology and Genetics, Johns Hopkins University School of Medicine, Baltimore, Maryland 21205, USA.

Salas, M. Centro de Biologia Molecular (CSIC-UAM), Universidad Autonoma, Canto Blanco, 28049 Madrid, SPAIN.

Sanchez, J. Aquiles. Department of Molecular and Cellular Biology, Roswell Park Cancer Institute, Buffalo, NY 14263, USA.

Santocanale, C. Dipartimento di Genetica et di Biologia dei Microrganismi, Università di Milano, Via Celoria 26, 20133 Milano, ITALY.

Santucci, Sabine. University of Nice, INSERM U.273, Centre de Biochimie, Parc Valrose, 06034 Nice Cédex, FRANCE.

Schaefer, Christoph. Max-Planck-Institut für Molekulare Genetik, Abt. Trautner, Ihnestrasse 63, D-1000 Berlin 33, GERMANY.

Schneider, Christine. Institute for Biochemistry, Karlstrasse 23, 8000 Munich 2, FEDERAL REPUBLIC OF GERMANY.

Serrano, M. Centro de Biologia Molecular (CSIC-UAM), Universidad Autonoma, Canto Blanco, 28049 Madrid, SPAIN.

Simancek, Pamela R. Department of Molecular Biology and Genetics, Johns Hopkins University School of Medicine, Baltimore, Maryland 21205, USA.

Simon, M. Institut Curie-Biologie, Bâtiment 110, Centre Universitaire, 91405 Orsay, FRANCE.

Skowyra, D. Division of Biophysics, Department of Molecular Biology, University of Gdansk, Kladki 24, 80-822 Gdansk, POLAND.

Sogo, José M. Department of Cell and Developmental Biology, Roche Institute of Molecular Biology, Roche Research Center, Nutley, NJ 07110, USA.

Stillman, Bruce. Cold Spring Harbor Laboratory, Cold Spring Harbor, NY 11724, USA.

Tazzi, R. Dipartimento di Genetica et di Biologia dei Microrganismi, Università di Milano, Via Celoria 26, 20133 Milano, ITALY.

Temperley, S.M. Department of Biochemistry and Microbiology, University of St. Andrews, St. Andrews, Fife KY169AL. SCOTLAND.

Theis, James F. Department of Microbiology and Molecular Genetics, UMDNJ-New Jersey Medical School, 185 South Orange Avenue, Newark, N.J. 07103, USA.

Tye, B.K. Section of Biochemistry, Molecular and Cell Biology, Cornell University, Ithaca, NY 14853, USA.

Umbricht, Christopher B. Department of Molecular Biology and Genetics, Johns Hopkins University School of Medicine, Baltimore, Maryland 21205, USA.

Van der Vliet, P.C. Laboratory for Physiological Chemistry, University of Utrecht, Vondellaan 24a, 3521 GG Utrecht, THE NETHERLANDS.

Van Driel, W. Laboratory for Physiological Chemistry, University of Utrecht, Vondellaan 24a, 3521 GG Utrecht, THE NETHERLANDS.

Van Oosterhout, J.A.W.M. Laboratory for Physiological Chemistry, University of Utrecht, Vondellaan 24a, 3521 GG Utrecht, THE NETHERLANDS.

Vassal, A. Institut Curie-Biologie, Bâtiment 110, Centre Universitaire, 91405 Orsay, FRANCE.

Vassilev, Lyubomir T. Department of Cell and Developmental Biology, Roche Institute of Molecular Biology, Roche Research Center, Nutley, NJ 07110, USA.

Verrijzer, C.P. Laboratory for Physiological Chemistry, University of Utrecht, Vondellaan 24a, 3521 GG Utrecht, THE NETHERLANDS.

Virshup, David M. Department of Molecular Biology and Genetics, Johns Hopkins University School of Medicine, Baltimore, Maryland 21205, USA.

Von Winkler, Dorothea. Institute for Biochemistry, Karlstrasse 23, 8000 Munich 2, FEDERAL REPUBLIC OF GERMANY.

Wang, S.S. Fred Hutchinson Cancer Research Center, 1124 Columbia St, Seattle, WA 98104, USA.

Weinberg, David H. Department of Molecular Biology and Genetics, Johns Hopkins University School of Medicine, Baltimore, Maryland 21205, USA.

Wellinger, R. Fred Hutchinson Cancer Research Center, 1124 Columbia St, Seattle, WA 98104, USA.

Wickner, Sue. Laboratory of Molecular Biology, National Cancer Institute, National Institutes of Health, Bethesda, MD 20892, USA.

Wiekowski, Maria. Department of Cell and Developmental Biology, Roche Institute of Molecular Biology, Roche Research Center, Nutley, NJ 07110, USA.

Wintersberger, Ulrike. Department of Molecular Genetics, Institute for Tumorbiology and Cancer Research, University of Vienna, Borschkegasse 8a, 1090 Wien, AUSTRIA.

Wold, Marc S. Department of Molecular Biology and Genetics, Johns Hopkins University School of Medicine, Baltimore, Maryland 21205, USA.

Wu, Jiarui. Department of Cell and Developmental Biology, Roche Institute of Molecular Biology, Roche Research Center, Nutley, NJ 07110, USA.

Yan, H. Section of Biochemistry, Molecular and Cell Biology, Cornell University, Ithaca, NY 14853, USA.

Yoon, Yeup. Department of Molecular and Cellular Biology, Roswell Park Cancer Institute, Buffalo, NY 14263, USA.

Yoshikawa, H. Department of Genetics, Osaka University Medical School, 2-2, Yamadaoka, Suita 565, JAPAN.

Zakian, V.A. Fred Hutchinson Cancer Research Center, 1124 Columbia St, Seattle, WA 98104, USA.

Zannis-Hadjopoulos, M. McGill Cancer Centre, McGill University, 3655 Drummond Street, Montreal,Québec H3G lY6, CANADA.

Zylicz, M. Division of Biophysics, Department of Molecular Biology, University of Gdansk, Kladki 24, 80-822 Gdansk, POLAND.

Regulation of Initiation of Chromosome Replication

Arthur Kornberg

Department of Biochemistry
Beckman Center
Stanford University
Stanford, CA94305-5307
U.S.A.

In opening this Symposium, I thought it appropriate to reflect on some broad issues of research on DNA replication before presenting a brief review of our *E. coli* work on the start and elongation of chains and on the initiation and termination of chromosomes. In this overview, I will offer the briefest glimpse of the physiological features of replication on the one hand and only hints of the structure-function relationships of the numerous proteins and enzymes and DNA on the other.

REFLECTIONS

I often think of how fortunate I have been to have spent my life in science, to have been part of this marvelous enterprise for almost fifty years. Beyond that, I feel especially lucky to have had these particular fifty years in which I could share in the most explosive advance in the history of biologic science.

When we say explosive, we think immediately of the revolutionary advances of genetic engineering and associated biotechnologies. To be sure, these advances are arguably the most impressive in the history of biologic science. Yet there is a significant advance, perhaps even more revolutionary, but unnoticed. It lacks a name and has no apparent commercial application. I refer to the remarkable coalescence of the previously discrete biosciences -from anatomy, at one time the most descriptive, to genetics, until recently the most abstract. The basis of this coalescence is that they can all be expressed in a single language, the language of chemistry. Chemistry is truly the international language, a language without dialects, a language for all of time.

In my attempts to understand the chemistry of the cell and organism, I have been driven to do battle with vitalism, a vitalism that goes back to what befell Pasteur. He believed and preached that alcoholic fermentation is intimately linked to the life of a yeast cell. We can also recall the vitalism of A.J. Kluyver, the most eminent microbiologist of his day, who generations later, stated in 1931 that fatty acid metabolism was too complex ever to be reduced to a cell-free system. Only a few years later, Luis Leloir demonstrated fatty acid oxidation in cell homogenates.

Vitalism dies hard. It really never does. I can sense a neovitalism that results from our mastery over DNA and the capacity we have to introduce and knock out genes. The consequences of such manipulations are observed either in intact cells in culture or in transgenic flies, worms and mice. Despite the impressive insights that such experiments provide, the last word about the molecular basis of physiologic events will never be spoken until these phenomena are observed in cell-free systems and then reconstituted from molecularly defined components.

In pursuing the reductionist approach, I have been guided, as most of this audience knows, by three articles of faith. The <u>first</u> is that all reactions in the cell are catalyzed and directed by enzymes. Chemists may squirm when they hear this. But as biochemists, we are reluctant to accept a spontaneous reaction in a metabolic pathway. We know that something as simple as the hydration of CO_2 is catalyzed by carbonic anhydrase. We know that the melting of duplex DNA requires helicases. So we are not surprised when we learn these days that the folding of proteins is also facilitated : by disulfide isomerases, by prolyl isomerases and by chaperonins.

The <u>second</u> faith is that biochemists devoted to enzymes should be able to reconstitute most cellular phenomena with defined molecules. Having succeeded in observing a cellular phenomenon in a cell-free system -whether it be that alcoholic fermentation in yeast juices, glycolysis in muscles extracts, firefly luminescence, or replication of DNA, the biochemist should be able to recreate the process as well as the cell does it, even better. After all, the cell is under great constraints, having to provide a consensus medium that supports thousands of diverse reactions, few of which operate under optimal conditions. By contrast, the biochemist enjoys the enormous advantage and freedom to saturate each enzyme with its substrate, provide the optimal pH, salt and metal concentrations, trap the products, etc. The biochemist can thus be far more creative and effective in analyzing the molecular basis of a reaction or pathway.

The third strong faith is in the universality of biochemistry. When the pathways of bioenergetics were defined early in this century, they revealed that the enzymes responsible for alcoholic fermentation in yeast were virtually identical to those that perform glycolysis in muscle. This astonishing fact, along with other such examples in biosynthetic pathways made it clear that mechanism and molecules have been preserved in nature virtually intact through billions of years of evolution. I regard this insight as one of the great revelations of our century.

Because of this universality of biochemistry, many of us working on replication focus on how prokaryotes replicate their own genomes and those of their phages and plasmids. This is where the light on replication is often the brightest. We make this choice with the confidence that these systems will be reliable prototypes for how the so-called higher organisms manage their DNA replications. Exciting advances have been made in discovering and characterizing eukaryotic helicases, topoisomerases, polymerases, primases and ligases. Although the variations are fascinating, these enzymes and mechanisms do adhere to the prokaryotic themes.

Before presenting some of our recent work I want to comment about the broad biologic significance of DNA replication. Even though DNA replication occupies a rather small area in the pathways and byways of cellular metabolism, it has been remarkably fertile in revealing patterns and mechanisms of great general significance. There are numerous, placental interdigitations with many components of cellular growth and maintenance. After tilling this area intensively for over 40 years, I have been pleased to observe another unanticipated bonus. This has been the clinical importance of replication in understanding and treating disease. As examples, the knowledge of DNA replication has been crucial to understanding mutagenesis and carcinogenesis,

DNA repair and recombination. In fact, most therapies for cancer, for autoimmune diseases, for AIDS and herpes are designed to interrupt replication.

This month saw the completion of the Second Edition of *DNA Replication* (Kornberg & Baker, 1991), this time with Tania Baker, a former student, as a coauthor. The book will appear this summer. In this Second Edition, I was impressed and distressed at how little we could salvage from the 1980 edition (Kornberg, 1980). Every chapter had to be revised extensively and five new ones were added to the original seventeen. Yet, the organization and philosophy of the book remain the same. We devote the first thirteen chapters to the enzymes relevant to replication. This constitutes the cast of the replication actors. The remaining nine chapters describe how each of these actors performs in a repertoire that includes the various acts of replication : the origin of chromosomes, chain starts, chain elongation and the termination of chromosomes ; the prokaryotic and eukaryotic growth cycles ; viral and plasmid replication ; recombination, repair and transformation ; inhibitors of replication and their use as drugs ; and so on.

Whatever impelled us undertake the daunting job of doing a Second Edition ? A major motive was to organize the vital new information that had accrued in the past ten years. Yet a more basic motive was to express a philosophy about how to solve a problem in biology. We wanted to show how the pursuit of enzymology, the reduction of a cellular phenomenon to a succession of molecularly defined reactions can, with the aid of genetics and inhibitors, contribute in a mighty was to understanding the physiology of cells and the pathology of organisms.

One further comment is appropriate in introducing this Conference. In addition to being grateful for being in science, we are also fortunate in the friendships and cooperation that are built over the years. These ties embrace several generations, transcend national boundaries and enable us to meet under attractive circumstances to be challenged, stimulated and refreshed.

A brief outline of our *E. coli* replication work now follows (Kornberg, 1988).

Start of DNA Chains

Inasmuch as no DNA polymerase can start a chain, the mechanism most generally used in nature depends on RNA priming (Kornberg, 1980). Primase, often facilitated by the helicase action of dnaB protein, lays down a transcript of a few nucleotides, the 3' terminus of which is extended by a DNA polymerase. However, the start of the complementary strand on phage øX174 proved more complex and disclosed the need for five additional proteins which together form a mobile primosome (Kornberg, 1982). What each of these proteins contributes to the priming of chains at the growing fork of the *E. coli* chromosome still remains uncertain.

Of these five proteins, dnaC and dnaT (formerly called protein i) (Masai *et al.*, 1986) are known from genetic studies to be essential. A role for dnaC protein is in forming a complex with dnaB protein to facilitate its interactions with DNA (Wahle *et al.*, 1989). The remaining three, proteins n, n' and n", had been purified and identified as essential in øX174 replication *in vitro*. Our discoveries of the genes for n' (*priA*) (Lee *et al.*, 1990) and n (*priB*) (Allen & Kornberg, 1991) now leaves only the gene for n" (*priC*) unknown. We anticipate that the regulated expression of these genes will reveal insights into their physiologic functions (Lee & Kornberg, 1991) and also provide substantial quantities of material for studies of the assembly and structure of the primosome.

Elongation of DNA Chains

DNA polymerase III holoenzyme is the replicative machine that elongates chains rapidly and processively with high fidelity (Kornberg, 1988). Its ten distinctive subunits are organized as an asymmetric dimer. The pair of polymerase cores has two arms which differ in their tenacity for the template, suggestive that one may serve in the continuous synthesis of the so-called "leading" strand and the other in discontinuous synthesis of the "lagging" strand.

Beyond the need to advance our understanding of the structural and functional features of the holoenzyme, is the uncertain nature of its interactions with the primosome which might create a "replisome" and thereby achieve the coordinated, nearly concurrent replication of both strands.

Initiation of the Chromosome

Plasmids with the minimal *E. coli* chromosome origin (*oriC*) behave as minichromosomes in the cell and are convenient chromosomal templates for replication by purified enzyme systems (Kaguni & Kornberg, 1984). Initiation depends on opening the supercoiled duplex wide enough to admit the dnaB helicase, thus setting the stage for priming and replication. First, the binding of the key initiator protein (dnaA) to its four 9-mer sequences (boxes) alters the DNA conformation which then facilitates the opening of an adjacent region of three tandem 13-mer (Bramhill & Kornberg, 1988).

The actions of dnaA protein depend on several accessory proteins : protein HU to facilitate DNA bending, RNase H and topoisomerase I to abort nonspecific starts, gyrase to maintain the proper superhelical density, and RNA polymerase to create an R-loop at a nearby promoter which activates an otherwise relaxed and inert 13-mer region.

The activity of dnaA protein depends on several additional conditions and factors : an aggregated inactive from of dnaA protein is convertible to the monomeric active form by dnaK protein or by phospholipase (Hwang *et al.*, 1990); a tightly bound ATP is essential, whereas ADP, generated by endogenous ATPase activity is ineffective (Sekimizu *et al.*, 1987) ; strong interaction with membranes, evidences by promoting exchange of the tightly held ATP (or ADP), is specific for acidic polar head groups of phosphatides in bilayers made fluid by unsaturated fatty acid side chains (Yung & Kornberg, 1988).

A negative effector has been discovered : the IciA protein (inhibitor of chromosome initiation) which blocks dnaA action at the initial stage but is no longer effective once the 13-mer region has been opened (Hwang & Kornberg, 1990). The gene (*iciA*) encoding the dimeric IciA protein is located at 62.8 min and bears an homology to the LysR group of transcriptional regulatory proteins (Thöny *et al.*, 1991). Modulated expression of the gene and assorted mutation of the 13-mers should provide insights into how the chromosome and minchromosomes are initiated.

Termination of the Chromosome

The genomes of *E. coli, B. subtilis* and some plasmids possess *ter* sequences which when oppositely oriented and bound by a protein (*tus* gene product), block the progress of the bidirectional replication forks (i.e., one orientation blocks the clockwise, the other the counterclockwise movement). Presumably, pausing provides time for topological actions needed to complete the daughter chromosomes, and averts rolling circles, catenanes and other aberrant forms.

Fork blockage, as observed *in vivo*, has been reproduced *in vitro* (Lee *et al.*, 1989). When *oriC* plasmids contain appropriately oriented and bound *ter* sequences, their replication by systems reconstituted with purified proteins is blocked precisely as anticipated. Blockage of the helicase action of dnaB protein by a Tus-protein-bound *ter* sequence appears to be most immediate. Studies with various helicases that differ in the polarity of their translocation confirm the strict orientation-specificity to the *ter* sequences.

Envoi

In the course of this lecture, I've mentioned over 30 distinctive polypeptides engaged in the events of replication. Even to an aficionado of replication enzymology, the multiple stages, the complexity of each of the stages and the large cast of actors may seem excessive. Yet, their number is fewer than those employed in ribosomal translation. Nor should initiation of a chromosome be expected to be simpler than the regulated expression of a gene. Nor have we reached the limit. We can anticipate that still more proteins will be discovered that either augment or antagonize those already known. Some day, when the balanced actions of these many proteins and factors, including nucleotides, membranes, chaperonins, and metals, are known, they will be recognized as the signal transmitted by cell mass or an external agent to operate the switch for chromosome initiation.

Having publicly expressed my love for enzymes and devotion to proteins, I must now confess to having seriously neglected DNA. We accept that biosynthetic and other metabolic pathways are governed by proteins acting on defined substrates. In the past when we regarded DNA, notwithstanding its extraordinary size and the intelligence it contains, we viewed it as a substrate, a rather fixed and rigid structure to be exploited by proteins.

My attitude toward DNA has changed profoundly. The rigid rod is now seen as a dynamic, plastic structure with a complex "personality". In addition to the sequences that orient proteins, studies of replication and other DNA transactions, now reveal unanticipated qualities of DNA, absolutely crucial to its functions. I have mentioned several of the features : the energy stored in negative supercoiling, the sequences that favor unwinding (or melting) in a thermodynamic sense (Kowalski & Eddy, 1989), the sequences which can transmit such kinetic instability, perhaps by translocation of a "minibubble", over long distances to perturb and activate another sequence to be unwound, and sequences which impede the transmission of kinetic instability.

Finally, a concluding remark about the origin of the *E. coli* chromosome. The minimal origin contains large stretches of highly conserved sequences separated by fixed lengths of random sequence (Zyskind & Smith, 1986). The conserved sequences include the sites for the binding of proteins : the key initiator (dnaA), the accessory bender (HU), and the inhibitor (IciA), and we can reasonably expect that others will be discovered. The fixed-length randomized sequences which, over a span of 100 million years of evolution, have not tolerated insertions or deletions may be regarded as spacers that give the entire origin a unique conformation. One day, when we know more about the *oriC* structure and the many proteins it binds and envelops, we will better understand the "orisome" and how it really functions.

Acknowledgments

I am grateful to Dr. Masamichi Kohiyama and Dr. Erik Boye for their efforts in organizing this Symposium and to the Human Frontier Science Program for their generous financial support. I also acknowledge grants from the National Institutes of Health and the National Science Foundation that made our research possible.

My thanks go to the many students and fellows for the research contributions which made this report possible, and particularly for the current efforts of Dr. Deog Su Hwang and Dr. Beat Thöny with the *iciA* gene and the IciA protein, of Dr. Elliott Crooke and Dr. Kirsten Skarstad with chromosome initiation, of Dr. Celina Castuma with the DnaA protein-membrane association, and of Dr. Eui Hum Lee and Rus Allen with chain initiation.

References

Allen GC Jr., Kornberg A (1991) The *priB* gene encoding the primosomal replication n protein of *Escherichia coli*. J Biol Chem, in press.

Baker TA, Kornberg A (1988) Transcriptional activation of initiation of replication from the *E. coli* chromosomal origin : an RNA-DNA hybrid near *oriC*. Cell 55:113-123.

Bramhill D, Kornberg A (1988) A model for initiation at origins of DNA replication. Cell 52:915-918.

Hwang DS, Crooke E, Kornberg A (1990) Aggregated dnaA protein is dissociated and activated for DNA replication by phospholipase or dnaK protein. J Biol Chem 265:19244-19248.

Hwang DS, Kornberg A (1990) A novel protein binds a key origin sequence to block replication of an *E. coli* minichromosome. Cell 63:325-331.

Kaguni JM, Kornberg A (1984) Replication initiated at the origin (*oriC*) of the *E. coli* chromosome reconstituted with purified enzymes. Cell 38:183-190.

Kornberg A (1980) DNA Replication. W.H. Freeman, New York.

Kornberg A (1982) Supplement to DNA Replication. W.H. Freeman, New York.

Kornberg A (1988) DNA Replication. J Biol Chem 263:1-4.

Kornberg A, Baker TA (1991) DNA Replication, Second Edition. W.H. Freeman, New York.

Kowalski D, Eddy MJ (1989) The DNA unwinding element : a novel, *cis*-acting component that facilitates opening of the *Escherichia coli* replication origin. EMBO J 8:4335-4344.

Lee EH, Kornberg A (1991) Replication deficiencies in *priA* mutants of *E. coli* lacking the primosomal replication n'protein. Proc Natl Acad Sci USA, in press.

Lee EH, Kornberg A, Hidaka M, Kobayashi T, Horiuchi T (1989) *Escherichia coli* replication termination protein impedes the action of helicases. Proc Natl Acad Sci USA 86:9104-9108.

Lee EH, Masai H, Allen GC Jr, Kornberg A (1990) The *priA* gene encoding the primosomal, replicatived n'protein of *E. coli*. Proc Natl Acad Sci USA 87:4620-4624.

Maki H, Maki S, Kornberg A (1988) DNA polymerase III holoenzyme of *Escherichia coli* : IV. The holoenzyme is an asymmetric dimer with twin active sites. J Biol Chem 263:6570-6578.

Masai H, Bond MW, Arai K (1986) Cloning of the *Escherichia coli* gene for primosomal protein i: The relationship to *dnaT*, essential for chromosomal DNA replication. Proc Natl Acad Sci USA 83:1256-1260.

Sekimizu K, Bramhill D, Kornberg A (1987) ATP activates dnaA protein in initiating replication of plasmids bearing the origin of the *E. coli* chromosome. Cell 50:259-275.

Skarstad K, Baker TA, Kornberg A (1990) Strand separation required for initiation of replication at the chromosomal origin of *E. coli* is facilitated by a distant RNA-DNA hybrid. EMBO J 9:2341-2348.

Thöny B, Hwang DS, Fradkin L, Kornberg A (1991) *IciA*, an *Escherichia coli* gene encoding a specific inhibitor of chromosomal initiation of replication *in vitro*. Proc Natl Acad Sci USA, in press.

Wahle E, Lasken RS, Kornberg A (1989) The dnaB-dnaC replication protein complex of *Escherichia coli*. II. Role of the complex in mobilizing dnaB functions. J Biol Chem 264:2469-2475.

Wahle E, Lasken RS, Kornberg A (1989) The dnaB-dnaC replication protein complex of *Escherichia coli*. I. Formation and properties. J Biol Chem 264:2463-2468.

Yung BY, Kornberg A (1988) Membrane attachment activates dnaA protein, the initiation protein of replication in *Escherichia coli*. Proc Natl Acad Sci USA 85:7202-7205.

Zyskind JW, Smith DW (1986) The Bacterial Origin of Replication, *oriC*. Cell 46:489-490.

Escherichia coli Strains in Which the Chromosome Is Replicated from Origins Other than *oriC*

Rolf Bernander, Santanu Dasgupta, and Kurt Nordström
Department of Microbiology
University of Uppsala
The Biomedical Center, Box 581
S-751 23 Uppsala
Sweden

The *Escherichia coli* chromosome is bidirectionally replicated from a unique origin, *oriC*. Various strains exist in which chromosome replication is controlled by other replicons, located at varying distance from *oriC*. The copy number determined by the alternative replicon may be different from that of the wild type, replication can be bidirectional or show varying degrees of unidirectionality, and initiation is usually randomly timed during the cell cycle. The more chromosome replication deviates from the normal situation with respect to the above characteristics, the more the cell cycle and growth of the strains are disturbed.

Introduction

In *Escherichia coli*, the configuration of the unique origin of replication, *oriC*, and the terminus region, *terC*, in the chromosome is symmetric (Fig. 1).Thus, the two replication forks that carry out the bidirectional replication will normally meet in the *terC* region even in the absence of the polar barriers to fork movement that

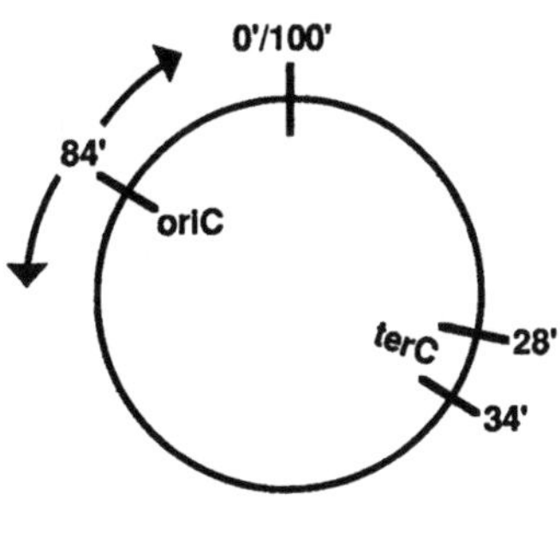

Figure 1. Map of the *E. coli* chromosome showing the locations of *oriC* and the *terC* region where initiation and termination of bidirectional replication occur, respectively.

are located at each border of the region, provided that the two forks travel at equal velocity. The consequences of a decreased symmetry of chromosome replication on various aspects of cell cycle control and coordination can be investigated *e. g.* by changing the location of the origin or the degree of bidirectionality. This can be achieved by letting replicons other than *oriC* control replication (Table 1).

Table 1. Examples of strains with alternative origins of replication

Description of systems	**Examples**
A plasmid replicon in addition to a functional *oriC*	Hfr; integratively suppressed strains at low temp.
Integrative suppression with a plasmid replicon with controlled copy number	*Low copy number*: ColB[1]; ColV2[2]; F/F'*lac*[3]; P1/P7[4]; R1*drd-19*[1]; R100/R100-1[2]; R144/R483[5]; RP1[6]; Rts1[7] *High copy number*: pKY31[8]; P1*cop*[9]; R6K[10]; R1162[11]; pUC13[12]
Integrative suppression with a phage replicon	P2[13]
Replication from other sites on the chromosome than *oriC*	SDR (Stable DNA Replication)

References: 1, Moody and Runge, 1972; 2, Nishimura *et al.*, 1973; 3, Nishimura *et al.*, 1971; 4, Chesney and Scott, 1978; 5, Datta and Barth, 1976; 6, Martin *et al.*, 1981; 7, Yoshimoto and Yoshikawa, 1975; 8, Yamaguchi and Tomizawa, 1980; 9, Froelich *et al.*, 1983; 10, Sotomura and Yoshikawa, 1975; 11, Brasch and Meyer, 1988; 12, Mao *et al.*, 1991; 13, Lindahl *et al.*, 1971

Hfr strains A classic example of strains in which alternative replicons are integrated into the chromosome at different locations is provided by so-called Hfr strains (Neidhardt *et al.*, 1987), which carry an integrated conjugative plasmid that can promote mobilization and transfer of the chromosome into suitable recipient strains. Usually an F plasmid is integrated but many other plasmids, of both low or high copy number classes, also give rise to Hfr strains (see *e. g.* references 1, 2, 5, 6, 7, and 10 in Table 1).

Integratively suppressed strains Strains that are temperature-sensitive for initiation of replication from *oriC* often show an increased frequency of temperature-resistant isolates if they contain plasmids, as compared to plasmid-free strains. This is due to the integration of a plasmid molecule into the chromosome which, therefore, can be replicated from the plasmid origin at high temperature. Plasmids ColV2, F, P1, and R100 are examples of low copy number plasmids that give rise to such *integratively suppressed* (Nishimura *et al.*, 1971) strains. Even high copy number plasmids such as R6K, R1162, and the ColE1-related pKY31 can rescue chromosome replication, but this requires that unintegrated plasmid copies also are present in the same cell (below).

Integrative suppression can also be obtained with prophage P2. The frequency of integrative suppression is low and appears to result from mutations in the replicon part of the phage (Lindahl *et al.*, 1971). The nature of the copy number control in these strains is not fully understood.

***Stable DNA replication and* oriJ** An additional example of replication from sites other than *oriC* is provided by SDR (Stable DNA Replication) strains (Kogoma, 1986). In such strains, in which replication from *oriC* has been inhibited, replication is instead initiated from alternative sites, collectively denoted *oriK*. Replication requires a mutated *rnh* (ribonuclease H) gene and is dependent on the RecA protein. In addition, a normally repressed origin, *oriJ*, is present in *E. coli* strains carrying a defective prophage integrated in the *terC* region (Diaz and Pritchard, 1978).

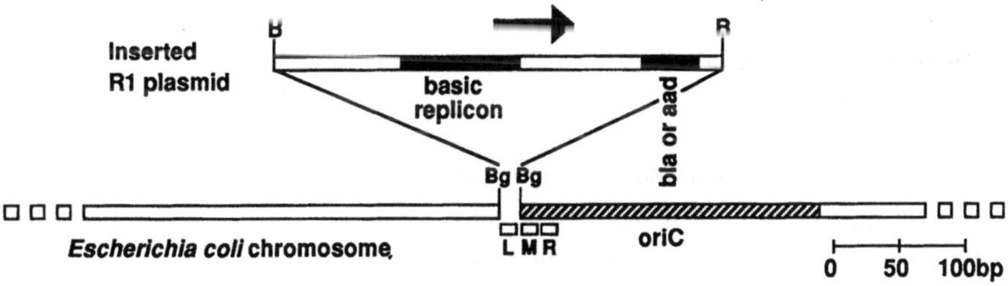

Figure 2. Map of *intR1* strains.
Open rectangles show the 13-mers involved in unwinding of *oriC* during initiation of replication; the leftmost 13-mer has been deleted in the *intR1* strains. The orientation that results in clockwise (*CW*) replication fork movement (arrow) with respect to the genetic map of the chromosome is illustrated; in the *CC* configuration the plasmid insert is inverted.

Targeted exchange of* oriC *for an R1 replicon;* intR1 *strains To study different aspects of chromosome replication and cell division in *E. coli*, we constructed strains in which chromosome replication is controlled by various derivatives of the low copy number plasmid R1. These *intR1* strains (Fig. 2 and Nordström *et al.*, 1991) carry a 16 bp deletion which inactivates *oriC* by removing the leftmost DNA unwinding element. Plasmid R1, which replicates unidirectionally, was inserted into the deletion site in two orientations, giving rise to clockwise (*CW*) and counter-clockwise (*CC*) fork movement with respect to the genetic map of *E. coli* (Fig. 2). The location of the R1 insertion ensures that replication starts close to the normal site. The R1 origin is the only functional origin in these strains and, furthermore, since the copy number of plasmid R1 is close to that of *oriC*, the DNA content is essentially normal. The copy number of the chromosome can be varied from normal to very high values by using a derivative of plasmid R1 whose copy number is normal at temperatures below 34°C and gradually increases at higher temperatures until replication becomes uncontrolled above 39°C (runaway replication).

Studies of the* E. coli *cell cycle using strains in which chromosome replication is controlled by alternative replicons In the various strains discussed above the alternative replicons display random timing of initiation of replication during the cell cycle, different copy numbers, varying degrees of uni- and bidirectionality, and may be inserted at different positions in the chromosome. By varying these and other replication parameters and studying the effects on growth and cell size distribution valuable information about cell cycle control and coordination can be obtained (below).

Effects of integrated replicons on DNA replication, growth and cell division

Strains in which the chromosome replicates from origins other than *oriC* deviate from wild-type bacteria in several important respects (Table 2). These differences generally result in two types of effects that are discussed below:

1. The growth characteristics may be altered such that *e. g.* growth rate and viability are affected, often accompanied by so-called rich medium sensitivity.

2. The cell size and DNA content distributions may be disturbed.

Timing and coordination of initiations Initiation of replication of the *E. coli* chromosome occurs at a defined cell mass per *oriC*, the *initiation mass*, which is independent of growth rate (Donachie, 1968). The inter-replication time, the time between consecutive initiations at the same origin, is consequently strictly one doubling time at all growth rates. Density-shift experiments with *intR1* and SDR strains show that in such strains the inter-replication time is variable (Koppes and

Nordström, 1986; von Meyenburg *et al.*, 1987), indicating that the timing of initiation of replication during the cell cycle is random.

Table 2. Important replication characteristics.

Case	Property	Alternatives
1.	*Timing and coordination*	
	Timing in cell cycle	Narrow interval or random
	Inter-replication time	One doubling time or random interval
	Replication of multiple origins	Coordinated or random
2.	*DNA per mass ratio*	Under-replication, normal replication or over-replication
3.	*Directionality of replication*	Uni- or bidirectional
4.	*Location of the replicon*	Varying distance from *oriC*

At moderate and high growth rates more than one origin is present in a wild-type cell at the time of initiation of replication; flow cytometry experiments show that initiation at these origins is coordinated such that all origins are initiated within a short time interval (Skarstad *et al.*, 1986). Similar experiments with *intR1* (unpublished) and SDR (von Meyenburg *et al.*, 1987) strains suggest uncoordinated initiations.

The results discussed above indicate that in these strains replication is uncoupled from normal cell cycle regulation. Thus, although replication in wild-type *E. coli* is tightly timed and coordinated during the cell cycle, less strict regulation does not appear to lead to major physiological consequences. The growth rate is usually somewhat reduced, and the cell size and DNA content distributions are broader than those of the wild type (Bernander *et al.*, 1989; von Meyenburg *et al.*, 1987), presumably as a consequence of a less precise coordination between growth, chromosome replication, and cell division.

DNA per mass ratio The copy number determined by the various integrated replicons may vary. Some display copy numbers similar to that resulting from *oriC*-controlled replication of the chromosome, whereas others have considerably higher copy numbers (Table 1). The copy number of certain *intR1* strains can be varied from very low to very high values (see Nordström *et al.*, 1991). Thus, the *intR1* and other systems offer an opportunity to study the effects of varying the chromosome copy number on growth characteristics and cell division.

At essentially wild-type chromosome copy number, growth and division of many strains in which replication is controlled by alternative replicons are similar to those of wild-type bacteria. The deviations observed can be explained by the randomness of initiation as discussed in the previous section.

Extensive over-replication of the chromosome is lethal to the bacteria, as illustrated by two examples: First, Yamaguchi and Tomizawa (1980) found that a ColE1-related plasmid, pKY31, can control chromosome replication only if free copies of the plasmid also are present, so that the combined copy number of the integrated and the free plasmid copies is that of pKY31. Thus, the chromosome copy number can be kept low despite the high copy number of pKY31. In support of this, the integrated ColE1 could not be transduced into a plasmid-free cell. Similar conclusions have been arrived at using other high copy number plasmids (Brasch and Meyer, 1988; François *et al.*, 1990; Froelich *et al.*, 1983). Second, in *intR1* strains extensive over-replication of the chromosome was shown to severely affect growth and division such that elongated cells, filaments, and DNA-less minicells were formed, and a two- to threefold increase in the DNA content of the cells was lethal (Bernander et al., 1989). A systematic study of the effects of a gradual increase in the chromosome copy number showed, however, that a moderate increase in the DNA content (about 50%) only slightly affected growth rate and cell size distribution. This led us to suggest that cell division is not directly triggered by chromosome replication but may be independently controlled (Bernander and Nordström, 1990).

Severe under-replication also causes a reduction in the growth rate, and results in the formation of elongated cells and filaments, indicating an inhibitory effect on cell division (Bernander *et al.*, 1991). In addition, DNA-less cells are formed. The quantitative correlations between under-replication and effects on growth rate and cell division are presently investigated in our laboratory using *intR1* strains.

Thus, both extensive under- and over-replication of the chromosome lead to severe disturbances in cell division. Elongated cells and filaments are abundant in both cases, whereas DNA-less cells and minicells predominantly only appear to be produced during under-replication and over-replication, respectively. Over-replication probably results in difficulties in segregation of complex daughter nucleoids which may lead to relocation of divisions to the cell poles. Severe under-replication results in inhibition of cell division, indicating that chromosome replication and nucleoid segregation have to be completed before cell division can take place.

Directionality of replication Replication from *oriC* is bidirectional and, through unknown mechanisms, many strains in which replication is controlled by integrated replicons also replicate bidirectionally despite unidirectional replication of the respective free replicons (Bird *et al.*, 1976; Kuempel *et al.*, 1977 and 1978; Louarn *et al.*, 1982). The *intR1* strains exhibit different degrees of bidirectionality depending on the orientation of the integrated plasmid. In the *CW* strains, in which the replisome assembled at the R1 origin replicates towards the inactivated *oriC*, replication is essentially bidirectional, whereas replication is biased towards unidirectionality in the *CC* strains (Nordström *et al.*, 1991). We have suggested that the approaching replication fork initiated at the R1 origin may exert unwinding activity on the remaining *oriC*, thereby inducing bidirectional replication through *replicational activation.* If correct, this may also apply to other strains in which an alternative replicon controls chromosome replication. Thus, it is possible that a replication fork may have the potential to induce bidirectional replication when it reaches *oriC* regardless of where on the chromosome, and from which origin, it was initiated.

The *intR1 CC* strains, exhibiting unidirectional replication, display a more severely disturbed cell morphology than the bidirectionally replicating *intR1 CW* strains, with an abundance of elongated cells, filaments and DNA-less minicells. An explanation for this may be the increased time required for chromosome replication during unidirectional replication, which causes a more severe disruption of the coordination between chromosome replication and cell division, due to that replication more often is not completed at the normal time for cell division. In addition to the disturbed coordination between replication and division, true unidirectional replication might impose great difficulties in initiating a new round of replication before the completion of the previous one. It is possible that true unidirectional replication of the chromosome cannot be sustained, and that all strains in which alternative replicons control chromosome replication exhibit some degree of bidirectionality (*cf.* discussion in Mao *et al.*, 1991). Accordingly, in strains in which chromosome replication is unidirectionally initiated from an integrated P2*sig5* prophage, marker frequency results indicate that after a time delay of about 40 min a second replication fork travelling in the opposite direction from the first one is initiated (Kuempel *et al.*, 1978).

Strains in which chromosome replication is controlled by replicons other than *oriC* often exhibit rich medium sensitivity, *i. e.* an inability to grow in rich culture media (Louarn *et al.*, 1982; Mao *et al.* 1991; Torrey *et al.*, 1984). This could be a consequence of an insufficient efficiency of chromosome replication at high growth rates, since the alternative replicons do not respond to an increased growth rate by adjusting the level of replication the way the normal, *oriC*-dependent, control system does. The rich medium sensitivity depends on how far from *oriC* the replicon is located, probably due to that the further away from *oriC* it is integrated the earlier one replisome will arrive at *terC*, where it is halted for an extensive time period resulting in delayed chromosome replication. This again suggests that the rich medium sensitivity is due to under-replication of the chromosome relative to mass increase.

Location of the inserted replicon Integratively suppressed strains that are selected as spontaneous temperature-resistant clones from *dnaA(ts)* strains preferentially carry the alternative replicon integrated in the *oriC*-containing half of the chromosome (Lindahl *et al.*, 1971; Martin *et al.*, 1981; Nishimura *et al.*, 1971; Nishimura *et al.*, 1973; Yoshimoto and Yoshikawa, 1975). The further away from *oriC* the replicon is located, the slower the growth and the more the cell morphology of the strain is affected. Accordingly, the slow growth and disturbed cell morphology of SDR strains might be due to the location of the majority of the SDR origins far from *oriC* (de Massy *et al.*, 1984). Several explanations for the effect of the location of the alternative replicon are possible. Initiation of replication far from *oriC* may lead to disturbed gene dosages that affect growth. Also, as discussed for rich medium sensitivity (above), the closer to *terC* the replicon is located the more chromosome replication will be delayed as a consequence of the early arrival and blockage of one replisome at *terC*. Furthermore, a greater part of the chromosome will be replicated in the "wrong" direction; it has been suggested that the replisome may be slowed down when it encounters intensively transcribed operons, which in wild-type strains are preferentially transcribed in the same direction as that of replication fork movement (Brewer, 1988). In several cases two or more functional origins may be simultaneously present in the chromosome, *e. g.* in integratively suppressed strains at permissive temperature or in SDR strains. It is not clear to what extent, or even if, multiple origins may be utilized during one round of replication in a single cell, although multiple origin usage in populations of growing cells has been reported (Bird *et al.*, 1976; de Massy *et al.*, 1984; Meyenburg *et al.*, 1987).

Effects of inactivation of host genes: *dnaA*, *tus*, and *recA*

Strains in which chromosome replication is controlled by alternative replicons allow a different perspective in the study of various host genes.

The *dnaA* gene is required for initiation of replication from *oriC*. Other replicons such as ColE1, R1, R100, and R1162 are able to function in the absence of this protein. We constructed *intR1* strains in which the *dnaA* gene was interrupted by a transposon insertion (Bernander *et al.*, 1991). These strains are viable, but the efficiency of initiation of chromosome replication from the R1 origin is reduced, resulting in a reduced growth rate and severe filamentation. Interestingly, increasing the production of the RepA protein, rate-limiting for initiation of R1 replication, resulted in an increased efficiency of initiation, and to a restored cell size distribution. It was concluded that efficient chromosome replication is necessary for normal cell division, and that the DnaA protein is required neither for initiation of replication nor for cell division in these strains.

The Tus protein is necessary for stalling of replication forks at *ter* sites in the chromosome. Inactivation of the *tus* gene had little effect on the cell size and DNA content distributions of wild-type and *intR1 CW* strains (Dasgupta *et al.*, 1991). In contrast, *intR1 CC* strains grew better and showed a reduced distortion in their cell size and DNA content distributions in the absence of Tus. Hence, at least part of the severe effects of unidirectional (or asymmetric bidirectional) replication appears to be due to the stalling of replisomes at *ter* sites.

The RecA protein is required for the viability of SDR strains. In contrast, wild-type and *intR1 CW* bacteria were viable also after inactivation of the *recA* gene. However, an *intR1 CC* strain could not be transduced with an inactivated *recA* gene (unpublished). Furthermore, Mao *et al.* (1991) recently described strains in which chromosome replication is initiated from plasmid pUC13 integrated at different sites. It was found that the closer to *terC* the plasmid was integrated, the more severe was the effect of inactivation of the *recA* gene on growth and viability. We have recently demonstrated that inactivation of the *tus* gene suppressed the lethality of an inactivated *recA* gene in an *intR1 CC* strain (unpublished). Taken together, these results indicate that strains in which chromosome replication is severely asymmetric in directionality or location may require the RecA protein to overcome the strong blocks exerted by the Tus protein at various *ter* sites.

Concluding remarks

Replication of the *E. coli* chromosome may be controlled by several alternative replicons. In such strains, replication is randomly initiated during the cell cycle and multiple initiations are not coordinated. This results in broadened cell size and DNA content distributions, and in a reduction in the growth rate of the strain.

The severity of the perturbations depends on several characteristics of the alternative replicon. The more the copy number deviates from that of a wild-type strain, the stronger are the negative effects on growth and survival of the cell. Both over- and under-replication of the chromosome interfere with cell division. Growth and viability are more affected the further away from *oriC* the replicon is integrated. Strains with bidirectionally replicated chromosomes exhibit less disturbed cell cycle characteristics than strains biased towards unidirectional replication. Unidirectional replication or integration close to *terC* appear to result in an increased dependence on the RecA protein for replication, possibly to overcome replication fork blocks at *ter* sequences.

Acknowledgements This work was supported by the Swedish Natural Science Research Council, The Swedish Cancer Society, and the Knut and Alice Wallenberg Foundation. The skillful technical assistance of Lena Möller is gratefully acknowledged.

References

Bernander R, Dasgupta S, and Nordström K (1991) The E. coli cell cycle and the plasmid R1 replication cycle in the absence of the DnaA protein. Cell 64: 1145-1153.

Bernander R, Merryweather A, and Nordström K (1989) Overinitiation of replication of the *Escherichia coli* chromosome from an integrated runaway-replication derivative of plasmid R1. J Bacteriol 171: 674-683.

Bernander R, and Nordström K (1990) Chromosome replication does not trigger cell division in E. coli. Cell 60: 365-374.

Bird RE, Chandler M, and Caro L (1976) Suppression of an *Escherichia coli dnaA* mutation by the integrated R factor R.100.1: Change of chromosome replication origin in synchronized cultures. J Bacteriol 126: 1215-1223.

Brasch MA, and Meyer RJ (1988) Integrative suppression of *dnaA46* by broad host-range plasmid R1162. Mol Gen Genet 215: 139-145.

Brewer BJ (1988) When polymerases collide: Replication and the transcriptional organization of the E. coli chromosome. Cell 53: 679-686.

Chesney RH, and Scott JR (1978) Suppression of a thermosensitive *dnaA* mutation of *Escherichia coli* by bacteriophage P1 and P7. Plasmid 1: 145-163.

Dasgupta S, Bernander R, and Nordström K (1991) *In vivo* effect of the *tus* mutation on cell division in an *Escherichia coli* strain where chromosome replication is under the control of plasmid R1. Res Microbiol 142: 177-180.

Datta N, and Barth PT (1976) Hfr formation by I pilus-determining plasmids in *Escherichia coli* K-12. J Bacteriol 125: 811-817.

de Massy B, Fayet O, and Kogoma T (1984) Multiple origin usage for DNA replication in *sdrA*(*rnh*) mutants of *Escherichia coli* K-12. Initiation in the absence of *oriC*. J Mol Biol 178: 227-236.

Diaz R, and Pritchard RH (1978) Cloning of replication origins from the *E. coli* K12 chromosome. Nature 275: 561-564.

Donachie WD (1968) Relationship between cell size and time of initiation of DNA replication. Nature 219: 1077-1079.

François V, Conter A, and Louarn J-M (1990) Properties of new *Escherichia coli* Hfr strains constructed by integration of pSC101-derived conjugative plasmids. J Bacteriol 172: 1436-1440.

Froehlich BJ, Tatti K, and Scott JR (1983) Evidence for positive regulation of plasmid prophage P1 replication: Integrative suppression by copy mutants. J Bacteriol 156: 205-211.

Kogoma T (1986) RNase H-defective mutants of *Escherichia coli*. J Bacteriol 166: 361-363.

Koppes L, and Nordström K (1986) Insertion of an R1 plasmid into the origin of replication of the E. coli chromosome: Random timing of replication of the hybrid chromosome. Cell 44: 117-124.

Kuempel PL, Duerr SA, and Maglothin PD (1978) Chromosome replication in an *Escherichia coli dnaA* mutant integratively suppressed by prophage P2. J Bacteriol 134: 902-912.

Kuempel PL, Duerr SA, and Seeley NR (1977) Terminus region of the chromosome in *Escherichia coli* inhibits replication forks. Proc Natl Acad Sci USA 74: 3927-3931.

Lindahl G, Hirota Y, and Jacob F (1971) On the process of cellular division in *Escherichia coli*: Replication of the bacterial chromosome under control of prophage P2. Proc Natl Acad Sci USA 68: 2407-2411.

Louarn J, Patte J, and Louarn J-M (1982) Suppression of *Escherichia coli dnaA46* mutations by integration of plasmid R100.1 derivatives: Constraints imposed by the replication terminus. J Bacteriol 151: 657-667.

Mao Y-M, Shi Q, Li Q-G, and Sheng Z-J (1991) *recA* gene dependence of replication of the *Escherichia coli* chromosome initiated by plasmid pUC13 integrated at predetermined sites. Mol Gen Genet 225: 234-240.

Martin RR, Thorlton CL, and Unger L (1981) Formation of *Escherichia coli* Hfr strains by integrative suppression with the P group plasmid RP1. J Bacteriol 145: 713-721.

Moody EEM, and Runge R (1972) The integration of autonomous transmissible plasmids into the chromosome of *Escherichia coli* K12. Genet Res 19: 181-186.

Neidhardt FC, Ingraham JL, Brooks Low K, Magasanik B, Schaechter M, and Umbarger HE (eds.) (1987) *Escherichia coli* and *Salmonella typhimurium*: Cellular and molecular biology. (Washington, DC: American Society for Microbiology) Chapters 65-67.

Nishimura Y, Caro L, Berg CM, and Hirota Y (1971) Chromosome replication in *Escherichia coli* IV. Control of chromosome replication and cell division by an integrated episome. J Mol Biol 55: 441-456.

Nishimura A, Nishimura Y, and Caro L (1973) Isolation of Hfr strains from R+ and ColV2+ strains of *Escherichia coli* and derivation of an R'*lac* factor by transduction. J Bacteriol 116: 1107-1112.

Nordström K, Bernander R, and Dasgupta S (1991) Analysis of the bacterial cell cycle using strains in which chromosome replication is controlled by plasmid R1. Res Microbiol 142: 181-188.

Skarstad K, Boye E, and Steen HB (1986) Timing of initiation of chromosome replication in individual *Escherichia coli* cells. EMBO J 5: 1711-1717.

Sotomura M, and Yoshikawa M (1975) Reinitiation of chromosome replication in the presence of chloramphenicol under an integratively suppressed state by R6K. J Bacteriol 122: 623-628.

Torrey TA, Atlung T, and Kogoma T (1984) *dnaA* suppressor (*dasF*) mutants of *Escherichia coli* are stable DNA replication (*sdrA/rnh*) mutants. Mol Gen Genet 196: 350-355.

von Meyenburg K, Boye E, Skarstad K, Koppes L, and Kogoma T (1987) Mode of initiation of constitutive stable DNA replication in RNase H-defective mutants of *Escherichia coli* K-12. J Bacteriol 169: 2650-2658.

Yamaguchi K, and Tomizawa J-I (1980) Establishment of *Escherichia coli* cells with an integrated high copy number plasmid. Mol Gen Genet 178: 525-533.

Yoshimoto H, and Yoshikawa M (1975) Chromosome-plasmid interaction in *Escherichia coli* K-12 carrying a thermosensitive plasmid, Rts1, in autonomous and in integrated states. J Bacteriol 124: 661-667.

Dam Methyltransferase in *Escherichia coli*: Effects of Different Enzymatic Levels on DNA Replication and Cell Growth

Erik Boye, Anita Lyngstadaas, and Anders Løbner-Olesen[1]
Department of Biophysics
Institute for Cancer Research
Montebello, 0310 Oslo
Norway

ABSTRACT

A certain level of *dam* methyltransferase is required for the cells to initiate simultaneously all chromosomal origins of replication. The *dam* gene of phage T4 was found to substitute for the chromosomal *dam* gene in this function, strongly suggesting that the effect is mediated by GATC methylation. A sudden overproduction of *dam* methyltransferase resulted in drastic reductions in the synthesis of RNA, protein, and DNA, by a mechanism that is yet unknown.

INTRODUCTION

Methylation of the DNA is instrumental in many regulatory processes, both in eukaryotes and prokaryotes. In *Escherichia coli* methylation of adenines in GATC sequences by *dam* methyltransferase is involved in the regulation of DNA replication and in the repair of mismatched bases (for reviews, see Modrich, 1987; Messer and Noyer-Weidner, 1988; Barras and Marinus, 1989).

Deviations from the wild type level of *dam* methyltransferase has been shown to affect both the time between successive initiations (Messer et al., 1985) as well as each individual

---

[1]Department of Microbiology, The Technical University of Denmark, 2800 Lyngby, Denmark.

cell's ability to initiate all origins simultaneously (Boye and Løbner-Olesen, 1990). In this work the effects of overproducing *dam* methyltransferase has been studied in detail. We have also investigated whether the *dam* gene encoded by phage T4 can substitute for the chromosomal *dam* gene in its DNA replication control functions.

MATERIALS AND METHODS

Bacterial strains and plasmids.

The *E. coli* K-12 strain ALO151 is strain W3110 containing the temperature inducible *dam* gene as described before (Boye and Løbner-Olesen, 1990). It harbours plasmid pALO91 carrying the wild type *dam* gene under the control of a λpL promoter. The promoter is inducible since the strain also harbours plasmid pNF2690 carrying the gene for the temperature sensitive cI857 repressor. The control strain ALO152 contains, in addition to plasmid pNF2690, the expression vector without a *dam* gene. EBO40 is MC1000 *dam*-3 (Boye and Løbner-Olesen, 1990) and harbors the repressor plasmid pNF2690 and plasmid pEBO5, which is the same λpL expression vector as above with the *Fnu*4H1 fragment containing the T4 *dam* gene (Macdonald and Mosig, 1984) cloned into the *Sma*I site downstream of the λpL promoter. Strain ALO482 is C600 *Δlac thr*$^+$ (Lene J. Rasmussen, unpublished) containing a plasmid-borne system for expressing *dam* from an IPTG-inducible promoter, as follows: the *dam*-carrying *Hind*III fragment from pALO91 was cloned in the *Hind*III site of the expression vector pUHE21-2 (Lanzer and Bujard, 1988). In addition, the cells contained plasmid pALO15 (Løbner-Olesen et al., 1989) carrying the *lac*I repressor gene.

Bacterial growth and flow cytometry.

The bacteria were grown in either Luria broth supplemented with 0.2% glucose or in AB medium (Clark and Maaløe, 1967) containing 0.5% casamino acids and 0.2% glucose. Bacterial growth was monitored by measuring the optical density at 600 nm (for LB) or at 450 nm (AB). For flow cytometry, the cells were incubated in the presence of rifampicin and cephalexin (Boye and Løbner-Olesen, Res. Microbiol., in press), fixed, stained and run in the flow cytometer as described before (Skarstad et al., 1983).

Radioactive labelling.

Proteins were labelled with a ^{14}C-amino acid mixture (Amersham, England). The cells were grown in AB medium without casamino acids to an OD of 0.1 and pulse labelled for 30-sec intervals before precipitation with 5% TCA and filtering onto fiberglass filters for liquid scintillation counting. For analysis on polyacrylamide gels, the cells were labelled for 1 min (42°C) or 2 min (31°C) before addition of 1% casamino acids and continued incubation for 5 min. The cells were mixed with the SDS-mercaptoethanol loading buffer before application on a 12% polyacrylamide gel (Laemmli, 1970). The gel was dried and subjected to autoradiography.

RNA was labelled with ^{3}H-uridine (Amersham). The cells were grown in AB medium with casamino acids and glucose and pulse labelled for 1-min intervals. Samples were precipitated directly in 5% TCA and filtered onto fiberglass filters before liquid scintillation counting.

DNA was labelled with ^{3}H-thymidine (Amersham). The cells were grown as for RNA labelling and pulse labelled for 1 min, followed by a 10-min chase in the presence of an excess of unlabelled thymidine. Samples were removed and processed as above.

RESULTS

When wild type *E. coli* cells grow rapidly, they contain several origins of DNA replication which are all initiated simultaneously (Skarstad et al., 1986). A high degree of synchrony is dependent upon *dam* methylation, so that only at a certain level of *dam* methyltransferase are initiations properly synchronized (Boye and Løbner-Olesen, 1990). This was shown by varying the intracellular level of the enzyme and subsequently analyzing initiation kinetics by flow cytometry. The basis for this analysis is that after treatment of the cells with rifampicin (to inhibit initiations) and cephalexin (to inhibit cell division) the cells contain a number of chromosomes which is equal to the number of replication origins present at the time of drug addition (Boye and Løbner-Olesen, Res. Microbiol., in press). By flow cytometry the fraction of cells with 1, 2, 3, 4, etc, chromosomes (origins) can be measured.

It is conceivable that *dam* methyltransferase has other functions in the cell beyond GATC methylation, and thus it is not proven that the effect of the enzyme on initiation synchrony is due to the methylating activity. To investigate this further, we have made a construct allowing graded induction of the T4 *dam* gene in the same vector and the same host cell as used for the studies of different levels of the *E. coli dam* methyltransferase (Boye and Løbner-Olesen, 1990). The two methyltransferases both methylate adenines in the sequence GATC, but structure- and sequencewise they appear to be unrelated.

Strain EBO40 containing the λpL-*dam*T4 construct was grown at temperatures between 32 and 42°C, and the methylation status of the chromosomal DNA was checked by restriction enzyme analyses. Exposure to *Dpn*I, which digests only fully methylated GATC sites, revealed that the DNA was fully methylated at 39°C and above. The cells were grown at the different temperatures, treated with rifampicin and cephalexin, and subjected to flow cytometric analyses. The DNA histograms of cells grown at the lowest temperatures (37°C and below) showed discrete peaks corresponding to 1, 2, 3, and 4 chromosomes (Fig 1, panels A to D). This is consistent with earlier observations that in *dam* deficient cells all origins present are not initiated simultaneously (Boye et al., 1988; Boye and Løbner-Olesen, 1990) and consistent with the DNA histogram of the corresponding *dam* deficient strain (panel I). The DNA histogram of the wild type control strain (MC1000 *dam*$^+$) when grown at 37°C and treated similarly with rifampicin and cephalexin, showed two prominent

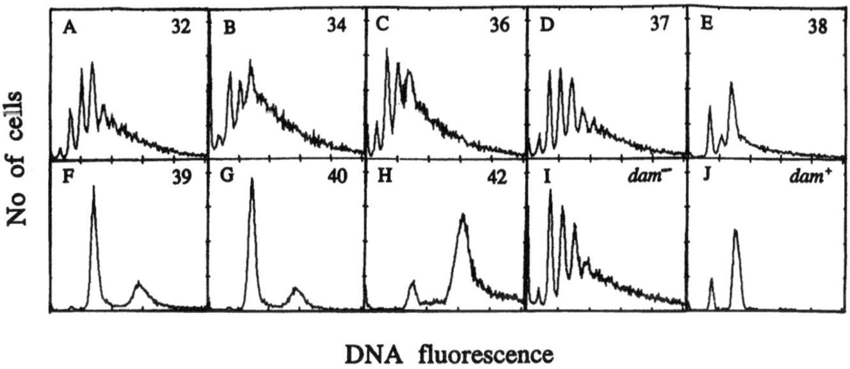

Figure 1. DNA histograms of strain EB040 (panels A to H) grown at the temperatures indicated in each panel, of strain MC1000 *dam*-3 (panel I) and strain MC1000 (panel J). All strains were grown to an OD of 0.1 before addition of rifampicin and cephalexin, fixation, staining and flow cytometric analyses. The two peaks in panel J represent cells with 2 and 4 fully replicated chromosomes.

peaks corresponding to 2 and 4 fully replicated chromosomes (Figure 1, panel J). At 38°C initiations in strain EB040 were almost perfectly synchronized (panel E), except for a minor peak representing 3-chromosome cells. The synchrony was even better at higher temperatures, although the number of origins was then significantly increased (panels F to G).

It has been reported that an increase in the cellular level of *dam* methyltransferase causes the interval between successive initiations to decrease (Messer et al., 1985). We have employed the *dam*-inducible system described above to study this subject further. When strain AL0151 was grown in AB medium with casamino acids and glucose and shifted from 30 to 42°C to allow induction of the plasmid-borne *dam* gene, the increase in cell growth was abruptly stopped, while growth was unimpeded in the control strain AL0152 (Fig. 2). Only after several hours of continued incubation did the optical density of strain AL0151 start to increase again, but the culture doubling time was significantly lower than at 30°C and than the control strain at either temperature. Also, growth of strain AL0151 at 42°C was extremely

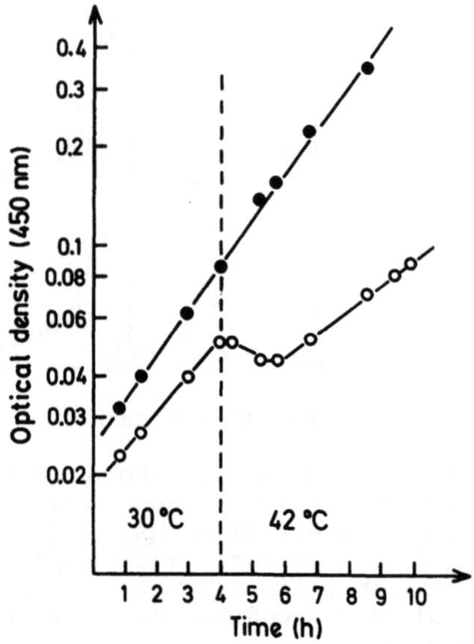

Figure 2. Growth of strain AL0151 (○) and strain AL0152 (●) after a shift from 30 to 42°C at 4 hours.

sensitive to even a slight further increase in the temperature. The inihibition of cell growth after overproduction of *dam* methyltransferase could not be relieved by addition of extra nutrients, e.g. LB medium, concentrated casamino acids, or extra methionine (data not shown). The growth kinetics of strain AL0482 after induction with 0.5mM IPTG at 37°C was similar to that shown in Fig. 2. However, when grown at 32°C the lag in growth after IPTG induction was greatly reduced, and at 42°C the OD decreased continuously for several hours after induction. Strain AL0482 was unable to form colonies at 42°C in the presence of IPTG. Thus, a sudden induction of *dam* methyltransferase does, at least transiently, inhibit cell growth, and the effect is potentiated at higher temperatures.

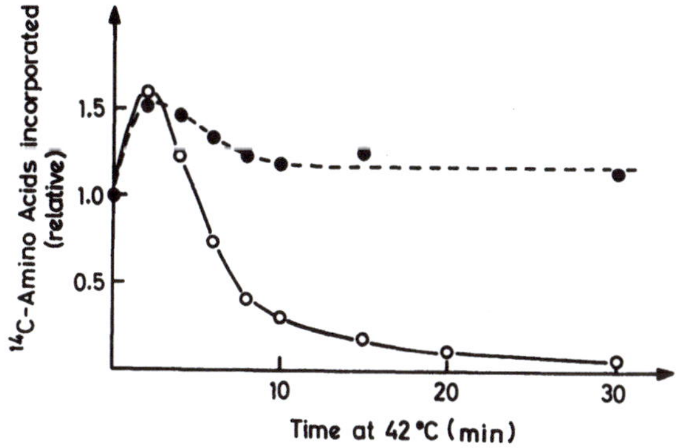

Figure 3. Protein synthesis in strain AL0151 (○) and strain AL0152 (•) after *dam* induction.

The growth inhibition was investigated by measuring the rate of macromolecular synthesis after temperature induction of strain AL0151. Protein synthesis was initially stimulated after a shift from 31 to 42°C, thereafter the rate of protein synthesis decreased sharply (Fig. 3). Ten min after the temperature shift the rate was at about 25% of the preshift value, and after 30 min there was hardly any protein synthesized. In contrast, in the control strain AL0152 protein synthesis first increased and was then stabilized at a higher level than before the temperature shift. Cells labelled at different times after the temperature shift were lysed and subjected to polyacrylamide gel electrophoresis. Autoradiography revealed that the synthesis of all proteins was inhibited (Fig. 4). Although heat-shock proteins could be observed at early times, there were hardly any protein bands visible 30 min after the temperature shift. As expected, little effects on protein synthesis was found for the control strain AL0152.

Total RNA synthesis decreased in much the same way as described for protein synthesis (Fig. 5). A slight post-shift increase was followed by a strong decline in strain AL0151, while synthesis continued at about the same level in strain AL0152. A similar picture could be found for DNA synthesis,

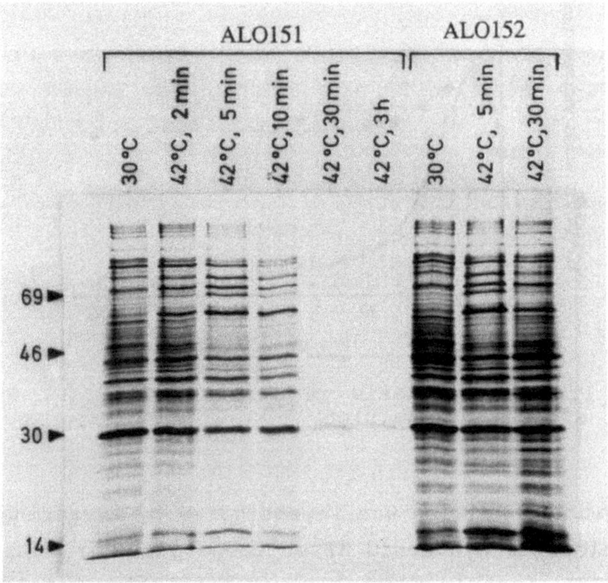

Figure 4. Pattern of protein synthesis in strains ALO151 and ALO152 after a shift from 30 to 42°C.

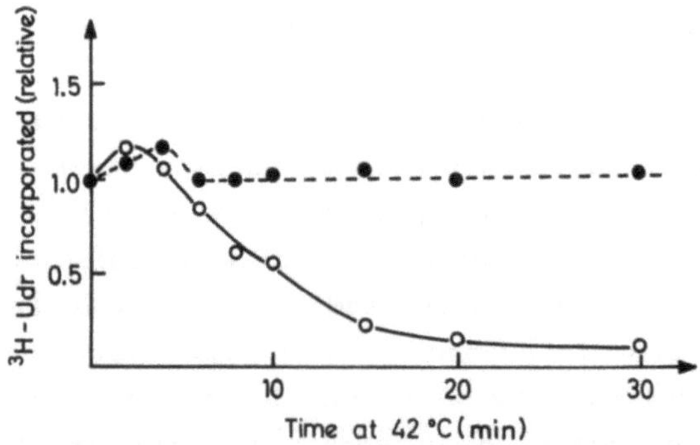

Figure 5. Synthesis of RNA in strain ALO151 (○) and strain ALO152 (•) after a shift to 42°C.

although in this case the time scale was changed (Fig. 6). Only after about 60 min was the rate of DNA synthesis decreased to 25% in strain AL0151, and even after 200 min the cells were still synthesizing DNA 10% as fast as before the shift.

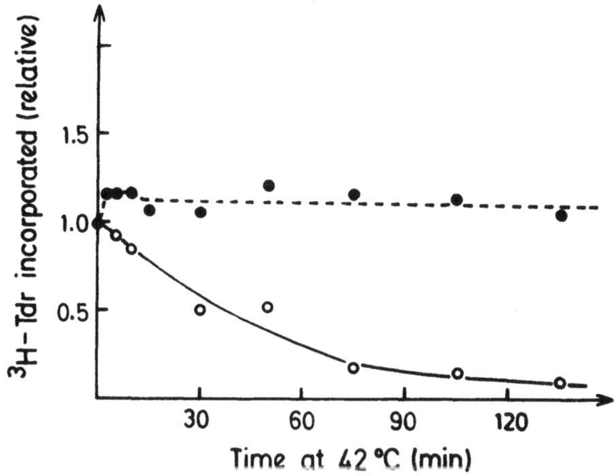

Figure 6. Synthesis of DNA in strain AL0151 (○) and strain AL0152 (●) after a shift to 42°C.

DISCUSSION

It has been shown here that the *dam* methyltransferase encoded in the genome of phage T4 can completely substitute for *E. coli*'s own enzyme, both concerning methylation of GATC sequences and in terms of coordinating initiations of multiple origins. A certain level of *dam* methyltransferase activity must be present in the cells in order to allow synchronized initiations to occur. Furthermore, on the basis of findings that *oriC* is transiently bound to the membrane in its hemimethylated form (Ogden et al., 1988, Landoulsi et al., 1990) and that *oriC* remains hemimethylated for a long time after initiation (Ogden et al., 1988, Campell and Kleckner, 1990) it may be concluded that it is the rate of methylation of GATC sites in *oriC* that is crucial for initiation synchrony.

In the present experiments with the T4 *dam* gene the cells were growing in steady state, and we can assume that the average rate of initiation was one per origin per generation. The finding that the number of origins present per cell was doubled from 38 to 42°C (Fig. 2, panels F to H) argues that the time of initiation has been moved forwards in the cell cycle. This is consistent with similar findings using *E. coli*'s own *dam* gene (Boye and Løbner-Olesen, 1990) and with *dam* methyltransferase being a stimulator of *oriC* initiation, as suggested earlier (Messer et al., 1985). Presumably, this occurs because the origins are methylated more rapidly and thus are more rapidly released from the cell membrane for a new round of initiations.

In contrast to the present results, we have earlier found that abnormally high levels of methyltransferase produces the asynchrony phenotype (Boye and Løbner-Olesen, 1990). Two lines of evidence indicate that the λpL-*dam*T4 construct yielded a less efficient expression of methyltransferase activity than the corresponding construct with the chromosomal *dam* gene: with the T4 gene full methylation was achieved at a higher temperature, and synchronous initiations were also observed at a higher temperature than with the chromosomal *dam* gene. Therefore, the latter construct presumably produced a higher level of enzymatic activity at 42°C. We propose that at very high enzyme levels *oriC* may become methylated before being sequestered in the membrane. This may result in multiple initiations of the same origin and DNA histograms showing cells containing all different numbers of origins, i.e. the asynchrony phenotype.

The data presented show that a sudden overproduction of *dam* methyltransferase rapidly inhibits RNA synthesis, an inhibition which is sufficient to explain the decreased rates of protein and DNA synthesis. The mechanism of this inhibitory action is not known. The efficiency of RNA polymerase in transcribing chromosomal DNA or plasmid DNA *in vitro* was found to be unaffected by high concentrations of *dam* methyltransferase (data not shown). Thus, there was no evidence of a direct interaction between RNA polymerase and *dam* methyltransferase.

The growth inhibition was not caused by the stringent

response, since the same results could be obtained in a *relA*⁻ host cell (data not shown). Also, it was not a heat shock effect, since the inhibition could be reproduced by using an IPTG-inducible *dam* gene. However, higher temperatures clearly potentiated the inhibition, since *dam* overproducers hardly grow at 42°C. It is possible that excess *dam* enzyme is localised to the cell membrane, thus contributing to cell envelope breakdown at higher temperatures. This possibility was investigated by immunoblotting. Cells of strain AL0482 were grown at 32°C, induced with IPTG at 42°C for 4 hours, sonicated, and the cell envelope was pelleted by centrifugation at 25000 x g for 20 min. The cytoplasmic and envelope fractions were assayed for *dam* methyltransferase by immunoblotting (Boye and Løbner-Olesen, 1990). No *dam* enzyme could be detected in the envelope fraction (data not shown).

The kinetics of DNA replication after *dam* induction is compatible with a "run-out", i.e. when initiations are inhibited and elongation is continuing. If this were the case, the cells would slowly accumulate fully replicated chromosomes, as is the case during treatment with rifampicin. Flow cytometry was employed to measure the DNA contents at different times after *dam* induction. The DNA histograms revealed no clear evidence of replication "run-out" (data not shown) and it may be concluded that initiations are still occurring. However, flow cytometry of rifampicin-treated cells revealed that the number of origins per cell was significantly reduced 30 min after *dam* induction, suggesting that both initiations and elongation were depressed.

The present work shows that a moderate overexpression of *dam* methyltransferase stimulates initiation of DNA replication, while a massive overexpression severely inhibits important cellular processes such as RNA, DNA and protein synthesis by an unidentified mechanism.

ACKNOWLEDGEMENTS

This work was supported by grant no GM39582 from the National Institutes of Health (to E.B.), by the Danish Center for Microbiology (to A.L.O.) and by a NATO collaborative Research Grant. A.L. is a fellow of the Norwegian Cancer Society.

LITERATURE REFERENCES

Barras F, Marinus MG (1989) The great GATC: DNA methylation in *E. coli*. Trends Genet 5:139-143

Boye E, Løbner-Olesen A, Skarstad K (1988) Timing of chromosomal replication in *Escherichia coli*. Biochim Biophys Acta 951:359-364

Boye E, Løbner-Olesen A (1990) The role of *dam* methyltransferase in the control of DNA replication in *E. coli*. Cell 62:981-989

Campbell JL, Kleckner N (1990) *E. coli oriC* and the *dnaA* gene promoter are sequestered from *dam* methyltransferase following the passage of the chromosomal replication fork. Cell 62:967-979

Clark DJ, Maaløe O (1967) DNA replication and the division cycle in *Escherichia coli*. J Mol Biol 23:99-112

Laemmli UK (1970) Cleavage of structural proteins during the assembly of head bacteriophage T4. Nature 227:680-685

Landoulsi A, Malki A, Kern R, Kohiyama M, Hughes P (1990) The *E. coli* surface specifically prevents the initiation of DNA replication at *oriC* on hemimethylated DNA templates. Cell 63:1053-1060

Lanzer M, Bujard H (1988) Promoters largely determine the efficiency of repressor action. Proc Natl Acad Sci USA 85:8973-8977

Løbner-Olesen A, Skarstad K, Hansen FG, von Meyenburg K, Boye E (1989) The DnaA protein determines the initiation mass of *Escherichia coli* K-12. Cell 57:881-889

Macdonald PM, Mosig G (1984) Regulation of a new bacteriophage T4 gene, *69*, that spans an origin of DNA replication. EMBO J 3:2863-2871

Messer W, Bellekes U, Lother H (1985) Effect of *dam* methylation on the activity of the *E. coli* replication origin, *oriC*. EMBO J 4:1327-1332

Messer W, Noyer-Weidner M (1988) Timing and targeting: the biological functions of Dam methylation in *E. coli*. Cell 54:735-737

Modrich P (1987) DNA mismatch correction. Ann Rev Biochem 56: 435-466

Ogden GB, Pratt MJ, Schaechter M (1988) The replicative origin of the *E. coli* chromosome binds to cell membranes only when hemimethylated. Cell 54:127-135

Skarstad K, Steen HB, Boye E (1983) Cell cycle parameters of slowly growing *Escherichia coli* B/r studied by flow cytometry. J Bacteriol 154:656-662

Skarstad K, Boye E, Steen HB (1986) Timing of initiation of chromosome replication in individual *Escherichia coli* cells. EMBO J 5:1711-1717

Further Characterization of a DNA Replication Inhibitor from the Cell Membrane of *Escherichia coli*

Patrick Hughes, Abderrahim Malki, John Herrick, Renée Kern and Masamichi Kohiyama.
Institut Jacques Monod
Tour 43
Université Paris VII
75251 Paris Cedex 05
France

SUMMARY

Hemimethylated DNA occurs briefly in *E.coli* after passage of the replication fork and correlates well with the inhibition of the functioning of a number of genes involved in DNA replication including initiation of DNA synthesis at *oriC* and pBR322 origins of replication and transcription of the initiator DnaA gene. Previously, we found that inhibition of initiation *in vitro* at *oriC* on a hemimethylated DNA template relies on strong affinity between (a) membrane component(s) and the hemimethylated *oriC* sequence. Here, we show that this affinity is also shared by hemimethylated DNA fragments carrying the *dnaA* promoter and the RNA II promoter region which furnishes primers for pBR322 DNA replication. Thus, sequestration of hemimethylated DNA by the cell membrane may represent a common mechanism for the inhibition of replication and transcription at these and other sites on newly replicated DNA. We discuss these findings with reference to the control of interinitiation times of DNA replication. Also, we further characterize the membrane fraction by identifying a 24kDa protein responsible for the specific affinity for hemimethylated *oriC* DNA and describe hemimethylated *oriC* - membrane binding sensitivity to the sulfhydryl (-SH) group blocking agent, N-ethylmaleimide.

INTRODUCTION

DNA replication in *E.coli* is a tightly controlled process regulated by the frequency of initiation at the chromosome origin, *oriC*. Initiation has been shown to occur once at a particular moment of the cell cycle when a certain ratio of cell mass to origin content is reached. Current models link this initiation mass to the cellular level of the DnaA initiator protein which is expected to accumulate in proportion to cell mass increase by autoregulation. By in large, many of the predictions of these models have been shown to hold true. Initiator DnaA protein binding to four recognition sites within the minimal *oriC* segment of the chromosome is a key event in initiation of DNA synthesis as originally predicted in the replicon hypothesis. DnaA protein binding to a recognition site in the *dnaA* gene promoter region represses transcriptional initiation from this site as predicted in the autorepressor model (reviewed in von Meyenburg & Hansen, 1987). Finally, increased DnaA protein cell mass ratios have been shown to result in over-initiation (Atlung et al., 1987) earlier in the cell cycle (Lobner-Olesen et al., 1989).
Despite these successes, these models, which attempt to describe initiation as a positively controlled event, have failed to explain satisfactorily what has been termed "switch off" (Pritchard, 1978) i.e., why are origins only initiated once and only once per cell cycle ? DnaA protein could effectively be inactivated after each initiation and a number of nucleotide binding sites (Sekimizu et al., 1987; Hughes et al., 1988) and a membrane binding domain (Sekimizu et al., 1988) on the protein have been suggested to play a role in this. However, even in this modified form, the model would fail to explain why multiple origins in fast growing cells are initiated synchronously, with a very small coefficient of variation (Skarstad et al., 1986).
An alternative explanation are recent results which link "switch off" of initiation to the state of postreplicative origin DNA rather than the initiator protein. The adenine residue of GATC sequences on the *E.coli* chromosome is methylated postreplicatively by the deoxyadenosine methylase (Dam). Thus, newly replicated (hemimethylated) DNA is distinguishable from unreplicated (methylated) DNA until remethylated. Remethylation rates have been shown

to vary depending on the region examined. The *oriC* region,*dnaA* and *mioC* promoters, all of which have been implicated in initiation and contain statistically elevated numbers of GATC sites, remain hemimethylated for periods ranging from 7 to 12 minutes after replication, a delay largely superior to that of other sites examined (less than one minute) (Ogden et al., 1988; Campbell & Kleckner,1990)

Transformation studies in *dam* mutants have shown that *oriC* plasmids do not replicate when hemimethylated (Smith et al., 1985; Messer et al., 1985; Russell & Zinder, 1987). Also, the hemimethylated state of the *dnaA* promoter correlates well *in vivo* with a period of transcriptional arrest of the *dnaA* gene (Campbell & Kleckner, 1990). *In vitro* studies have shown that hemimethylated *oriC* DNA has preferential affinity for an outer cell membrane fraction (Ogden et al., 1988) and that the addition of this membrane fraction to soluble replication assays specifically inhibits DNA synthesis on hemimethylated *oriC* plasmids (Landoulsi et al., 1989). The results were interpreted to mean that the inhibition of initiation *in vivo* on hemimethylated DNA could be the result of sequestration of this post-replicative form of DNA by the cell membrane.

Here, we extend this study to two other sites besides *oriC*, the *dnaA* promoter region and the a DNA fragment containing the RNA primer promoter region for pBR322 DNA replication. We also further characterize the replication inhibitor present in the OM fraction.

EXPERIMENTAL PROCEDURES

DNA Binding and Replication Assays.

Replication and DNA binding assays were as described previously by Landoulsi et al. (1990).

South-Western Blots.

These were realized essentially according to the original method of Bowen et al. (1980) with the following modifications. Radiolabelled hemimethylated or unmethylated *oriC* probes were prepared by primer (5'GGGCCG3') extension from the *Sma* I side on denatured *Sma* I - *Xho* I restriction fragments purified from respectively methylated and unmethylated pOC42 DNA. Modified T7

DNA polymerase (Biolabs) was used to extend the primer and the radiolabelled (^{32}P dATP) products of the reaction were examined on polyacrylamide gels to ensure that greater than 90% corresponded to complete 463bp *Sma* I - *Xho* I DNA fragments. Cytosol, total membrane, and outer membrane fractions were obtained from a synchronized culture of *E.coli* strain PC2 according to Ogden et al. (1988). Proteins from cytosol and total membrane fractions were concentrated by 5% tricloroacetic acid precipitation, washed with acetone and resuspended in loading buffer (4% sodium dodecyl sulfate (SDS), 8 mM 2-mercaptoethanol, 20 mM Tris pH 6.8 and 20% glycerol). Outer membrane fraction was simply centrifuged at 15,000 g for 15 min and the pellet resuspended directly in loading buffer. Samples were applied directly to the SDS-15% polyacrylamide (1% bis-acrylamide) gels (Laemmli, 1970), without boiling, along with prestained SDS-PAGE low-range molecular weight standards (Bio-Rad). After electrophoresis, proteins were transferred at 4°C in Tris-glycine transfer buffer (Towbin et al. (1979) to Hybond (0.45 µm) nylon filters according to the manufacturers (Amersham) instructions. Filters were then blocked by incubation in binding buffer (10 mM Hepes/KOH (pH 7.4), 100 mM KCl, 5 mM $MgCl_2$ and 1 mM EDTA) containing 5% skimmed milk for one hour at ambient temperature. After brief washing, the filters were incubated overnight with either hemimethylated or unmethylated *oriC* DNA fragment (10^{-11} M) as ^{32}P labelled probe (approximately 2.10^8 cpm per µg) and a 10,000 fold excess of cold calf thymus DNA in binding buffer at 4°C. After extensive washing in binding buffer to reduce background radioactivity, the filters were autoradiographed.

RESULTS

An OM Fraction Specifically Inhibits Initiation of DNA Replication on Hemimethylated *oriC* Plasmids.

The isolation of OM (OM) fraction - origin DNA complexes in which the DNA was shown to be largely hemimethylated has argued in favor of a tight association of the origin with one or more proteins of the cell envelope after each cycle of initiation (Ogden et al. 1988). This has led to two different but not necessarily conflicting viewpoints; sequestration of the origin after initiation so as to ensure correct segregation of daughter chromosomes and inhibition of premature reinitiation of new cycles of DNA replication on newly replicated DNA. Although experimental evidence for the former hypothesis is lacking, the latter hypothesis is supported by experimental evidence presented in an earlier report (Landoulsi et al. 1990) and in Figure 1. Here, it can be seen that the *in vitro* replication of a hemimethylated *oriC* plasmid compared to a methylated or unmethylated plasmid is preferentially inhibited by the OM fraction.

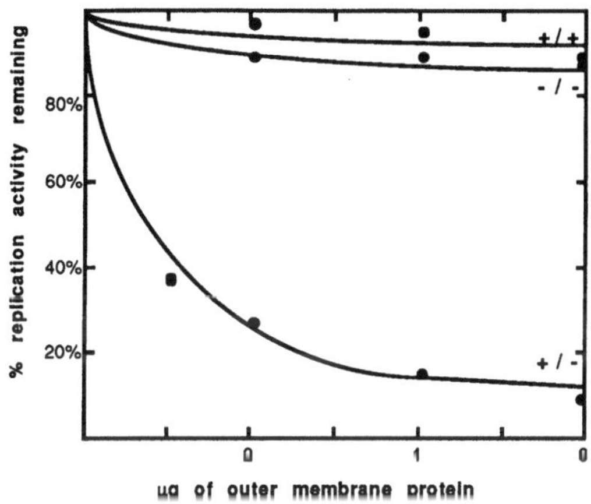

Fig.1. Outer membrane fraction inhibits preferentially initiation of DNA replication at *oriC* on hemimethylated DNA templates (+/-) compared to methylated (+/+) or unmethylated (-/-) DNA templates. Reactions were carried out as described in Landoulsi et al. (1990).

An involvement of methylation in initiation is also supported by flow cytometry data (Boye & Lobner-Olesen, 1990; Boye et al., this volume) and density shift experiments in *dam* mutants which lack hemimethylated DNA (Bakker & Smith, 1989). Interpretation of these latter results suggests that initiation probably occurs more than once within a short period of time (less than one cell cycle time interval) on the same chromosome with consequent bad synchronization and timing of initiation events within the cell cycle. Also, the average interinitiation time has been shown to be reduced in Dam methylase overproducers (Messer et al., 1985) with a probable consequent increase in origin to cell mass ratios (Boye & Lobner-Olesen, 1990). However, origin DNA has been shown to remain entirely hemimethylated only for 8 minutes at most (Campbell & Kleckner, 1990) after initiation, an insufficient period of time to explain the entire interinitiation time even at fast growth rates if the time of hemimethylation corresponds to the residence time of origin attachment to the membrane. This suggests that hemimethylated origin - membrane attachment mechanism may exist only to ensure against over-initiation during a time window when initiation is particularly favorable, i.e., at the time when the critical initiation mass is reached. Why then, once this period has passed, is initiation no longer probable until after another doubling in cell mass ?

OM Protein Fraction Recognizes Other Hemimethylated Regions Besides *oriC*.

(i) The *dnaA* Promoter.

One possible response to the above question has come from Campbell & Kleckner (1990) who have recently shown that the quantity of the *dnaA* gene transcript is reduced during a time period which corresponds to the hemimethylation of its promoter region. It appears likely that transcription of the *dnaA* gene, like initiation of DNA synthesis at *oriC*, is shut down at this time. *In vitro*, we have found that, as with hemimethylated *oriC* DNA for initiation, hemimethylated *dnaA* gene promoter DNA is active for transcription with purified RNA polymerase (unpublished results). Is hemimethylated *dnaA* promoter inactive for transcription *in vivo* because of heightened affinity of this form for the cell membrane ? To test this possibility, we subcloned the DnaA protein gene promoter region into the *oriC* plasmid pOC42 to make the new plasmid pOC43 (Fig.2A). After restriction endonuclease digestion and end labelling with Klenow fragment and ^{32}P dATP, the DNA was incubated with OM protein and filtered through nitrocellulose membranes as described previously (Landoulsi et al. 1990). As can be seen in Figure 2B, three DNA fragments from hemimethylated pOC43 are bound to the nitrocellulose membrane. No detectable retention of methylated or unmethylated DNA was observed. In agreement with previous data, one of these fragments contains *oriC*. The two other DNA fragments correspond to DNA from the *dnaA* promoter region and the RNA primer promoter region of pBR322. Both are retained less well (especially the pBR322 DNA fragment) compared to *oriC* indicating preferential affinity of *oriC* for the membrane. Binding of the *dnaA* gene promoter DNA fragment when hemimethylated by the membrane fraction suggests that inhibition of transcriptional initiation at this site *in vivo* is the result of sequestration by the cell membrane. The results also predict that DnaA protein concentration in the cell should fall nearly immediately after initiation as the *dnaA* gene is located only about 30 seconds in replication time from the origin. Therefore, one of the reasons why reinitiation might not occur immediately after remethylation and detachment of *oriC* from the membrane is because DnaA protein concentration has become limiting at this time. The period until the next initiation event would then represent the period required for *dnaA* gene transcription to catch up with cell growth.

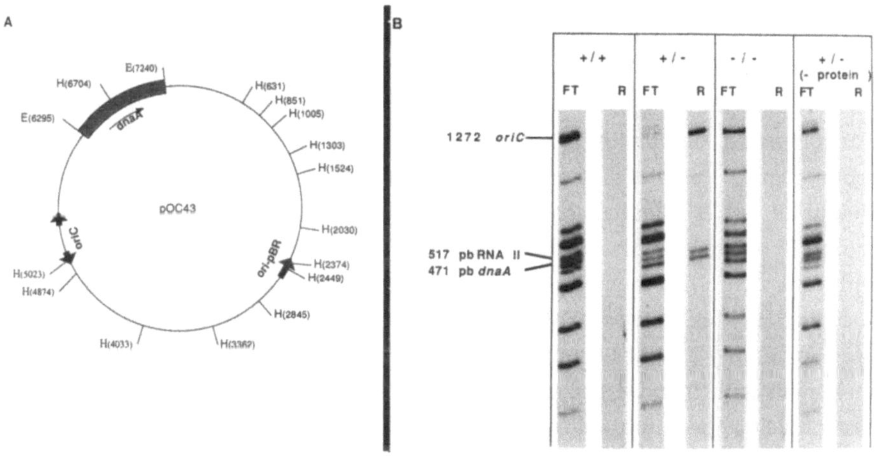

Fig. 2 (A) Pertinent regulatory regions and restriction endonuclease sites on the plasmid pOC43. (B) Binding affinity of outer membrane fraction for *Hinf* 1 & *Eco* R1 restriction fragments from methylated (+/+), hemimethylated (+/-) and unmethylated (-/-) pOC43 DNA using the nitrocellulose binding assay. Assays were as described previously by Landoulsi et al. (1990). DNA-bound (R) and flow-through(FT) fractions were subjected to polyacrylamide gel electrophoresis.

<u>(ii) The Primer Promoter Region for pBR322 DNA Replication.</u>

Binding of the pBR322 RNA primer promoter region by the OM fraction is in support of previous *in vivo* data which showed that hemimethylated pBR322 DNA does not allow initiation of DNA replication (Smith et al., 1985; Messer et al., 1985; Russell & Zinder, 1987). In fact, we have recently found *in vitro* that the OM fraction specifically inhibits hemimethylated pBR322 DNA replication but at concentrations higher than those used for the inhibition of hemimethylated *oriC* directed DNA replication (unpublished results). Also, the higher concentrations of OM fraction used probably explains why we previously failed to observe this DNA fragment in membrane binding assays (Landoulsi et al., 1990). pBR322 DNA replication is apparently asynchronous within the cell cycle and thus independent of cell mass (Leonard & Helmstetter, 1986). In this case, attachment of the hemimethylated origin to the membrane probably uniquely represents a way to prevent too many cycles of DNA replication on a restricted subset of the cellular pBR322 population.

Specific Affinity for Hemimethylated *oriC* DNA can be attributed to the presence of a 24KD protein in the OM Fraction.

In order to define better the nature and identity of the membrane elements involved in the inhibition of DNA replication on hemimethylated templates we carried out South-Western blot analysis (Bowen et al. 1980) of the OM fraction using hemimethylated or unmethylated *oriC* DNA as radioactive probe. Figure 3 shows that one protein band with an approximate molecular weight of 24KD interacts specifically with the hemimethylated *oriC* DNA probe. This band is only weakly visible when unmethylated DNA is employed and, consistent with membrane location, it is absent from the cytosol fraction. A coomassie blue stained polyacrylamide gel shows the corresponding protein band to be only a minor component of this fast sedimenting - OM fraction. In agreement with this, we could not detect a 24KD protein in a total membrane fraction using the South - Western blot technique.

N-ethylmaleimide Sensitivity of Membrane protein - Hemimethylated DNA Interactions

In Figure 4 we show the sulfhydryl blocking agent, N-ethylmaleimide (NEM) at a concentration of 2 to 4 mM severely inhibits the retention of OM protein - hemimethylated complexes on nitrocellulose filters. DNA - membrane complexes can be observed, however, if NEM is added after DNA binding has been allowed to occur. This suggests that the specific DNA binding protein contains one or more essential sulfhydryl (-SH) group(s) which become(s) inaccessible to NEM once the protein is complexed with DNA.

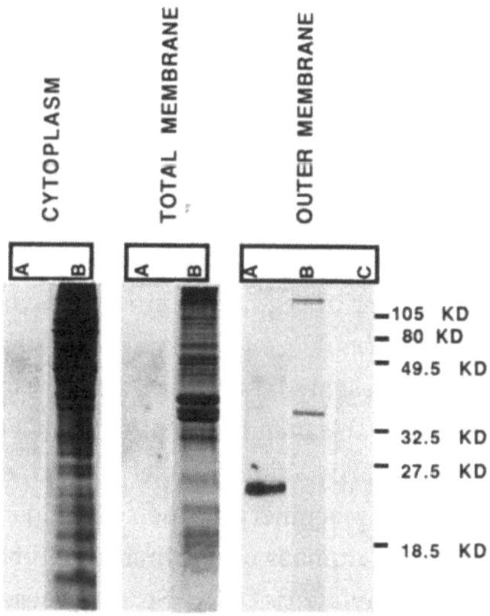

Fig. 3 South - Western blots of outer membrane fractions using either hemimethylated (A) or unmethylated (C) *oriC* DNA as labelled probe and corresponding coomassie brilliant blue stained gels (B). The quantities of protein applied to the gels were as follows ; 60 µg of cytoplasm (A) & (B); 30 µg of total membrane (A) & (B); 8 µg (A,C) & 16 µg (B) of outer membrane fraction.

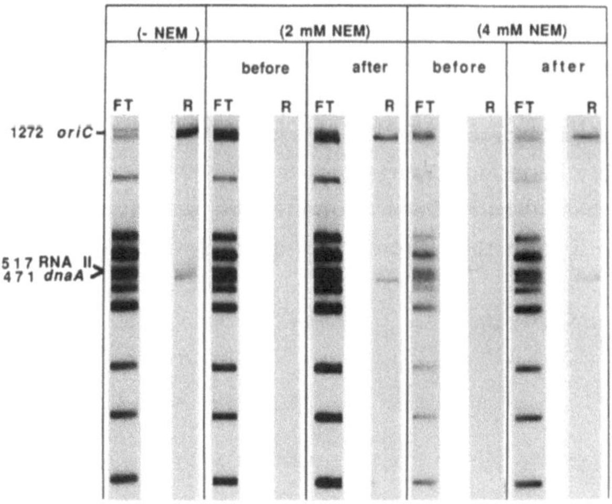

Fig. 4. N-ethylmaleimide (NEM) sensitivity of outer membrane protein binding to labelled restriction fragments from pOC43 (Fig. 2 (A)). NEM at 2mM & 4mM was incubated for 30 min at 37°C in binding buffer with the membrane fraction either before or after the addition of labelled DNA fragments. RNAII & *dnaA* refer respectively to the RNAII primer promoter for pBR322 DNA replication and the *dnaA* gene promoter.

CONCLUSION

The results presented here confirm and extend the likelyhood that a protein found in an OM fraction is important in the control of initiation of DNA synthesis in *E.coli*. The protein with an approximate molecular weight of 24KD (see above) binds to a number of regulatory sites involved in DNA replication including *oriC* and the *dnaA* gene promoter when these sites are hemimethylated. Results presented here and elsewhere suggest that it can sequester these regions into the membrane after each DNA replication cycle and renders them temporarily inaccessible to the soluble methylation, transcription and replication machinery of the cell. Thus, the protein probably functions as an inhibitor of a number of enzymatic activities at specific sites on newly replicated (hemimethylated) DNA. So far, these sites have all been shown to be involved in initiation of chromosomal DNA synthesis and as discussed above, the sequestration after replication of these sites i.e., *oriC* and the *dnaA* promoter, by the membrane may be an important element in the strict timing between replication cycles observed at different growth rates in *E.coli*.

Finally, it would not be surprising if this type of mechanism also functions for the replication control of some other replicons and to couple the transcription of other genes to the cell cycle. An example of the former is suggested by our discovery of an OM protein binding site in the RNA II primer promoter region of pBR322 (see above). Possible examples of the latter will probably be found amongst those genes whose transcriptional efficiency *in vitro* has already been found to be weakly inhibited by unmethylated GATC sites in their promoter sequences (Barras & Marrinus, 1989).

REFERENCES

Atlung T, Lobner-Olesen A, Hansen F (1987) Overproduction of DnaA protein stimulates initiation of chromosome and minichromosome replication in Escherichia coli. Mol Gen Genet 206 : 51-59

Bakker A, Smith DW (1989) Methylation of GATC sites is required for precise timing between rounds of DNA replication in *Escherichia coli.* J Bacteriol 171 : 5738-5742.

Barras F, Marinus MG (1989) The great GATC : DNA methylation in *E. coli.* Trends Genet 5 : 139-143

Boye E, Lobner-Olesen A (1990) The role of *dam* methyltransferase in the control of DNA replication in *E. coli.* Cell 62 : 981-989

Bowen B, Steinberg J, Laemmli UK, Weintraub H (1980) The detection of DNA-binding proteins by immunoblotting. Nucleic Acids Res 8 : 1-20

Campbell JL, Kleckner N (1990) *E. coli* and the *dnaA* gene promoter are sequestered from dam methyltransferase following the passage of the chromosomal replication fork. Cell 62 : 967-979

Hughes P, Landoulsi A, Kohiyama M (1988) A novel role for cAMP in the control of the activity of the *E. coli* chromosome replication initiator protein, DnaA. Cell 55 : 343-350

Laemmli UK (1970) Cleavage of structural proteins during the assembly of head bacteriophage T4. Nature 227 : 680-685

Landoulsi A, Malki A, Kern R, Kohiyama M, Hughes P (1990) The *E. coli* cell surface specifically prevents the initiation of DNA replication at *oriC* on hemimethylated DNA templates. Cell 63 : 1053-1060

Leonard AC, Helmstetter CE (1986) Cell cycle specific replication of *Escherichia coli* minichromosomes. Proc Natl Acad Sci USA 83 : 5101-5105.

Lobner-Olesen A, Skarstad K, Hansen FG, von Meyenburg K, Boye E (1989) The DnaA protein determines the initiation mass of *Escherichia coli.* Cell 57 : 881-889

Messer W, Bellekes U, Lother H (1985) Effect of *dam* methylation on the activity of the *E. coli* replication origin, *oriC.* EMBO J 4: 1327-1332

Ogden GB, Prat MJ, Schaechter M (1988) The replicative origin of the *E. coli* chromosome binds to cell membranes only when hemimethylated. Cell 54 : 127-135

Pritchard RH (1978) Control of DNA replication in bacteria. In DNA synthesis present and future, I. Molineux and M. kohiyama (eds). (New York: Plenum Publishing Corp.), pp. 1-26

Russell DW, Zinder ND (1987) Hemimethylation prevents DNA replication in *E. coli.* Cell 50 : 1071-1079

Sekimizu K, Bramhill D, Kornberg A (1987) ATP activates DnaA protein in initiating replication of plasmids bearing the origin of replication of the *E. coli* chromosome. Cell 50 : 259-265

Sekimizu K, Yat-Ming Yung B, Kornberg A (1988) The DnaA protein of *Escherichia coli:* abundance, improved purification and membrane binding. J Biol Chem 263 : 7136-7140

Skarstad K, Boye E, Steen HB (1986) Timing of initiation of chromosome replication in individual *Escherichia coli* cells. EMBO J 5 : 1711-1717

Smith DW, Garland AM, Hermann G, Enns RE, Baker TA, Zyskind JW (1985) Importance of state of *oriC* GATC sites in initiation of DNA replication in *Escherichia coli.* EMBO J 4 : 1319-1327

Towbin H, Staehelin T, Gordon J (1979) Electrophoresis transfer of proteins from polyacrylamide gels to nitrocellulose sheets : procedure and some applications. Proc Natl Acad Sci USA 76 : 4350-4354

von Meyenburg K, Hansen FG (1987) Regulation of chromosome replication. In *Escherichia coli* and *Salmonella typhimurium*: Cellular and Molecular Biology. Neidhart FC, Ingraham JL, Low KB, Magnasnik B, Schaechter M and Umbarger HE (eds). (Washington, DC: American Society of Microbiology), pp. 1555-1577

The Involvement of cdc2 in Cell Cycle Control of DNA Replication in Xenopus Egg Extracts

J. Julian Blow and Paul Nurse
ICRF Cell Cycle Group
Microbiology Unit
Department of Biochemistry
University of Oxford
South Parks Road
Oxford OX1 3QU

DNA replication must be strictly controlled in order to coordinate it with the other major events of the cell division cycle. Two main strategies could be employed to investigate this problem. The first is to generate temperature-sensitive (cdc) mutations in yeast that are specifically defective in steps that normally control chromosomal DNA replication. Although this approach has identified a range of genes that are required for DNA replication, the absence of cell-free systems from yeast means that the precise function of these gene products often remains unknown. The second approach is to study cell-free systems that are capable of supporting the initiation of DNA replication from chromosomal (non-viral) origins of replication under cell cycle control. This has been possible following the development of a cell-free extract of Xenopus eggs (Lohka and Masui,1983) that can support the efficient initiation and completion of chromosomal DNA replication in vitro (Blow and Laskey, 1986; Blow and Watson, 1987). However, the dependence of chromosome replication on a range of basic cellular functions, such as nuclear assembly, severely complicates biochemical fractionation of replication proteins from this system .

We have attempted to exploit the strengths of both the genetic and biochemical methods with a mixed approach. By studying Xenopus homologues of yeast proteins involved in the control of DNA replication, we hope to focus biochemical attention on those proteins playing key roles in replication control in the Xenopus cell-free system. Our first target has been the cdc2 protein, which is a major regulator of the fundamental events of the yeast cell division cycle. We have shown that egg extracts depleted of a Xenopus cdc2 protein are specifically unable to initiate chromosome replication (Blow and Nurse, 1990). Depleted extracts display an arrest similar to that observed in yeast cells containing a loss-of-function mutation in this gene.

The Xenopus cell-free system

The Xenopus cell-free system that we have used is based upon the original procedure of Lohka and Masui (1983). It supports a range of cell cycle events including nuclear assembly (Lohka and Masui, 1983), DNA replication (Blow and Laskey, 1986) and progression into mitosis (Lohka and Masui, 1983, 1985; Hutchison et al, 1988). DNA replication in this system appears to be similar to that occurring in vivo (reviewed in Blow et al, 1989). Sperm chromatin, the natural substrate for DNA replication in egg cytoplasm, is replicated efficiently, with more than 70% of added molecules being replicated fully (Blow and Laskey, 1986; Blow and Watson, 1987). Purified DNA is also replicated in vitro, but with an efficiency varying between 3-30% (Blow and Laskey, 1986; Blow and Sleeman, 1990). This is similar to efficiencies that are obtained on microinjection of similar DNA templates into intact eggs (Harland and Laskey, 1980; Mechali and Kearsey, 1984). DNA undergoes a single, complete round of semiconservative replication during each cell cycle in vitro (Blow and Laskey, 1986; Blow and Watson, 1987; Blow and Sleeman, 1990). Replicated DNA must experience an early mitotic event (nuclear envelope permeabilisation may be sufficient) before a further round of replication can be initiated (Blow and Laskey, 1988; Minshull et al., 1989).

However, the initiation of DNA replication can only occur in interphase, and not in extracts arrested in mitosis (Blow and Sleeman, 1990). This is illustrated in Fig 1, which shows density substitution profiles of plasmid DNA incubated in the cell free system in either interphase (open squares), or in arrested in metaphase (closed diamonds). Replicated DNA is labelled with α^{32}P-dATP and a density marker, BrdUTP, and has been separated on CsCl density equilibrium gradients. In the interphase extract, approximately 30% of the template DNA has undergone a single complete round of semiconservative replication, resulting in a single peak of heavy/light (HL) DNA. No DNA has re-replicated (which would give a heavy/heavy peak); nor is there any sign of DNA repair, which would give a smear of label extending down from the light/light position. In contrast, no significant DNA replication is seen in the metaphase-arrested extract. Even when very high

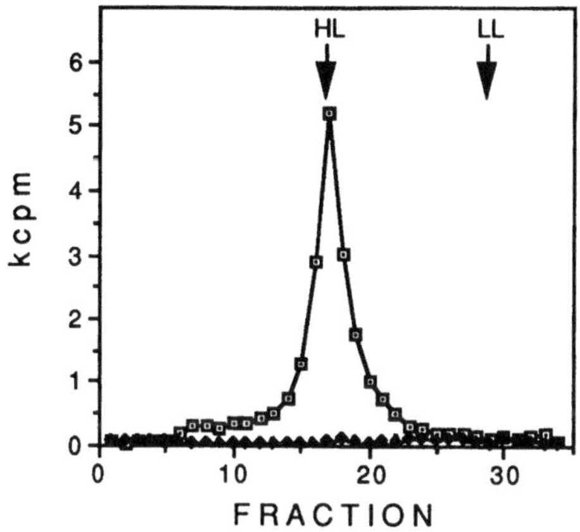

Fig. 1. Replication of purified DNA in mitotic and interphase extracts. Plasmid pUCm54 was incubated in egg extract supplemented with [α-^{32}P]dATP and BrdUTP, after which DNA was isolated and fractionated on CsCl equilibrium density gradients. (◆——◆) Extract blocked in metaphase; <1% of the plasmid DNA was replicated. (▫——▫) Extract released into interphase by the addition of 0.3mM $CaCl_2$; approximately 29% of the plasmid DNA was replicated. Vertical arrows show the expected densities of heavy/light (1.75) and light/light DNA (1.71). Reproduced from Blow and Sleeman, 1989.

concentrations of DNA are used, metaphase extracts produce no heavy/light DNA, and show only incorporation consistent with DNA repair-type nucleotide turnover (Blow and Sleeman, 1990). Although incorporation of labelled dATP is observed when very large quantities of plasmid DNA are microinjected into metaphase-arrested embryos (Newport and Kirschner,1984), this is probably due to repair-type nucleotide turnover, not DNA replication (Blow and Sleeman, 1990).

The inability of metaphase extracts to support the initiation of DNA replication can be explained at least in part by their inability to assemble DNA into interphase nuclei at this stage of the cycle. The Xenopus extract must assemble template DNA into an interphase nucleus before the initiation of DNA replication can occur (Blow and Laskey, 1986; Newport, 1987; Sheehan et al., 1988; Blow and Sleeman, 1990). A second reason is that $p34^{cdc2}$ cannot perform its essential replication function until after exit from metaphase (Blow and Nurse, 1990; see below).

The cdc2 gene function

In yeast it has been possible to identify different stages in the cell cycle by the use of temperature-sensitive mutations in genes required for specific cell cycle transitions. The point of commitment to DNA replication is called 'Start', when the yeast cell becomes unable to pursue its alternative developmental fates of conjugation, sporulation or entry into stationary phase (reviewed in Bartlett and Nurse, 1990; Nasmyth, 1990). Several genes required for performing Start have been defined in both fission yeast (*Schizosaccharomyces pombe*) and budding yeast (*Saccharomyces cerevisiae*). Interestingly two of these Start genes from the different yeasts, the *cdc2* gene from S. pombe and the *CDC28* gene from S. cerevisiae, show strong sequence homology. These two genes are also functionally homologous as they can substitute for one another in the different yeasts (Beach et al, 1982). *cdc2* has subsequently been identified in a range of eukaryotic organisms, including humans (Lee and Nurse, 1987; Norbury and Nurse, 1989). The human gene, *CDC2Hs*, can functionally substitute for the yeast gene, showing that its function has been remarkably well conserved throughout evolution. *cdc2* and *CDC28* are also unique in being required for two separate transitions in the yeast cell cycle: firstly at Start just prior to the initiation of DNA replication, and secondly at the G2/M boundary just prior to the initiation of mitosis (Nurse and Bisset, 1981; Piggot et al, 1982; Reed and Wittenberg, 1990). Its involvement in both these crucial transitions suggests that $p34^{cdc2}$, the cdc2 gene product, plays a key role in controlling both these major events of the cell cycle.

The central role of cdc2 in cell cycle control has been confirmed by the study of MPF, an activity originally identified in frog eggs that is capable of inducing all the major events of mitosis. MPF activity can be identified in a wide range of eukaryotic cells whilst they are in mitosis, and appears to be the key activator of this process in vivo (reviewed in Nurse, 1990). Purified MPF consists of $p34^{cdc2}$ and cyclin B (Lohka et al, 1988; Dunphy et al, 1988; Gautier et al, 1988; Draetta et al, 1989; Labbe et al, 1989; Gautier et al, 1990). Cyclins are rapidly degraded as cells exit from metaphase (Evans et al, 1983), and therefore must be synthesised anew during interphase of each cell cycle in order to allow cells to generate MPF and progress into mitosis (Minshull et al, 1989; Murray and Kirschner, 1989). $p34^{cdc2}$ also complexes with another protein, $p13^{suc1}$ (Hayles et al, 1986; Brizuela et al, 1987). Addition of $p13^{suc1}$ to Xenopus egg extracts inhibits MPF activity (Dunphy et al, 1988), and agarose beads with $p13^{suc1}$ covalently linked to them (suc1 beads) can deplete Xenopus egg extracts of MPF activity (Dunphy et al., 1988). This shows that $p34^{cdc2}$ plays an important role in controlling the G2/M transition in higher eukaryotes.

Depletion of $p34^{cdc2}$ from Xenopus egg extracts

We were interested to determine whether $p34^{cdc2}$ has a role in the control of DNA replication in the Xenopus cell-free system (Blow and Nurse, 1990). To do this extracts were incubated with suc1 beads to deplete them of $p34^{cdc2}$, and they were then assayed for their ability to replicate sperm chromatin (Figure 2). Addition of 30 - 40% vol:vol of suc1 beads to the extract caused an almost complete inhibition of subsequent DNA synthesis (open squares). Treatment with control beads (agarose beads without suc1 crosslinked to them) did not cause this inhibition (filled diamonds). Inhibition of DNA synthesis was not simply due to the addition of suc1 protein, since addition of high concentrations of free $p13^{suc1}$ (up to 3mg/ml) did not significantly reduce DNA synthesis (open circles). This provides a biochemical distinction between the G1/S and G2/M transitions involving cdc2, since the MPF activity of Xenopus egg extracts can be completely inhibited by addition of

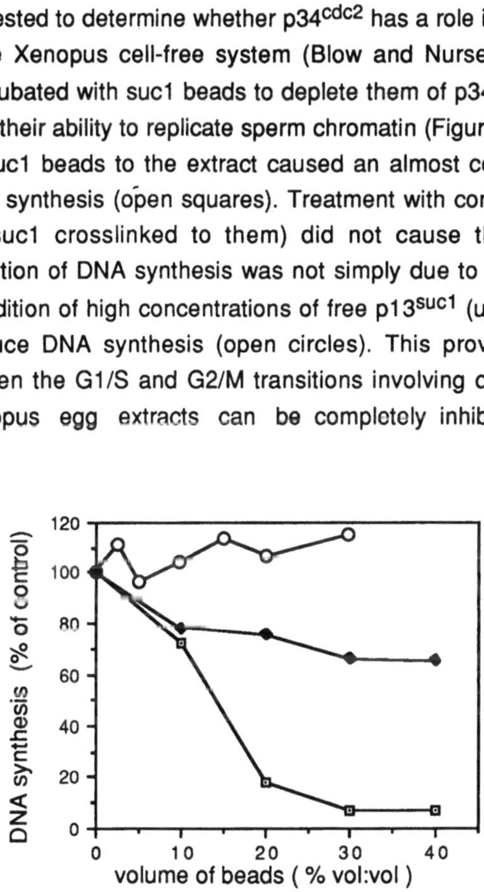

Fig. 2. Inhibition of DNA synthesis by depletion of extracts with suc1 beads. Mitotic extracts were treated with either suc1 beads (□——□), control beads (◆——◆) or 0.1 mg/ml suc1 protein (○——○), and then released into interphase by addition of Ca++. Extracts were then assayed for their ability to replicate sperm chromatin. The volume of beads added is expressed as a percentage of the extract volume to which they were subsequently added. DNA synthesis is expressed as a percentage of that obtained in an untreated control. Reproduced from Blow and Nurse, 1990.

0.03mg/ml suc1 protein (Dunphy et al, 1988). In contrast, DNA synthesis is unaffected by 100 times this concentration of suc1 (Blow and Nurse, 1990; Fig 1).

To confirm the involvement of $p34^{cdc2}$ in DNA replication in this system, we used an antiserum raised against $p34^{cdc2}$ to immunodeplete the Xenopus egg extract. The antiserum immunoprecipitates a single band of the size expected of $p34^{cdc2}$ on silver stained gels. Immunodepletion of $p34^{cdc2}$ renders the egg extract incapable of supporting DNA replication (Fig 3, open squares). In contrast, treatment with pre-immune IgG from the same rabbit only slightly inhibits subsequent DNA replication (Fig 3, filled diamonds). Mixing experiments, involving extracts affinity depleted with suc1 beads or immunodepleted extracts, demonstrate that these treatments remove the same essential replication factor from Xenopus egg extracts (Blow and Nurse, 1990). Given the specificity of both these reagents, this factor is almost certainly $p34^{cdc2}$, although the possibility that it is another protein closely related to $p34^{cdc2}$ cannot be completely ruled out at present.

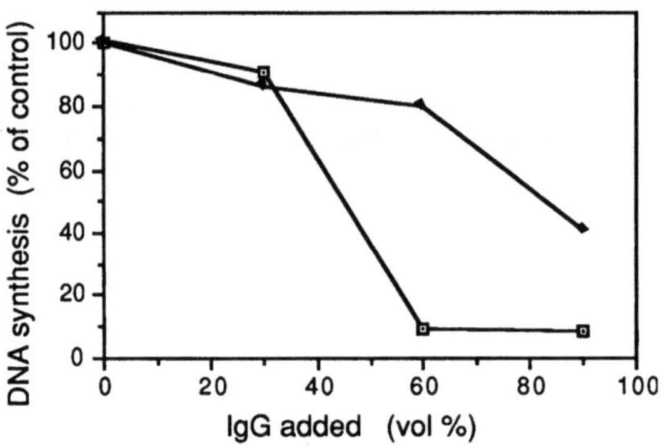

Fig. 3. Inhibition of DNA synthesis by immunodepletion of $p34^{cdc2}$.
Extracts were immunodepleted with either pre-immune IgG (◆——◆) or anti-$p34^{cdc2}$ IgG (▫——▫). Extracts were assayed for their subsequent ability to replicate sperm chromatin. DNA synthesis is expressed as a percentage of that attained when buffer was added instead of serum. The volume of IgG added is expressed as a percentage of the volume of extract to which it was added. Reproduced from Blow and Nurse, 1990.

cdc2 is required for initiation but acts soon after exit from metaphase

In yeast, the G1/S function of $p34^{cdc2}$ is performed as cells pass through Start and become committed to another round of DNA replication. Although this function may be performed early in the cycle (Novak and Mitchison, 1989), it is not required for exit from mitosis and nuclear assembly. A similar conclusion can be drawn in the Xenopus system, since depletion of $p34^{cdc2}$ does not prevent the extract from assembling interphase nuclei (Blow and Nurse, 1990). In addition, egg extracts depleted of $p34^{cdc2}$ can still support the elongation of replication forks that have already been initiated. This shows that in Xenopus, as in yeast, the $p34^{cdc2}$ function is required for the initiation of DNA replication itself, but not for the elongation of previously initiated forks (Blow and Nurse, 1990).

Fig 4 shows that $p34^{cdc2}$ performs its essential G1/S function soon after exit from metaphase. When Ca^{++} is added to a metaphase extract, it enters interphase and starts the series of steps leading to the initiation of DNA replication. The extract can

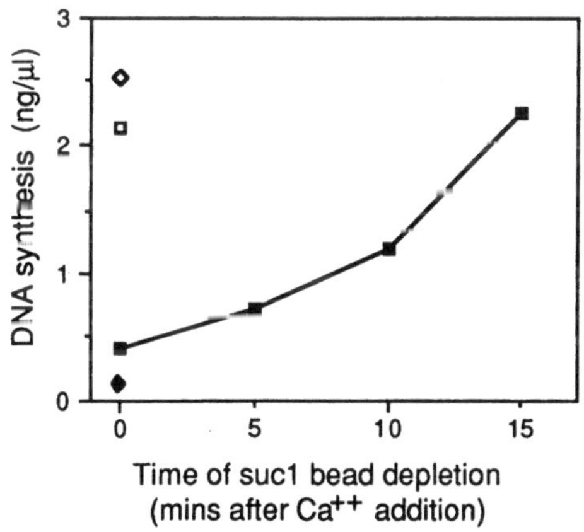

Fig. 4. Suc1 bead depletion at different times after exit from metaphase.
Sperm nuclei were incubated with α^{32}P-dATP in extracts for various times after Ca++ release from metaphase. Suc1 beads (■ , ◆) or control beads (□ , ◇) were added at various times after this, and allowed to deplete the extract at 4°C before the incubation was resumed at 25°C for 3hr. The final extent of DNA synthesis was then assessed. (■ , □): DNA and Ca++ were added to extract at the same time. (◆ , ◇): DNA was preincubated in mitotic extract for 20mins at 25°C before Ca++ addition and treatment with beads. Reproduced from Blow and Nurse, 1990.

be depleted with suc1 beads at various times after Ca^{++} release in order to determine the time at which the cdc2-requiring step has been completed. Fig 4 (filled squares) shows that after only 15mins in interphase, suc1 bead depletion has become totally incapable of inhibiting subsequent DNA synthesis. 15mins is well before the initiation of DNA replication, which does not start until about 45min after Ca^{++} release (Blow and Laskey, 1986; data not shown). It is also before the first signs of nuclear envelope formation, which are not visible until about 30mins after Ca^{++} release.

However, the cdc2-requiring step cannot be performed during metaphase. If DNA is incubated in mitotic extract for 15min before depletion with suc1 beads and Ca^{++} release, subsequent DNA synthesis is efficiently inhibited (Fig 4, filled diamond). These results demonstrate that the cdc2-requiring step is normally executed very soon after the Xenopus egg exits from meiotic metaphase into the first mitotic interphase.

As a component of MPF, $p34^{cdc2}$ plays a central role in the induction of mitosis (Dunphy et al, 1988; Gautier et al, 1988). Our results also show that it has an important role in the initiation of DNA replication (Blow and Nurse, 1990). The

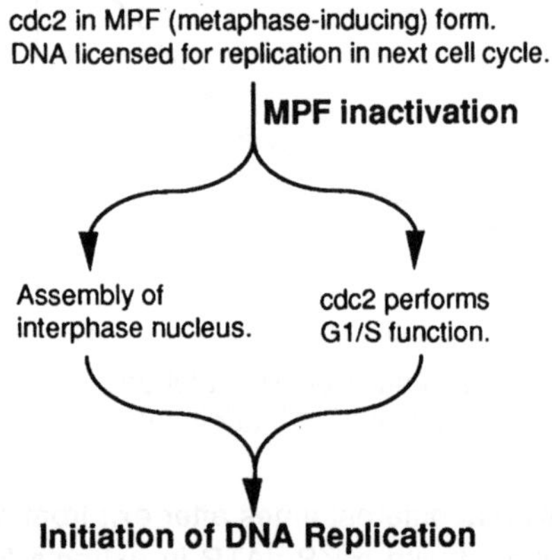

Fig. 5. Cartoon of the involvement of cdc2 in the initiation of DNA replication. When MPF is inactivated at the end of mitosis, two distinct events occur that are both required for the subsequent initiation of DNA replication: a) chromosomes decondense and are assembled into an interphase nucleus, and b) cdc2 can execute its G1/S function. Reproduced from Blow and Nurse, 1990.

involvement of cdc2 in these two different cell cycle transitions suggests that it controls DNA replication in the Xenopus egg extract in at least two distinct ways (Fig 5). Firstly, replicated nuclei must be disassembled in mitosis under the action of MPF in order for them to be licensed for a single round of replication in the next cell cycle (Blow and Laskey, 1988; Minshull et al., 1989). However, the initiation of replication depends on the subsequent assembly of template DNA into an interphase nucleus (Blow and Sleeman, 1990). This requires that the metaphase-inducing MPF activity of $p34^{cdc2}$ be turned off. Secondly, we demonstrate here that after release from metaphase, $p34^{cdc2}$ or a very closely related protein is required prior to the initiation of DNA replication, even in the presence of apparently normal nuclear assembly. The nature of this interphase activity of $p34^{cdc2}$ is currently unknown. Although the two different pathways leading to nuclear assembly and the execution of the G1/S function of $p34^{cdc2}$ can apparently occur independently of one another, both are required before DNA replication can be initiated (Fig 5).

References

Bartlett, R. and Nurse, P. (1990). Yeast as a model system for understanding the control of DNA replication in eukaryotes. Bioessays 12, 457-463.

Beach, D., Durkacz, B. and Nurse, P. (1982). Functionally homologous cell cycle control genes in budding and fission yeast. Nature 300, 682-687.

Blow, J.J. and Laskey, R.A. (1986). Initiation of DNA replication in nuclei and purified DNA by a cell-free extract of Xenopus eggs. Cell 47, 577-587.

Blow, J.J. and Laskey, R.A. (1988). A role for the nuclear envelope in controlling DNA replication within the cell cycle. Nature 332, 546-548.

Blow, J.J. and Nurse, P. (1990). A cdc2-like protein is involved in the initiation of DNA replication in Xenopus egg extracts. Cell 62, 855-862.

Blow, J.J., Sheehan, M.A., Watson, J.V. and Laskey, R.A. (1989). Nuclear structure and the control of DNA replication in the Xenopus embryo. J. Cell Sci. suppl. 12, 183-195.

Blow, J.J. and Sleeman, A.M. (1990). Replication of purified DNA in Xenopus egg extract is dependent on nuclear assembly. J. Cell Sci. 95, 383-391.

Blow, J.J. and Watson, J.V. (1987). Nuclei act as independent and integrated units of replication in a Xenopus cell-free system. EMBO J. 6, 1997-2002.

Brizuela, L., Draetta, G. and Beach, D. (1987). $p13^{suc1}$ acts in the fission yeast cell cycle as a component of the $p34^{cdc2}$ protein kinase. EMBO J. 6, 3507-3514.

Draetta, G., Luca, F., Westendorf, J., Brizuela, L., Ruderman, J. and Beach, D. (1989). cdc2 protein kinase is complexed with both cyclin A and B: evidence for proteolytic inactivation of MPF. Cell 56, 829-838.

Dunphy, W.G., Brizuela, L., Beach, D. and Newport, J. (1988). The Xenopus cdc2 protein is a component of MPF, a cytoplasmic regulator of mitosis. Cell 54, 423-431.

Evans, T., Rosenthal, E.T., Youngblom, J., Distel, D. and Hunt, T. (1983). Cyclin: a protein specified by maternal mRNA in sea urchin eggs that is destroyed at each cleavage division. Cell 33, 389-396.

Gautier, J., Norbury, C., Lohka, M., Nurse, P. and Maller, J. (1988). Purified

maturation-promoting factor contains the product of a Xenopus homolog of the fission yeast cell cycle control gene cdc2+. Cell 54, 433-439.
Harland, R.M. and Laskey, R.A. (1980). Regulated DNA replication of DNA microinjected into eggs of Xenopus laevis. Cell 21,
Hayles, J., Beach, D., Durkacz, B. and Nurse, P. (1986). The fission yeast cell cycle control gene cdc2: isolation of a sequence suc1 that suppresses cdc2 mutant function. Mol. gen. Genet. 202, 291-293.
Hutchison, C.J., Cox, R. and Ford, C.C. (1988). The control of DNA replication in a cell-free extract that recapitulates a basic cell-cycle in vitro. Development 103, 553-566.
Labbe, J., Capony, J., Caput, D., Cavadore, J., Derancourt, J., Kaghad, M., Lelias, J., Picard, A. and Doree, M. (1989). MPF from starfish oocytes at first meiotic metaphase is a heterodimer containing one molecule of cdc2 and one molecule of cyclin B. EMBO J. 8, 3053-3058.
Lee, M.G. and Nurse, P. (1987). Complementation used to clone a human homologue of the fission yeast cell cycle control gene cdc2. Nature 327, 31-35.
Lohka, M.J., Hayes, M.K. and Maller, J.L. (1988). Purification of maturation-promoting factor, an intracellular regulator of early mitotic events. Proc. Natl. Acad. Sci. USA 85, 3009-3013.
Lohka, M.J. and Masui, Y. (1983). Formation in vitro of sperm pronuclei and mitotic chromosomes induced by amphibian ooplasmic contents. Science 220, 719-721.
Lohka, M.J. and Masui, Y. (1985). Effect of Ca2+ ions on the formation of metaphase chromosomes and sperm pronuclei in cell-free preparations from unactivated Rana pipiens eggs. Devl. Biol. 103, 434-442.
Mechali, M. and Kearsey, S. (1984). Lack of specific sequence requirement for DNA replication in Xenopus eggs compared with high sequence specificity in yeast. Cell 38, 55-64.
Minshull, J., Blow, J. and Hunt, T. (1989). Translation of cyclin mRNA is necessary for extracts of activated Xenopus eggs to enter mitosis. Cell 56, 947-956.
Murray, A.W. and Kirschner, M.W. (1989). Cyclin synthesis drives the early embryonic cell cycle. Nature 339, 275-280.
Nasmyth, K. (1990). FAR-reaching discoveries about the regulation of START. Cell 63, 1117-1120.
Newport, J. (1987). Nuclear reconstitution in vitro: stages of assembly around protein-free DNA. Cell 48, 205-217.
Newport, J.W. and Kirschner, J.W. (1984). Regulation of the cell cycle during early Xenopus development. Cell 37, 731-741.
Norbury, C.J. and Nurse, P. (1989). Control of the higher eukaryote cell cycle by p34cdc2 homologues. Biochim. Biophys. Acta 989, 85-95.
Novak, B. and Mitchison, J.M. (1989). The first transition point of the mutant cdc2.33 in the fission yeast Schizosaccharomyces pombe. J. Cell Sci. 94, 657-662.
Nurse, P. (1990). Universal control mechanism regulating onset of M-phase. Nature 344, 503-508.
Nurse, P. and Bissett, Y. (1981). Gene required in G1 for commitment to the cell cycle and in G2 for control of mitosis in fission yeast. Nature 292, 558-560.
Piggott, J.R., Rai, R. and Carter, B.L.A. (1982). A bifunctional gene product involved in two phases of the yeast cell cycle. Nature 298, 391-394.
Reed, S.I. and Wittenberg, C. (1990). Mitotic role for the CDC28 protein kinase of Saccharomyces cerevisiae. Proc. Natl. Acad. Sci. USA 87, 5697-5701.
Sheehan, M.A., Mills, A.D., Sleeman, A.M., Laskey, R.A. and Blow, J.J. (1988). Steps in the assembly of replication-competent nuclei in a cell-free system from Xenopus eggs. J. Cell Biol. 106, 1-12.

A Gene of *Saccharomyces cerevisiae* Probably Involved in the Coordination of Chromatin Replication and Cell Division

Ulrike Wintersberger
Anneliese Karwan
Department of Molecular Genetics
Institute for Tumorbiology
and Cancer Research
University of Vienna
Borschkegasse 8a
1090 Wien
Austria

Introduction

The exact duplication of the genetic material and its perfect segregation between the two products of cell division has to be tightly coordinated during the cell cycle. A number of genes involved in the orderly progression of events necessary for cell division have been identified in the yeast, Saccharomyces cerevisiae, by screening for mutants which under certain restrictive conditions (e.g. high temperature) stop to proliferate. Such cell division cycle (cdc) mutants arrest at certain stages of the cell cycle and show a "terminal phenotype" indicating which of the sequencial steps the mutated gene might be resposible for (for review see Pringle and Hartwell, 1981). In addition to genes and proteins absolutely essential for genome and cell duplication, the existence of functions bringing about valuable but not vital improvements for the precision of the process has to be assumed. Such genes will not be picked up by screening for conditionally lethal mutations.

We have fortuitously isolated and sequenced a gene which, as disclosed by the genetic study described below, is involved in the maintenance of the exact ploidy of S. cerevisiae cells. The gene, which was named SCP(160) for S. cerevisiae gene for control of ploidy, codes for a polypeptide visible in immunoblots as a band of molecular weight around 160.000 (manuscript in preparation). The deduced amino acid sequence of the SCP(160) protein does not reveal any significant homologies to proteins of known functions except for several weak similarities over short stretches. Therefore no conclusions for its biological role could be drawn from the sequence data and analyses of the properties of deletion mutants were undertaken in order to get information about the function of SCP(160).

Materials and Methods

Yeast strains : All strains used in the genetic crosses descend from the two diploids, HRI 525 (a/α, ura3-52/ura3-52, leu2-3, 112/leu2-3,112, his3/his3, trp1/trp1) and JRY (a/α, ura3-52/ura3-52, his4-519/his4-519, leu2/leu2) originally obtained from H. Altmann (Basel). As will be described in detail elsewhere (manuscript in preparation) different parts of the SCP(160) gene were deleted and replaced by marker gene insertions in one allele each of the diploid strains. Thus, four heterozygous deletion mutants were obtained, called in the following HRI-M1(ΔC) [like HRI 525, but SCP(160)/ scp(160)::LEU2], HRI-M2(ΔN) [like HRI 525, but SCP(160)/ scp(160)::URA3], HRI-M4(ΔT) [like HRI 525, but SCP(160)/ scp(160)::LEU2], and JRY-M1(ΔC) [like JRY, but SCP(160)/ scp(160):LEU2]. The designations (ΔC), (ΔN) and (ΔT) indicate that part of the C-terminal half or of the N-terminal half, or the total gene was deleted, respectively. The haploid strains MZ3 α, pep4-3, trp1, leu2 (Δ400), ura3 (Δ200), and MZ3/K4 (ΔC) [like MZ3, but Δ scp(160)::LEU2] were constructed starting from strain 20-B12 (Yeast Stock Center) by Markus Zwickl in our laboratory. Strain RC43 (a, his1-7, hom3-10, ade2-1, trp5-48, arg4-17, lys1-5) was obtained from F. Fabre (Paris).

Genetic crosses, sporulation, and dissection of asci were carried out according to standard methods. Growth curves and fractions of colony forming cells were determined as described previously (Klein et al., 1989), and pedigree analyses were done exactly as in earlier studies (Wintersberger and Karwan, 1987; Klein et al., 1990). The amount of DNA was determined according to Burton (1968).

Results and Discussion

Deletion of the SCP(160) gene is not lethal but vitality of mutants is decreased

Three kinds of deletions of the SCP(160) gene were constructed and used to transform two different diploid yeast strains (manuscript in preparation). According to Southern blot analyses the transformants contained one truncated and one wild type allele. In order to find out whether the SCP(160) gene was essential for survival the diploids were sporulated and the expected haploid descendents, now possessing only one allele of the SCP(160) gene, namely a wild type or a deleted one, were tested for viability. As seen from Table 1 several asci containing four viable spores were obtained from all mutants tested, the fraction of viable descendents was found, however, somewhat decreased compared to that from the untransformed cells, especially in the genetic background of the strain HRI 525. Thus, deletion of parts of or the total SCP(160) gene is not lethal under laboratory conditions.

Deletion-mutants are hypersensitive to starvation but not to heat shock

The clones descending from the spores were tested for segregation of markers and (by Southern blotting) for the presence of the wild type or the mutated allele, designated SCP(160) and scp(160), respectively. While handling these clones we observed that the viability of the mutants was considerably lower than that of their siblings containing the wild type gene. This property was most obvious when the cells were starved during prolonged stationary phase. In Fig.1 growth

Table 1: **VIABILITY OF SPORES DESCENDING FROM DIPLOID CELLS CONTAINING ONE DELETED ALLELE OF THE SCP(160) GENE**

Diploid strain[1]		Asci[2]	Viable spores per ascus 4	3	2	1	0	Total number of spores viable/isolated
Original strains								
	SCP(160)							
HRI525	+/+	10	7	1	2	0	0	33/40
JRY	+/+	10	9	0	1	0	0	38/40
Trans-formants								
	SCP(160)							
HRI-M1(ΔC)	+/-	28	16	2	8	2	0	88/112
JRY-M1(ΔC)	+/-	10	9	0	1	0	0	38/40
HRI-M2(ΔN)	+/-	20	4	2	10	2	2	44/80
HRI-M4(ΔT)	+/-	11	6	4	1	0	0	38/44

1 For detailed description of the original diploid strains and the heterozygous transformants see Materials and Methods.

2 Numbers of asci dissected.

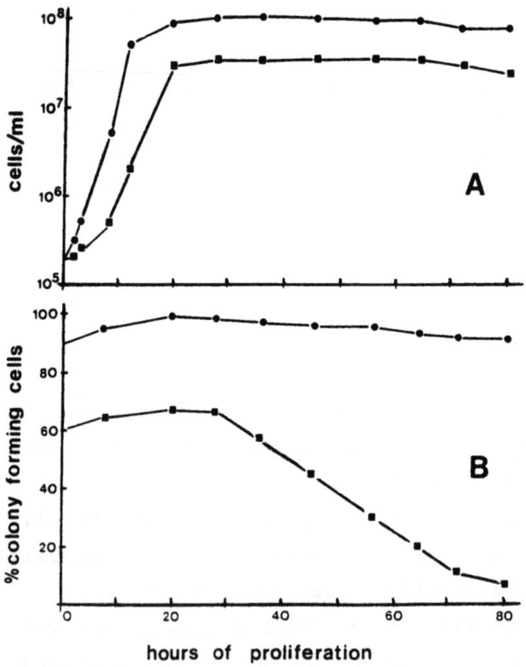

Fig 1. Growth curves (A) and colony forming abilities (B) of spore clones descending from strain HRI M1(ΔC), (●) SCP(160), and (■) scp(160), as determined on full medium.

curves together with colony forming abilites of two spore clones, both descending from HRI-M1 (ΔC), one scp(160) and the other SCP(160), are depicted. Whereas during logarithmic growth about 60%, and somewhat more in other experiments, of the cells from the mutant clone were able to develop into visible colonies the fraction of colony forming individuals dropped dramatically during the period of starvation. As this was not the case for the SCP(160) clone, and similar results were obtained also for other pairs of spore clones we had to assume that the hypersensitivity to starvation was caused by the loss of a function exhibited by the SCP(160) protein.

Hypersensitivity to starvation is typical for cells having lost a function necessary for entering a resting state, the G_1 or G_0 phase of the cell cycle. Studies of others have shown that such cells do not become heat shock resistent when reaching stationary phase, as it is the case for healthy S. cerevisiae cells. A well studied example for a mutation rendering yeast cells hypersensitive to starvation and at the same time to heat shock is the $RAS2^{Val19}$ mutation (Sass et al., 1986). The RAS2 protein is part of the pathway transporting information about external nutritional conditions to cellular functions regulating growth and division. We thought to get information of whether the SCP(160) protein might be involved in the same pathway by testing our mutants for heat shock sensitivity. We observed, however, that during early stationary phase, the viable fraction of a population of scp(160)-mutants was not more sensitive to heat shock than wild type cells. For example, sibling populations descending from strain HRI Mutant 4(ΔT) survived a 20 min period of incubation at 50°C up to around 30% if taken from an early stationary phase culture, but only to less than 0.1% if taken from a logarithmically growing culture, irrespectively of whether the cells contained an intact or a truncated copy of the SCP(160) gene (data not shown). Hence, functions responsible for saving non-growing cells from killing by elevated temperature are not defective in scp(160)-mutants, and unability to cope with starvation in our mutants is not coupled with heat shock sensitivity.

Deletion of the SCP(160) gene destroyes the control of ploidy

Diploid cells unable to endure starvation often were found defective also in sporulation. We have crossed mutated as well as normal cells descending from the spore clones with each other in order to get different kinds of diploid progeny. The clones resulting from the crosses were able to sporulate irrespectively of whether they did or did not contain a wild type allele of the SCP(160) gene, although the viability of spores descending from mutants was much lower than of those resulting from the crosses between two SCP(160) positive cells (Table 2). To our surprise, in asci descending from SCP(160)/scp(160) as well as from scp(160)/scp(160) zygotes, the mating types of the spore clones did not segregate regularely and many clones unable to mate were obtained.

Table 2: **VIABILITY AND MATING BEHAVIOUR OF CLONES ORIGINATING FROM CROSSES BETWEEN THE DESCENDENTS OF THE SCP(160)/scp(160) STRAINS**

Mutant strain from which clones originate[1]		Asci[2]	Viable spores per ascus					Total number of spores viable/ isolated	Mating behaviour of spore clones		
			4	3	2	1	0		a	alpha	nm[3]
HRI-M1 (ΔC)											
SCP(160)	+/+	10	9	0	1	0	0	38/40	19	19	0
	+/-	20	8	2	1	2	7	42/80	19	23	0
	-/-	40	7	3	7	10	13	61/160	20	24	17
JRY-M1 (ΔC)											
SCP(160)	+/+	10	8	2	0	0	0	38/40	19	19	0
	+/-	60	11	7	12	13	17	102/240	49	42	11
	-/-	40	23	8	4	5	0	129/160	37	34	58
HRI-M2 (ΔN)											
SCP(160)	+/+	10	8	2	0	0	0	38/40	19	19	0
	+/-	10	7	1	2	0	0	35/40	17	18	0
	-/-	30	2	1	6	9	12	32/120	17	1	14

[1] Clones descending from spores obtained by dissecting the transformed strains HRI-M1(ΔC), JRY-M1(ΔC) and HRI-M2(ΔN) were crossed among each other.
[2] Numbers of asci dissected.
[3] nm, nonmaters: cells unable to mate, many of which sporulate.

Furthermore a number of these nonmaters were able to sporulate and even to produce viable spores themselves (data not shown), indicating that in these cells not only a- as well as α-information was expressed but that the deletion of the SCP(160) gene had entailed a defect in the control of the ploidy of the cells. It is important to note that the patterns of mating type segregation were totally irregular and different from zygote to zygote and from ascus to ascus. Thus, some of the expectedly haploid scp(160) spore clones employed in the crosses must have been diploid and others probably aneuploid with respect to various parts of the genome. In crosses between scp(160) mutants and unrelated strains low spore viability and irregular segregation also of genetic markers other than mating types was observed (data not shown).

In the light of the genetic data we determined the DNA content per cell of a number of different expectedly haploid and diploid mutant cells. In Table 3 results for clones descending from the originally transformed heterozygotes are shown. It is obvious that the DNA content of the scp(160) clones and even that of the originally transformed SCP(160)/scp(160) heterozygotes is increased compared to normal cells. To be sure that this defect was independent of meiosis we constructed a deletion mutant starting from a haploid strain (MZ3), and we again obtained a clone with an abnormally high DNA content (MZ3/K4, last line of Table 3).

What kind of defect could be responsible for the observed mutant phenotype? (i) The increased amount of DNA per cell could result from overinitiation of chromosomal replication during the S-phase of the cell cycle leading to DNA amplification. (ii) A defect in chomosomal segregation or, in S. cerevisiae, in the division of the nuclei containing the replicated chromosomes, could result in cell pairs of which one individual might contain part of the chromosomes in duplicate whereas the other might miss several chromosomes or parts thereof. (iii) Another possibility would be that partitioning of the divided nuclei between the mother and the daughter cell

Table 3: **DNA CONTENT OF YEAST STRAINS CONTAINING TRUNCATED SCP(160) GENES**

Strain[1]		SCP(160)	DNA content of early stationary phase cells ($\mu g/10^8$ cells)
Original strain HRI 525, a/alpha		+/+	5.2
HRI-M1(ΔC), a/alpha		+/-	7.7
Tetrade(R)	A(R) a	+	2.3
	B(R) alpha	+	2.3
	C(R) a	-	3.1
	D(R) alpha	-	5.5
HRI-M2(ΔN), a/alpha		+/-	6.6
Tetrade(P)	A(P) a	+	2.8
	B(P) alpha	+	2.6
	C(P) a	-	10.1
	D(P) alpha	-	10.9
HRI-M4(ΔT), a/alpha		+/-	5.5
Tetrade(II)	A (II) alpha	-	3.5
	B (II) a	-	5.7
	C (II) alpha	+	2.5
	D (II) a	+	2.8
Crosses between descendents of mutants			
HRI-M1	A(R)xB(R)	+/+	4.9
HRI-M1	B(R)xC(R)	+/-	9.4
HRI-M1	C(R)xD(R)	-/-	10.6
HRI-M2	A(P)xB(P)	+/+	4.8, 5.3
HRI-M2	B(P)xC(P)	+/-	6.8
HRI-M2	C(P)xD(P)	-/-	13.6, 12.2
HRI-M1,D(R) x RC43, a/alpha		-/+	6.8
MZ3 (pep4-3, trp1, Δura3, Δleu2) alpha		+	2.6
MZ3/K4(ΔC) alpha		-	8.7

[1] For description of strains see Materials and Methods, and Tables 1 and 2.

would not work properly. (iv) The phenotype could also result from a defective timing of cell division such, that the next round of DNA replication is underway before the preceeding cell separation was completed.

Considering possibility (ii) one could explain the large fraction of mutant cells not able to develop into viable colonies (see Fig.1) as consisting of individuals which during cell division did not receive the full set of chromosomes. To get further information of whether missegregation of chromosomes might be the cause of the mutant phenotype we tested the mutants for hypersensitivity against caffeine, because mutants of Schizosaccharomyces pombe defective in chromosome segregation were recently shown to be hypersensitive to caffeine (Ohkura et al., 1988). In a preliminary spot-test on solid medium, 4mM caffeine indeed inhibited growth of the scp(160)-mutants. This phenomenon was studied in more detail by pedigree analyses.

SCP(160) deletion mutants are hypersensitive to killing but not to growth retardation by caffeine

In previous studies we successfully used pedigree analyses to follow the fate of progeny of individual cells treated with DNA damaging agents (Klein et al., 1990). We showed that pedigree analyses (see Fig.2 for our mode of documentation of results) allow discrimination between mother and daughter cells concerning (i) their viability and (ii) the time periods elapsing between two landmarks of the S. cerevisiae cell cycle: the time needed for bud emergence (including the G1 phase and the initiation of DNA replication, period (a) of Fig.2), and the period between bud emergence and cell separation (including progression and termination of DNA replication and mitosis, period (b) of Fig.2). Therefore, by subjecting scp(160)-mutants (and wild type cells as controls) to pedigree analyses we expected answers to the following questions: (i) Is there a difference in viability between mothers and daughters? Do,

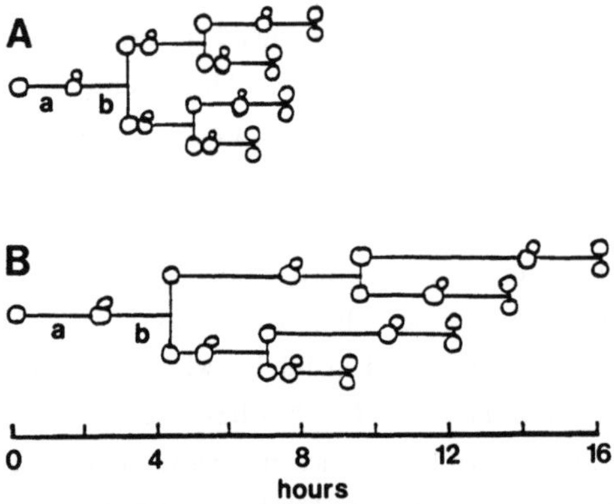

Fig.2. Pedigrees of individual cells observed for three generations. Large circles, mother-cells; small circles, daughter-cells; they were separated by micromanipulation as described earlier (Wintersberger and Karwan, 1987); (a), period elapsing until bud emergence; (b), period between bud emergence and cell separation. The time periods depicted represent the average of ten pedigrees each for SCP(160) cells [strain HRI-M2(ΔN)A(P), SCP(160)] growing in absence (A), or in presence (B) of 4mM caffeine.

Table 4: **SYNERGISM BETWEEN KILLING EFFECTS OF LOSS OF THE SCP(160) GENE AND GROWTH IN THE PRESENCE OF CAFFEINE, AS OBSERVED BY PEDIGREE ANALYSES.**

Strain[1]	Caffeine concentration	Number of cells employed in the pedigree analyses		Colony forming cells after 3 generations obtained/ expected
		cells at start	cells producing viable progeny	
HRI-M2 (ΔN),A(P) SCP(160) clone	0	20	20	156/160
	4mM	20	16	98/160
HRI-M2 (ΔN),C(P) scp(160) clone	0	10	7	32/80
	2mM	30	7	26/240
	4mM	20	2	6/160

[1] Strain designation as in Table 3

e.g., daughter cells die preferentially because transfer of nuclear material to the developing buds might be defective? (ii) Is one of the two periods of the cell cycle which we can determine by the pedigree analyses prolonged, as mutant cells might have difficulties to perform a certain step properly on time?

One of our assumptions was that a defect in chromosome segregation might show off as a retardation of mitosis. However period (b) (Fig.2) which includes the time cells need for mitosis was not extended during the cell cycle of the scp(160)-mutants compared to that of the wild type cells. Interestingly, period (b) was also not extended for normal cells growing on medium containing 4mM caffeine. The drug, in our experiments did, however, cause a considerable prolongation of period (a) exclusively for daughter cells (Fig.2). Growth-time of scp(160) daughter cells did not differ from that of normal cells, irrespectively of whether they grew in presence or absence of caffeine. Hence, the time needed for progression through the cell cycle is equal for scp(160)-mutants and wild type cells. The reason for the daughter specific retardation effect of caffeine could be that the growth of the new born cells to the size necessary for starting the next cell cycle might be prolonged in presence of the drug. As mothers have the right size already at the moment of cell division they obviously are insensitive against the retardation effect.

Table 4 summerizes the results concerning the viability of wild type and mutant cells observed with the pedigree analyses. If we follow cell divisions over three generations we expect for a healthy cell to produce 8 third generation colonies (see Fig.2). This was found to be the case for most cell families emerging from haploid SCP(160) positive individuals (first line of Table 4). Cells containing a truncated SCP(160) gene produced a considerable fraction of progeny not able to divide further (third line of Table 4) as to be anticipated (Fig.1), and less than half of the expected number of third generation colonies was obtained. It should be noted that inviable cells occurred in every generation such that some family trees ended up with a single third generation colony! No preference of dying was observed, neither for mothers nor for daughters. Those cells which were unable to divide further quickly lysed, a behaviour different from that of many well studied conditionally lethal cell cycle mutants. In the presence of 4mM caffeine SCP(160) positive cells produced similar family trees as scp(160) mutants in the absence of the drug (second line of Table 4). As on 4mM caffeine containing medium two of 20 scp(160)-mutants only, started to divide, in total producing 6 third generation colonies (fifth line of Table 4) we analyzed the divisions of mutants in presence of 2mM caffeine (forth line of Table 4). In every generation a high fraction of cells (mothers as well as daughters) did not divide further but lysed. Thus, the deletion of the SCP(160) gene and caffeine affect cells synergistically.

Concluding Remarks

Although the results reported support the idea that the SCP(160) protein might be involved in the control of chromosome distribution between mother and daughter cells, chromosome missegregation alone cannot fully explain the phenotype of the scp(160) deletion mutants. Why should cells be more prone to chromosome missegregation during starvation than during logarithmic growth? Could the synergism between caffeine (which is known to inhibit cellular phosphodiesterases) and loss of SCP(160) function indicate an involvement of the SCP(160) protein in a cyclic-nucleotide dependent signal transduction pathway, as we originally assumed from the starvation sensitivity of our mutants? May such a pathway regulate the coupling between segregation of the duplicated chromatin and cell division? We have recently isolated a mutant from an scp(160) strain in which the ploidy defect is suppressed. We hope that studying this mutant may help to find the regulatory pathway in which the SCP(160) protein participates.

Acknowledgements

We are grateful to Gabriela Kretschmer for carrying out the experiment of Fig.1 and to Aloisia Kopp for help with the preparation of the manuscript. The work was supported by the Fonds zur Förderung der wissenschaftlichen Forschung in Österreich (Project number 7209med).

References

Burton K (1968) Determination of DNA concentration with diphenylamin. Meth Enzymol 12:163-166

Klein F, Karwan A, Wintersberger U (1989) After a single treatment with EMS the number of non-colony-forming cells increases for many generations in yeast populations. Mutat Res 210:157-164

Klein F, Karwan A, Wintersberger U (1990) Pedigree analysis of yeast cells recovering from DNA damage allow assignment of lethal events to individual post-treatment generations. Genetics 124:57-65

Ohkura H, Adachi Y, Kinoshita N, Niwa O, Toda T, Yanagida M (1988) Cold-sensitive and caffeine-supersensitive mutants of the Schizosaccharomyces pombe dis genes implicated in sister chromatid separation during mitosis. EMBO J 7:1465-1473

Pringle JR, Hartwell LH (1981) The Saccharomyces cerevisiae cell cycle. In: Strathern JN, Jones EW, Broach JR (eds) The molecular biology of the yeast Saccharomyces. Cold Spring Harbor Laboratory, Cold Spring Harbor, NY pp97-142

Sass P, Field J, Nikawa J, Toda T, Wigler M (1986) Cloning and characterization of the high-affinity cAMP phosphodiesterase of Saccharomyces cerevisiae. Proc Natl Acad Sci USA 83:9303-9307

Wintersberger U, Karwan A (1987) Retardation of cell cycle progression in yeast cells recovering from DNA damage: A study at the single cell level. Mol Gen Genet 207:320-327

Origins of DNA Replication and Nucleosome Segregation in Mammalian Chromosomes

William C. Burhans, Lyubomir T. Vassilev, Jiarui Wu[1], José M. Sogo[1],
Ferez S. Nallaseth and Melvin L. DePamphilis
Department of Cell and Developmental Biology
Roche Institute of Molecular Biology
Roche Research Center, Nutley, NJ 07110, USA.

Initiation of DNA replication in bacterial, plasmid, bacteriophage, yeast, mitochondrial and mammalian viral genomes occurs at specific cis-acting DNA sequences referred to as origins of replication (Umek et al., 1989). Nevertheless, some studies in higher eukaryotes suggest that initiation of replication in cellular chromosomes may not require origin of replication sequences. Replication bubbles in the early cleavage embryos of flies and newts are at least 5 times more frequent than in the differentiated cells of the same animal, and most DNA molecules replicate when injected into non-mammalian eggs (reviewed in Kaiserman et al., 1989; DePamphilis, 1991). Replication of DNA transfected into cultured mammalian cells has been found, in some cases, to depend on DNA length rather than DNA sequence (Heinzel et al., 1991; Krysan and Calos, 1991). Furthermore, the same 2-dimensional gel electrophoresis techniques that identified unique origins of replication in yeast plasmids and chromosomes have detected what appears to be random initiation of bidirectional replication throughout a 30 kb region downstream of the dihydrofolate reductase (DHFR) gene in Chinese hamster ovary (CHO) cells (Vaughn et al., 1990).

On the other hand, these conclusions are contradicted by other studies. Injection of DNA into mouse embryos does not replicate unless it contains a functional origin (Wirak et al., 1985; Martinez-Salas et al., 1988; DePamphilis et al., 1988). In some cases, replication of DNA transfected into cultured mammalian cells also has been found to depend on DNA sequence (Frappier and Zannis-Hadjopoulos, 1987; McWhinney and Leffak, 1990), and, as discussed below, nascent DNA synthesis has been observed to begin at specific chromosomal loci. Thus, the question remains: do the chromosomes of higher eukaryotic organisms initiate replication at specific DNA sequences?

[1]Institute of Cell Biology, ETH-Hönggerberg, 8093 Zürich, Switzerland.

One approach to resolving this problem has been to map the genomic locations for initiation of replication in the chromosomes of higher eukaryotes. Many studies of the region downstream of the DHFR gene in CHO cells concluded that DNA synthesis began bidirectionally approximately 17 kb downstream from the DHFR gene (Fig. 1). However, interpretation of these results in terms of a discrete origin of replication is complicated by the fact that these studies used metabolic inhibitors to synchronize the cells, and were carried out with CHOC 400 cells, a derivative of CHO cells in which the DHFR gene region is amplified 1000-fold. In addition, none of them established the mechanism by which replication occurred.

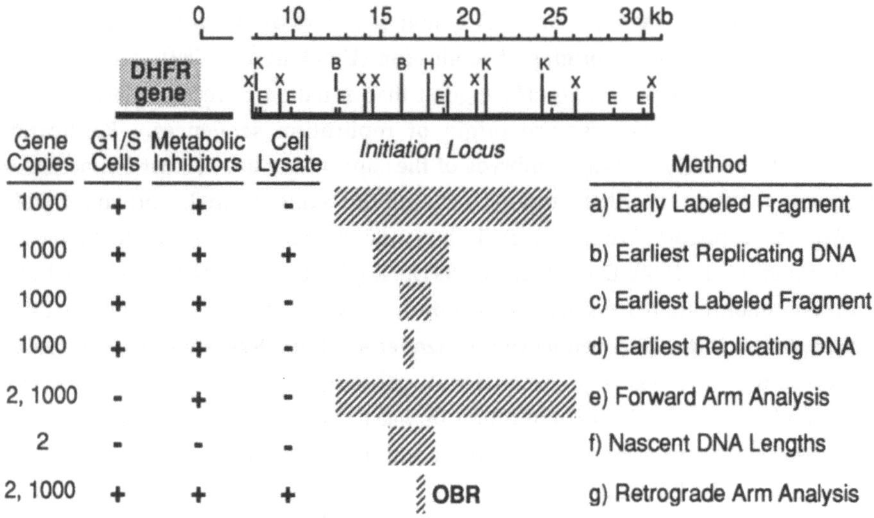

Figure 1. Summary of experimental results that identify DHFR-1, an origin of bidirectional DNA replication downstream of the dihydrofolate reductase gene in CHO cells. Genomic position is indicated by appropriate restriction enzyme cleavage sites, and distance is in kilobases. Number of gene copies is given for a diploid cell. The use of cells synchronized at their G1/S interphase, metabolic inhibitors such as aphidicolin or emetine, or nascent DNA radiolabeled in cell lysates is indicated by "+". Methods are found in (a) Heintz and Hamlin (1982); Heintz and Stillman (1988), (b) Burhans et al. (1986a, 1986b), (c) Leu and Hamlin (1989), Ma et al. (1990), (d) Anachkova and Hamlin (1989), (e) Handeli et al. (1989); this study, (f) Vassilev et al. (1990), and (g) Burhans et al. (1990).

Method I (Nascent DNA Lengths)

To overcome these problems, three independent techniques for locating origins of DNA replication in mammalian chromosomes have been developed and applied to the DHFR locus in CHO cells. The first method determines the distance from specific DNA sequence markers to a nearby origin by identifying the shortest nascent DNA chains that have traversed specific DNA sequence markers on either side of the origin. This method avoids the use of metabolic inhibitors, the requirement for cell synchronization, the need for multiple copies of the origin sequence. When applied to simian virus 40 (SV40), this method correctly identified the viral origin of replication (Vassilev and Johnson, 1989). When applied to exponentially growing CHO cells containing the nonamplified, single-copy DHFR gene locus (Vassilev et al., 1990), this method revealed a specific initiation locus for bidirectional DNA replication that lies within a 2.5 kb long sequence centered about 17 kb downstream of the DHFR gene (Fig. 1f).

Method II (Retrograde Arm Analysis)

The second method identifies the functional origin of bidirectional replication (OBR) by the transition from discontinuous to continuous DNA synthesis that must occur on each template strand at the site where replication forks originate (Fig. 2). This transition, which can be identified by hybridizing radiolabeled RNA-primed nascent DNA fragments ("Okazaki fragments") to unique strand-specific sequences spanning the putative origin region, identified the OBR at single nucleotide resolution in SV40 and polyomavirus chromosomes, coincident with the genetically defined origin of replication (Hay and DePamphilis, 1982; Hendrickson et. al., 1987a; 1987b, DePamphilis et al., 1988). When this method was applied to the DHFR gene locus in CHO and CHOC 400 cells, an OBR was identified within a specific 0.45 kb sequence approximately 17 kb downstream from the 3'-end of the DHFR gene (Burhans et al., 1990). At least 80% of the replication forks in a 27 kb region emanated from this OBR, which was located within the initiation locus identified by Vassilev et al. (1990). Since more than one OBR may exist downstream of the DHFR gene (Leu and Hamlin, 1989; Anachkova and Hamlin, 1989), we refer to this OBR as "DHFR OBR-1". In these experiments, Okazaki fragments were identified as nascent DNA fragments with a mean size of 105 nucleotides that were rapidly converted into long nascent DNA strands. We have recently determined that at least 90% of these nascent

DNA fragments were covalently attached to RNA with a mean length of ~10 bases ("RNA primer"), confirming their identification as Okazaki fragments.

Method III (Forward Arm analysis)

The third method induces an asymmetry at replication forks by treating exponentially proliferating cells with emetine, a potent inhibitor of protein synthesis. Based on previous studies (Riley and Weintraub, 1979; Seidman et al., 1979); Roufa and Marchionni (1982) and Handeli et al. (1989) assumed that, in the absence of histone synthesis, histone octamers in front of replication forks would segregate exclusively to forward arms of replication forks, leaving nascent DNA on retrograde arms unprotected by histones and therefore sensitive to non-specific endonucleases (Fig. 2, "conservative nucleosome segregation"). An OBR could then be recognized by the transition from nuclease protected to nuclease sensitive nascent DNA on each template strand, analogous to measuring the transition from continuous to discontinuous DNA synthesis. When Handeli et al. (1989) applied this approach to the DHFR locus, they concluded that the bulk of replication forks downstream of the DHFR gene in CHO cells emanated from an OBR somewhere within a 14 kb region downstream of the gene (Fig. 1), consistent with the results of Burhans et al. (1990) and Vassilev et al. (1990).

Neither Handeli et al. (1989) nor Roufa and Marchionni (1982), however, demonstrated directly that preferential protection of one strand of nascent DNA resulted from nucleosome protection by examining nascent DNA prior to digestion of chromatin with nuclease. Furthermore, the results of Seidman et al. (1979) could not be reproduced; under the same conditions in the absence of protein synthesis, Cusick et al., (1984), Sogo et al. (1986) and Jackson (1988; cited in Discussion) found that old histones were distributed evenly to both arms of SV40 replication forks. Thus, the mechanistic basis for this origin mapping protocol was not clear, raising doubts as to its validity. Therefore, to confirm the previous studies that led to identification of a specific origin of replication downstream of the DHFR gene ("DHFR OBR-1"), we attempted to reproduce the results of Handeli et al. (1989), and to determine its underlying mechanism.

In their protocol (Fig. 2, "conservative nucleosome segregation"), emetine was used to block total protein synthesis. Nascent DNA was then density-labeled with BrdU so that it could be isolated from unreplicated DNA. To destroy non-nucleosomal DNA, nuclei were isolated and digested with micrococcal nuclease. Nascent nucleosomal DNA was then isolated by isopycnic centrifugation in alkaline Cs_2SO_4 gradients, immobilized on nitrocellulose membranes, and annealed to unique radiolabeled probes that were complementary to either the forward or retrograde templates in order to identify its template specificity. We have modified their protocol to improve hybridization efficiency, strand specificity, and reproducibility.

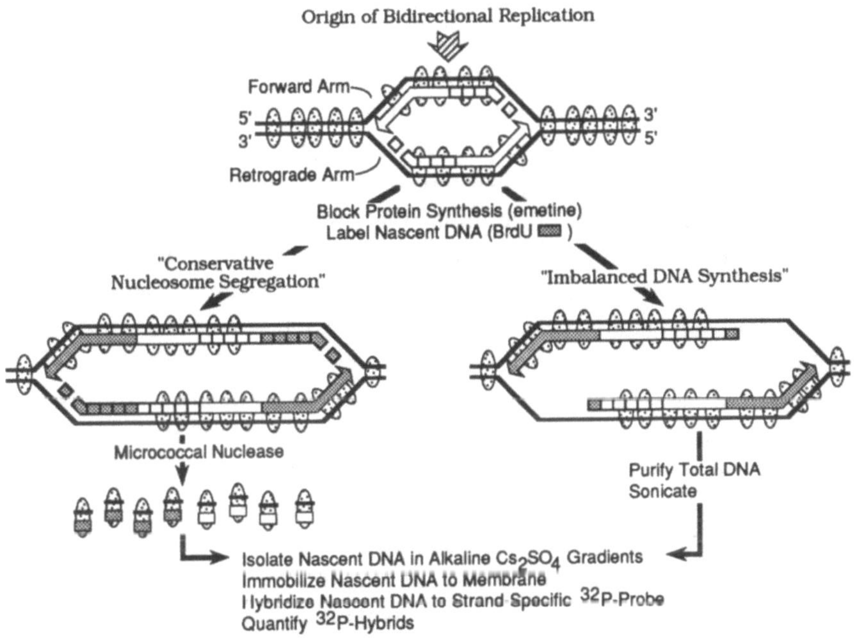

Figure 2. Mapping origins of replication by analysis of forward arm DNA synthesis in the presence of emetine.

In agreement with the results of Handeli et al. (1989), emetine induced a 5 to 7-fold accumulation of nascent DNA on forward arms of replication forks relative to retrograde arms, revealing that at least 85% of replication forks in this region emanated from OBR-1. However, this bias did not result from conservative nucleosome segregation to forward arms, as previously assumed, but from preferential inhibition of Okazaki fragment synthesis on retrograde arms.

Inhibition of protein synthesis causes imbalanced DNA synthesis, not conservative nucleosome segregation

Our results reveal that emetine induces imbalanced DNA synthesis by preferentially inhibiting DNA synthesis on the retrograde arm of replication forks (Fig. 2). First, hybridization of strand-specific probes to nascent DNA synthesized in the presence of emetine revealed that this DNA was enriched for sequences from the forward arm to the same extent as reported by Handeli et al. (1989). However, digestion of non-nucleosomal DNA by non-specific endonucleases was not required to observe this enrichment. Therefore, this enrichment was not dependent upon selective protection of the forward arm by nucleosomes. Second, although emetine eventually reduced the overall rate of DNA synthesis to <10% of controls, the proportion of total synthesis on retrograde arms was reduced more rapidly than synthesis on forward arms. The fraction of nascent DNA appearing as RNA-primed Okazaki fragments, which originate predominantly if not exclusively from retrograde arms, was reduced 4-fold within the first 1 hour of emetine treatment. Roufa (1982) reported that emetine had no effect on discontinuous DNA synthesis. However, his conclusion was based on DNA products labeled during a 20 min period *in vivo* that were greater than 10 kb long; he did not measure synthesis of RNA-primed Okazaki fragments. Third, electron microscopic analysis of SV40 DNA replicating *in vivo* revealed that emetine caused the appearance of ssDNA at replication forks in 90% of the replicating intermediates observed. These regions most often occurred in *trans* at both branch points of replication bubbles and represented the average length of an Okazaki fragment "initiation zone" (DePamphilis et al., 1988), consistent with preferential inhibition of Okazaki fragment synthesis by emetine.

When protein synthesis in mammalian cells was blocked by cycloheximide, Seidman et al. (1979) reported that essentially all prefork histones in SV40 chromosomes passed to forward arms of replication forks. However, Cusick et al. (1984) observed the opposite result; prefork histones were distributed equally to both arms of replication forks. Cusick pointed out that one explanation for this discrepancy: hybridization patterns were not reproducible when separated strands of DNA fragments were blotted to nitrocellulose membranes (conditions used by Seidman et al., 1979). Ironically, the faster migrating DNA band, representing retrograde DNA templates occasionally peeled off the paper during the hybridization reaction in some, but not all, lanes of the same blot. In some instances, only part of a band was missing. Their loss would have been misinterpreted as conservative nucleosome segregation to the forward arm. This problem was solved when DNA strands were covalently attached to membranes. Subsequently, the results of Cusick and coworkers were confirmed by psoralen-crosslinking of replicating SV40 DNA in virus-infected cells treated either with cycloheximide (Sogo et al., 1986) or emetine (unpublished data). Nucleosome segregation was random, and nucleosomes on newly replicated DNA tended to be grouped in clusters, as previously proposed by Pospelov et al. (1982).

The ability to induce imbalanced DNA synthesis reveals that synthesis on the retrograde arm of replication forks can be uncoupled from synthesis on the forward arm. Since synthesis on the forward arm template is carried out by DNA polymerase–δ while synthesis on the retrograde arm template is carried out by DNA primase-DNA polymerase–α, one possibility is that one or more components of the DNA primase-DNA polymerase–α complex is rapidly lost in the absence of protein synthesis. Alternatively, emetine as well as other protein synthesis inhibitors may interfere directly with DNA replication rather than indirectly through their effect on protein synthesis.

Summary

The results of eleven independent studies employing five different origin mapping methods are consistent with DHFR OBR-1 acting as a primary origin of replication in CHO cells (Fig. 1). These studies include both synchronized and exponentially proliferating cells, cells containing single copies or amplified copies of the DHFR locus, cells treated with metabolic inhibitors as well as untreated cells, and nascent DNA labeled in whole cells or in cell extracts. Our recent results demonstrate three important points. First, they confirm that at least 85% of replication forks in the 27 kb region downstream of the DHFR gene in both CHO and CHOC 400 cells emanate from DHFR OBR-1. Second, they confirm the utility of the mapping protocol described by Handeli et al. (1989). Finally, and perhaps most importantly, they provide it with a sound mechanistic basis by showing that the replication fork asymmetry induced by emetine results from imbalanced DNA synthesis (Fig. 2), not conservative nucleosome segregation, as previously believed. Moreover, our data, as well as those of Handeli et al. (1989), further support the conclusions of Burhans et al. (1990) and Vassilev et al. (1990) that at least 80% of replication forks within the 27 kb region downstream of the DHFR gene emanated from DHFR OBR-1 in both CHO and CHOC 400 cells.

Nascent DNA strand analysis compared with total DNA strand analysis

In apparent contradiction to the studies cited in Figure 1, 2-dimensional gel electrophoretic analysis of DNA structures produced during replication of the DHFR region led to the conclusion that replication bubbles were initiated randomly and replication forks proceeded in both directions throughout the 30 kb region downstream of the DHFR gene (Vaughn et al., 1990; Dijkwel et al., this volume). If this were true, then normal Okazaki fragments as well as long nascent DNA chains produced during imbalanced DNA synthesis would anneal equally well to both DNA strands throughout this region, and OBR-1 should not have been identified in the studies cited above. Three explanations come to mind. First, the 2-D gel method searches for structures that migrate as replication intermediates regardless of whether or not they contain nascent DNA, while the three methods described above observe only newly synthesized DNA. Thus, it is possible that replication begins at many sites throughout this region, but that most of the events outside the OBR are either aborted completely or continue only in the direction away from the OBR (Linskens and Huberman, 1990). Second, the ratio of replication bubbles to replication forks in the 2-D gel data appears to vary from ~1% to ~30%, suggesting that it is difficult to quantify the number of initiation events in any given region. Thus, the OBR may be a highly preferred initiation site, but not the only initiation site (Kaiserman et al., 1989), consistent with studies of phage T7 and T4 (see S. Rabkin, this volume). Third, some putative replication bubbles may actually be replication forks in which the DNA remains bound to cellular material, resulting in anomalously slower migration during gel electrophoresis. Resolution of this paradox will require an assay for origin activity that is independent of any of these mapping protocols.

REFERENCES

Anachkova, B. and Hamlin, J.L. (1989) Replication in the amplified dihydrofolate reductase domain in CHO cells may initiate at two distinct sites, one of which is a repetitive sequence element. *Mol. Cell. Biol.* **9**, 532-540.

Burhans, W.C., J.E. Selegue and N.H. Heintz (1986a) Isolation of the origin of replication associated with the amplified Chinese hamster dihydrofolate reductase domain. Proc. Natl. Acad. Sci. USA **83**: 7790-7794.

Burhans, W. C., J. E. Selegue and N. H. Heintz (1986b) Replication intermediates formed duiring initiation of DNA synthesis in methotrexate-resistant CHOC 400 cells are enriched for sequences derived from a specific, amplified restriction fragment. Biochemistry **25**: 441-449.

Burhans, W.C., L.T. Vassilev, M.S. Caddle, N.H. Heintz, and M.L. DePamphilis (1990) Identification of an origin of bidirectional DNA replication in mammalian chromosomes, Cell **62**: 955 - 965.

Cusick, M.E., M.L. DePamphilis and P.M. Wassarman (1984) Dispersive Segregation of Nucleosomes During Replication of SV40 Chromosomes, J. Mol. Biol. **178**: 249-271.

DePamphilis, M.L., Martínez-Salas, E., Cupo, D.Y., Hendrickson, E.A., Fritze, C.E., Folk, W.R. and U. Heine (1988) Initiation of Polyomavirus and SV40 DNA Replication, and the Requirements for DNA Replication During Mammalian Development, p. 165-175 *In* Eukaryotic DNA Replication (B. Stillman and T. Kelly, eds), Cancer Cells, vol. 6, Cold Spring Harbor Laboratory, N.Y.

DePamphilis, M.L. (1991) DNA Replication and the Role of Transcriptional Elements During Animal Development, *in* Advances in Developmental Biochemistry (P.M. Wassarman, ed.) JAI Press, Greenwich Conn., in press.

Frappier, L. and M. Zannis-Hadjopoulos. (1987) Autonomous replication of plasmids bearing monkey DNA origin-enriched sequences. *Proc. Natl. Acad. Sci. USA* **84**, 6668-6672.

Handeli, S., A. Klar, M. Meuth and H. Cedar (1989) Mapping replication units in animal cells. *Cell* **57**, 909-920.

Hay, R.T. and DePamphilis, M.L. (1982) Initiation of Simian Virus 40 DNA Replication in Vivo: Location and Structure of 5'-Ends of DNA Synthesized in the *Ori* Region, *Cell* **28**, 767-779.

Hendrickson, E.A., C.E. Fritze, W.R. Folk, and M.L. DePamphilis (1987a) The Origin of Bidirectional DNA Replication in Polyoma Virus, *EMBO J.* **6**, 2011-2018.

Hendrickson, E.A., C.E. Fritze, W.R. Folk, and M.L. DePamphilis (1987b) Polyoma Virus DNA Replication is Semi-Discontinuous, Nuc. Acids. Res. **15**: 6369-6385.

Heintz, N.H. and J.L. Hamlin (1982) An amplified chromosomal sequence that includes the gene for dihydrofolate reductase initiates replication within specific restriction fragments. *Proc. Nat. Acad. Sci. USA* **79**: 4083-4087.

Heintz, N.H. and Stillman, B.W. (1988) Nuclear DNA synthesis in vitro is mediated via stable replication forks assembled in a temporally specific fashion in vivo. *Mol. Cell. Biol.* **8**, 1923-1931.

Heinzel, S.S., Krysan, P.J., Tran, C.T. and Calos, M.P. (1991). Autonomous DNA replication in human cells is affected byu the size and source of the DNA. *Mol. Cell. Biol.* **11**: 2263-2272.

Kaiserman, H.B., N.J. Marini, E.H.A. Poll and R.M. Benbow (1989) Molecular, biochemical and sequence requirements for DNA replication in *Xenopus laevis* embryos, p. 354-383 *In* P. Strauss and S. Wilson (ed.), The Eukaryotic Nucleus: Molecular Structures and Macromolecular Assemblies. Telford Press, West Caldwell, N.J.

Kaufmann, G., S. Anderson and M.L. DePamphilis. (1977) RNA Primers in Simian Virus 40 DNA Replication: Distribution of 5'-Terminal Oligoribonucleotides in Nascent DNA. J. Mol. Biol. **116**: 549-567.

Krysan, P.J. and Calos, M.P. (1991) Replication initiates at multiple locations on an autonomously replicating plasmid in human cells. Mol. Cell. Biol. **11**: 1464-1472.

Leu, T-H. and J.L. Hamlin (1989) High-resolution mapping of replication fork movement through the amplified dihydrofolate reductase domain in CHO cells by in-gel renaturation analysis. Mol. Cell. Biol. **9**: 523-531.

Linskens, M.H.K. and Huberman, J.A. (1990) The two faces of higher eukaryotic DNA replication origins. *Cell* **62**, 845-847.

Ma, C., T-H. Leu, and J.L. Hamlin. (1990) Multiple origins of replication in the dihydrofolate reductase amplicons of a methotrexate resistant Chinese hamster cell line. *Mol. Cell. Biol.* **10**, 1338-1346.

Martínez-Salas, E., D.Y. Cupo and M.L. DePamphilis. (1988) The need for enhancers is acquired upon formation of a diploid nucleus during early mouse development, *Genes and Development* **2**: 1115-1126.

McWhinney, C. and M. Leffak. (1990) Autonomous replication of a DNA fragment containing the chromosomal replication origin of the human c-myc gene. *Nucl. Acids Res.* **18**, 1233-1242.

Riley, D. and Weintraub, H. (1979) Conservative segregation of parental histones during replication in the presence of cycloheximide, *Proc. Natl. Acad. Sci. USA* **76**, 328-332.

Roufa, D.J. (1978) Replication of a mammalian genome: the role of de novo protein biosynthesis during S phase. *Cell*, **13**, 129-138.

Seidman, M.M., Levine, A.J. and Weintraub, H. (1979) The asymmetric segregation of parental nucleosomes during chromosome replication. Cell **18**: 439-449.

Sogo, J. M., H. Stahl, Th. Koller and R. Knippers (1986) Structure of replicating SV40 minichromosomes: the replication fork, core histone segregation and terminal structures. J. Mol. Biol. **189**: 189-204.

Umek, R. M., M. H. K. Linskens, D. Kowalski and J. A. Huberman (1989) New beginnings in studies of eukaryotic DNA replication origins. Biochim. Biophys. Acta.**1007**: 1-14.

Vassilev, L. and E. M. Johnson (1989) Mapping initiation sites of DNA replication *in vivo* using polymerase chain reaction amplification of nascent strand segments. Nucl. Acids Res. **17**: 7693-7705.

Vassilev, L.T., W.C. Burhans and M.L. DePamphilis (1990) Mapping an origin of DNA replication at a single copy locus in exponentially proliferating mammalian cells, Mol. Cell. Biol., **10**, 4685-4689.

Vaughn, J. P., P. A. Dijkwel, and J. L. Hamlin (1990) Replication initiates in a broad zone in the amplified CHO dihydrofolate reductase domain. Cell **61**, 1075-1087.

Wirak, D.O., L.E. Chalifour, P.M. Wassarman, W.J. Muller, J.A. Hassell, and M.L. DePamphilis. (1985) Sequence-Dependent DNA Replication in Preimplantation Mouse Embryos, *Mol. Cell. Biol.* **5**, 2924-2935 .

Directions of DNA Replication in Yeast and Mammalian Cells

Joel A. Huberman, Dharani D. Dubey[1], Kevin A. Nawotka, George Russev[2], J. Aquiles Sanchez, Yeup Yoon, and Maarten H. K. Linskens[3]
Department of Molecular and Cellular Biology
Roswell Park Cancer Institute
Buffalo, NY 14263
U. S. A.

INTRODUCTION

Importance of determining directions of DNA replication. Because replication forks move outwards from chromosomal origins, and because replication terminates when adjacent converging replication forks collide, locations of origins and termini can be deduced when direction(s) of replication are measured through long stretches of DNA. In addition, measurement of direction(s) of both replication and transcription through a chromosomal region permits testing the hypothesis that replication and transcription are likely to proceed in the same direction (Seidman et al., 1979; Brewer, 1988).

Evidence that replication and transcription frequently proceed in the same direction. The first investigation of possible correlations between directions of replication and transcription was carried out by Seidman et al. (1979). To measure replication direction, they used an early version of the technique now called "leading strand polarity analysis" (see below). They pulse-labeled mammalian cells with ^{3}H-dThd while protein synthesis was inhibited. Their measurements suggested that, under these conditions, no new histones would be synthesized and all parental histones would segregate conservatively to the leading sides of replication forks, thereby protecting the leading sides from attack by micrococcal nuclease. However, when the micrococcal-nuclease-resistant ^{3}H-labelled DNA was purified and enriched for single copy sequences, it hybridized normally to total cellular DNA but poorly to nuclear RNA. This finding suggested that ^{3}H-labelled nascent leading strands have the same polarity as nuclear RNA, i.e., that nuclear RNA is transcribed in the same direction as replication. To test this suggestion, Seidman et al. (1979) attempted to purify ^{3}H-DNA from lagging strands by taking advantage of the presumed greater sensitivity of lagging strands to restriction nuclease digestion after inhibition of protein synthesis. ^{3}H-DNA from presumed lagging strands hybridized prefe-

[1]Current address: Department of Zoology, Kutir Mahavidyalaya, Chakkey, Jaunpur-222146, U.P., India
[2]Current address: Institute of Molecular Biology, Bulgarian Academy of Sciences, Sofia 1113, Bulgaria
[3]Current address: Cold Spring Harbor Laboratory, Cold Spring Harbor, NY 11724

rentially to nuclear RNA, again suggesting that replication forks move in the same direction as RNA polymerases.

Since the direction of DNA replication is known in every portion of the *E. coli* genome and the directions of *E. coli* gene transcription are known in many cases, Brewer (1988) was able to conclude, after an extensive literature review, that directions of replication and transcription are frequently correlated even in this prokaryotic organism. Specifically, Brewer (1988) found that (i) long, heavily transcribed genes (like the rDNA genes) are always transcribed in the direction of replication, (ii) short, lightly transcribed genes are randomly oriented, and (iii) genes of intermediate length and/or transcriptional frequency display intermediate levels of co-orientation with replication. These observations suggest an evolutionary equilibrium between the negative selective pressure generated by collisions between RNA polymerases and replication forks and the normal tendency of genes to rearrange and reorient with time.

Methods available for measuring directions of replication in eukaryotic chromosomes.

A. Replication timing. Measurements of the times in S phase when neighboring DNA segments replicate can provide suggestive information about directions of DNA replication: origin-proximal segments should replicate earlier than origin-distal segments. Information obtained by replication timing measurements cannot be interpreted unambiguously, however. If two closely spaced (compared to the resolution of the timing measurements) DNA segments both contain origins, but one of them initiates slightly later than the other, replication timing measurements would probably not reveal the existence of the later origin or of the short stretch of DNA replicated toward the earlier origin.

B. Two dimensional (2D) gel electrophoresis. Two independent 2D gel methods (Brewer and Fangman, 1987; Nawotka and Huberman, 1988) have proved capable of mapping replication origins and termination sites in yeast chromosomal DNA (Huberman et al., 1988; Linskens and Huberman, 1988). One of these methods, the neutral/alkaline method (Nawotka and Huberman, 1988), provides direct information about the direction(s) of replication through restriction fragments. This information is obtained by measuring the minimum nascent strand sizes detectable by short probes located at various positions throughout the restriction fragment of interest. The smaller the nascent strands detectable by a probe, the closer that probe must be to the origin. The other method, the neutral/neutral method (Brewer and Fangman, 1987), can also provide information about direction(s) of replication if it is modified by the addition of a restriction digestion between the first and second dimensions (B. Brewer, personal communication). Although these methods appear to provide mutually consistent information when applied to either yeast or mammalian chromosomes, the 2D gel data obtained with yeast chromosomes suggest that yeast origins are localizable within a few hundred bp and are experimentally indistinguishable from yeast *ARS* elements (Huberman et al., 1988; Linskens and

Huberman, 1988), whereas the data obtained with mammalian chromosomes suggest that replication may initiate within broad regions of 30 kb or more (Vaughn et al., 1990). Although these discrepancies are most likely due to real differences between yeast and mammalian replication, it is possible that information obtained with 2D gel techniques can be misinterpreted. In particular, one must bear in mind that the 2D gel techniques provide information about nonlinear DNA molecules present in the population. Replicating DNA molecules are nonlinear, but so are recombining molecules. Nonlinear structures may arise in additional ways of which we are not yet aware. Therefore, particularly in the case of mammalian cells, alternative, independent techniques for measuring direction of replication would be desirable. Several are available, as described in the next paragraphs.

C. In vitro runoff. James and Leffak (1986) suggested that the direction in which replication forks in isolated nuclei proceed through a stretch of DNA can be determined by cutting the nuclear DNA with a restriction enzyme and then permitting continued replication fork movement in the presence of BrdUTP. Because the double-strand cuts produced by the restriction enzyme should block fork movement, each restriction fragment should become denser at its origin-distal end than at its origin-proximal end. Results obtained with this method have been consistent with the hypothesis that mammalian replication generally proceeds in the direction of transcription (James and Leffak, 1986; Trempe et al., 1988; Leffak and James, 1989). However, the method depends on the assumption that replication in isolated nuclei proceeds in the same directions as *in vivo*.

D. Leading strand polarity analysis. The suggestion by Seidman et al. (1979) that, in the absence of protein synthesis, parental histones segregate conservatively to the leading sides of replication forks, provided the rationale for development of a general method for measuring directions of replication fork movement, based on the presumed relative nuclease resistance of the leading sides of forks (Roufa and Marchioni, 1982). This method was later applied by Handeli et al. (1989) to analysis of directions of replication near the DHFR gene in Chinese hamster cells. The results obtained were consistent with results obtained independently by replication timing analysis (Leu and Hamlin, 1989) and by Okazaki fragment polarity analysis (next paragraph; Burhans et al., 1990), but only partially consistent with the results of 2D gel analyses (Vaughn et al., 1990).

A major problem with leading strand polarity analysis is that the mechanism by which it works is unknown. Although the method was based on evidence suggesting that histones segregate conservatively to the leading sides of forks, that evidence has been questioned (Cusick et al., 1984). Furthermore, Burhans and DePamphilis (personal communication) have shown that omitting micrococcal nuclease digestion does not alter the results obtained; therefore the results obtained with the leading strand polarity method must be independent of nucleosome segregation patterns. Until the mechanism of leading strand polarity analysis is understood, the results obtained with it should be interpreted with caution.

E. Okazaki fragment polarity analysis. Okazaki fragments are intermediates in the discontinuous synthesis of lagging strands. Because eukaryotic Okazaki fragments are known to be very short (<300 b), one should be able to purify them on the basis of size and BrdU content from cells pulse-labeled with BrdUrd (or from nuclei pulse-labeled with BrdUTP). Burhans et al. (1990) used this approach to isolate Okazaki fragments from Chinese hamster cell nuclei and found that the purified Okazaki fragments hybridized primarily to just one of the two DNA strands in the tested regions near the DHFR gene. The results for all the tested regions suggested that replication forks moved outwards from an origin localizable to <450 nucleotides, in a region downstream of the DHFR gene that both replication timing measurements and leading strand polarity analysis had also implicated as containing an origin (see above). Only 2D gel measurements seemed to disagree with this conclusion (see above).

The Okazaki fragment polarity analysis conducted by Burhans et al. (1990) depended on *in vitro* DNA synthesis in broken cells, and may, therefore, have provided misleading results. An improved approach would be to purify Okazaki fragments from intact cells. Such "*in vivo*" Okazaki fragments could not be radioactively labeled to the same high specific activity obtained by Burhans et al. (1990), but sufficient sensitivity for analysis of "*in vivo*" Okazaki fragments could probably be obtained by use of the polymerase chain reaction (PCR).

F. Nascent strand size analysis. As discussed above for the neutral/alkaline 2D gel technique, the minimum size nascent strand detectable by a short genomic probe sequence provides an estimate of the distance between the probe sequence and the origin. We (see below) and Vassilev and Johnson (1989) have independently developed techniques that permit such measurements in mammalian cells. Because these techniques can take advantage of PCR amplification, they can be used to measure directions of DNA replication through single-copy genomic regions in non-synchronized, intact mammalian cells. A more extensive description of the advantages, disadvantages, and techniques of nascent strand size analysis will be presented in the last portion of this paper.

In this paper we shall summarize results obtained in yeast, with neutral/alkaline 2D gel electrophoresis, which suggest that most yeast genes are randomly oriented with respect to direction of replication. We shall also present new procedures for nascent strand size analysis in mammalian cells.

RESULTS AND DISCUSSION

<u>Directions of DNA replication in the A6C replicon on chromosome III.</u> The directions of DNA replication in the leftmost ≈60 kb of *S. cerevisiae* chromosome III have been measured by Huberman et al. (1988) and by Dubey, Davis, Newlon and Huberman (manuscript in prepa-

ration). Within this region, just a single origin, the A6C origin, was detected, and this origin corresponds to an *ARS* element, *ARS305* (Fig. 1; Huberman et al., 1988). Four additional *ARS* elements previously mapped in this region (Feldman et al., 1984; Newlon et al., 1986) have proved to be inactive as chromosomal origins (Dubey et al., submitted for publication; Dubey, Davis, Newlon, and Huberman, manuscript in preparation). We have attempted to determine whether or not there is a correlation between replication and transcription within this region by mapping the locations and directions of transcripts detectable by Northern blotting of total cellular RNA. To date, 14 transcribed regions have been detected within the leftmost 47 kb of this region, and the orientations of 8 of these transcripts have been determined, as shown in Fig. 1:

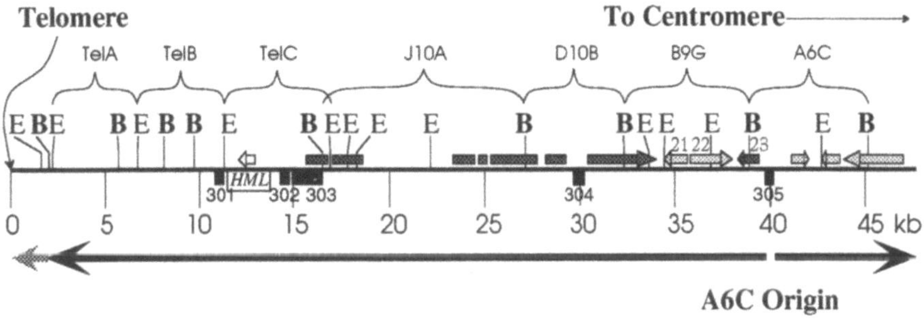

Fig. 1. Directions of DNA replication and transcription in the leftmost 47 kb of yeast chromosome III. The top horizontal line shows a restriction map of the left end of yeast chromosome III, determined by Newlon et al. (1986) and Synn and Newlon (unpublished). E=*Eco*RI; **B**=*Bam*HI. The brackets above the restriction maps indicate named *Eco*RI or *Bam*HI restriction fragments cloned by the Newlon laboratory. The black boxes below the map, numbered 301 to 305, show the locations of *ARS* elements. The solid black arrows at the bottom of the diagram indicate directions of DNA replication mapped by neutral/alkaline 2D gel electrophoresis. The shaded arrow at the left telomere indicates that evidence for this region is incomplete, but is suggestive of fork movement from right to left all the way to the telomere. Transcripts are indicated by thick boxes (some of which have arrowheads) just above the restriction map line. These transcripts were localized by probing Northern blots of total *S. cerevisiae* RNA (strain 4910-3-3; Huberman et al., 1988) first with large and then with smaller DNA restriction fragments from the region shown. Polarities were determined by use of restriction fragments cloned into pBluescript vectors (Stratagene), then transcribed into RNA probes by either T7 or T3 RNA polymerase. Where determined, polarities are shown by arrowheads. The 3 intensities of shading of the transcript boxes indicate (from light to dark) weak, moderate, and strong relative signal intensities when probes of comparable specific activity were used. The numbers 21, 22, and 23 over certain transcripts indicate correspondence with transcripts 21, 22, and 23 of Yoshikawa and Isono (1990).

We did not detect any transcripts to the left of the *HML* locus, and weak transcription from *HML* could be detected only in *sir*⁻ strains (Dubey et al., submitted for publication).

However, starting at the right end of the "TelC" restriction fragment, transcripts were detected at frequent intervals along the chromosome, consistent with the known high density of genes in yeast chromosomes. The results presented in Fig. 1 are in general agreement with the data of Yoshikawa and Isono (1990), who have roughly mapped transcripts throughout chromosome III. Although Yoshikawa and Isono (1990) detected transcription to the left of *HML*, the level of detected transcription is very low. Their use of poly(A)$^+$ RNA rather than total RNA probably permitted them to detect lower levels of transcription. There is minor disagreement about the locations, sizes, and abundances of transcripts in some other parts of the region mapped by both groups, probably due to the use of different strains and growth conditions. However, in the region immediately to the left of the A6C origin, agreement between the two sets of results appears to be very good: the transcripts numbered 21, 22, and 23 in Fig. 1 correspond nearly exactly in location, size, and abundance to transcripts 21, 22, and 23 of Yoshikawa and Isono (1990). Transcripts 21 and 22 were determined by Yoshikawa and Isono (1990) to be of average abundance among poly(A)$^+$ RNAs, and transcript 23 was more abundant than average, but not one of the most abundant transcripts from chromosome III. The data on transcription polarity in Fig. 1 suggest that there is no strong correlation between directions of transcription and replication for the short, relatively lightly transcribed genes surrounding the A6C origin.

Directions of DNA replication near the *GAL7, 10, 1* gene cluster on chromosome II. Three inducible genes encoding enzymes in the galactose utilization pathway are located in a cluster on *S. cerevisiae* chromosome II and are divergently transcribed as shown in Fig. 2:

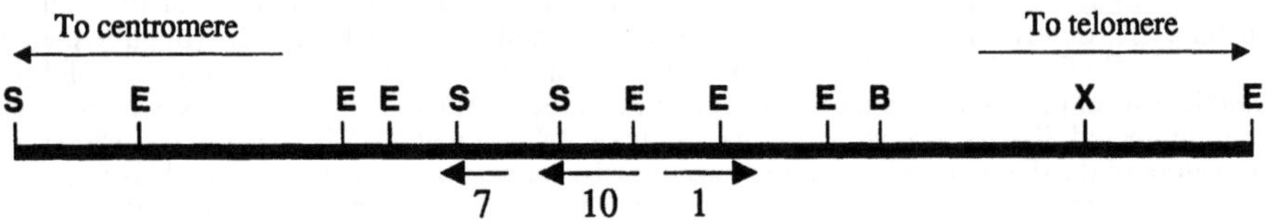

Fig. 2. The *GAL7, GAL10,* and *GAL1* genes of *S. cerevisiae.* The diagram shows a 28 kbp stretch from chromosome II. S=*Sal*I; E=*Eco*RI; B=*Bam*HI; X=*Xho*I. The locations of the genes and restriction sites were determined by St John and Davis (1981). DNA clones covering this entire region were provided to us by Ronald Davis.

No transcription can be detected from these genes when cells are grown in glucose, but when glucose is replaced by galactose, these genes are among the most abundantly transcribed single-copy genes in the genome. If the hypothesis that abundantly transcribed genes are replicated in the direction of transcription is correct, then one would predict that a replication origin should be located between the *GAL10* and *GAL1* genes, at least when cells are grown in galactose. Using neutral/alkaline 2D gel electrophoresis we have found, however, that replication forks move through the region shown in Fig. 2 from the left to the right, regardless of whether the cells are grown in glucose or galactose (data to be published elsewhere). Thus yeast replication forks appear to be capable of moving through even the most heavily transcribed single-copy, short yeast genes counter to the direction of transcription.

Replication of yeast rDNA genes. In contrast to the apparently random relationship between yeast replication forks and single-copy genes, we (Linskens and Huberman, 1988) and Brewer and Fangman (1988) have shown that, in the tandemly repeated array of *S. cerevisiae* rDNA genes on chromosome XII, there are blocks to counter-transcriptional replication fork movement at the 3' ends of all rDNA genes. The consequence of these blocks is that replication forks can move through the rDNA genes only in the direction of transcription.

Relationship between replication and transcription in *S. cerevisiae*. Although directions of replication and transcription have been measured in only a tiny fraction of the yeast genome, the data obtained so far show that some typical yeast genes — short and single-copy — are oriented so that DNA polymerases and RNA polymerases move through them in the same direction, while other short, single-copy genes are oriented so that polymerases move through in opposite directions. The limited available data are consistent with the possibility that typical yeast genes are oriented completely randomly with respect to direction of replication. In contrast, yeast rDNA genes are invariably replicated in the direction of transcription. However, the rDNA genes are the most abundantly transcribed in the genome. Furthermore, the rDNA transcribed region (over 6 kb) is probably the longest stretch of transcribed DNA in the genome. Consequently, the relationship between transcription and replication in *S. cerevisiae* appears qualitatively similar to that in *E. coli*, as described by Brewer (1988): replication forks must travel through the longest, most heavily transcribed regions in the same direction as transcription, but there is no obligatory relationship between directions of replication fork movement and orientation of shorter, more lightly transcribed genes.

The relationship between yeast genes and replication forks is apparently not as constrained as the data of Seidman et al. (1979) suggest for mammalian genes and replication forks. It will be important to determine actual directions of replication and compare them with directions of transcription in multiple locations in mammalian genomes, to test whether the data of Seidman et al. (1979) provide an accurate picture of reality.

Nascent strand size analysis as a method to measure directions of replication fork movement in mammalian cells. As described in the Introduction, several methods are now available for determining directions of replication in mammalian cells. Among these methods, nascent strand size analysis (NSSA) offers two potential advantages: (i) the rationale for the method is thoroughly understood (see below), and (ii) the method has sufficient sensitivity to permit measurements in single-copy DNA regions in experimentally feasible numbers of unsynchronized cells.

The rationale of NSSA is very similar to that of the neutral/alkaline 2D gel technique (see Introduction): the smaller the nascent strands detectable by a probe, the closer that probe must be to an origin.

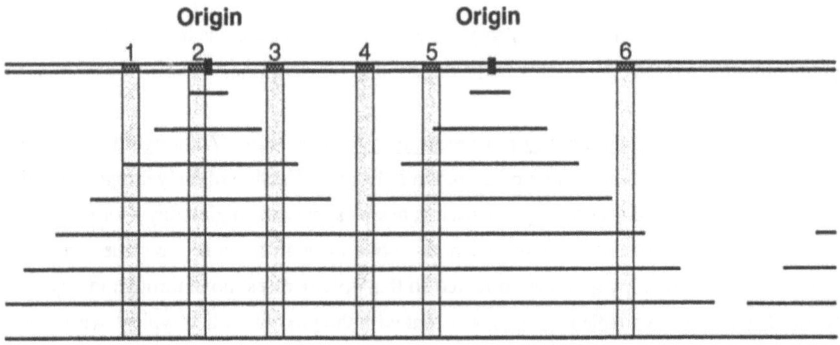

Fig. 3. Determination of directions of DNA replication and origin and terminus locations by nascent strand size analysis (NSSA). The pair of horizontal lines at the top represents a hypothetical stretch of chromosomal DNA containing two replication origins. Bidirectional DNA replication from each origin in a population of unsynchronized cells generates a family of nascent strands of increasing sizes. If these nascent strands are purified and fractionated according to size by the methods described in the text, then each of the hybridization probes labelled 1-6 will detect all indicated nascent strands longer than a minimum size. The minimum nascent strand size detectable by each probe is different. For each pair of neighboring probes, replication forks proceed from the probe detecting shorter nascent strands towards the probe detecting only longer nascent strands.

In neutral/alkaline 2D gel analysis, however, sensitivity is limited by (i) background due to the presence of parental strands and non-replicating DNA on the same membrane and (ii) the use of conventional hybridization technology, which requires at least femtogram amounts of DNA to provide a signal. As will be described below, with NSSA it is possible to purify the nascent strands away from virtually all parental and non-replicating strands; it is also possible to employ the polymerase chain reaction (PCR) to obtain significant signal amplification.

Purification of nascent strands is based on the incorporation of BrdUrd into the nascent strands during a brief labeling period. The nascent BrdUrd-labeled strands can then be released from non-replicating DNA either in single-stranded form by alkaline denaturation (Fig. 4A) or in double-stranded form by use of a single-strand-specific endonuclease to cut the DNA specifically at replication forks (Herman et al., 1979; Tsubota et al., 1982; Hanley et al., 1987; Fig. 4B).

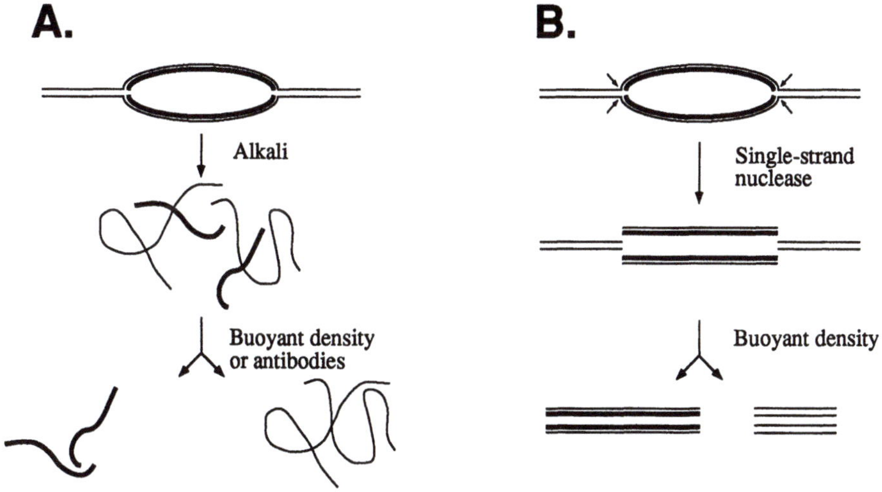

Fig. 4. Methods for purification of BrdUrd-labeled nascent DNA. The thin lines represent parental DNA strands, and the thick lines represent nascent strands into which BrdUrd has been incorporated. The BrdUrd-labeled strands can be separated from nonreplicating strands in either of two ways: **A.** All of the DNA can be denatured at high pH, and the nascent strands can then be separated from the nonreplicating strands by isopycnic centrifugation or by immunoprecipitation with anti-BrdUrd antibodies; or **B.** the parental DNA strands can be cut at the replication forks by a single-strand specific endonuclease (small arrows), releasing the nascent DNA in double-stranded form. Although the double-stranded nascent DNA is not reactive with anti-BrdUrd antibodies, it can be separated from nonreplicating DNA on the basis of its density.

There are advantages and disadvantages to each of the approaches presented in Fig. 4. The alkaline approach (Fig. 4A) is simple and efficient. The resulting single strands can be purified in either of two ways, use of anti-BrdUrd antibodies (Vassilev and Russev, 1988; Vassilev and Johnson, 1989) or isopycnic centrifugation. However, Vassilev and Johnson (1990) have noted that hybridization between repetitive sequences in nonreplicating DNA and similar sequences in replicating DNA under the neutral conditions required for antibody purification may lead to contamination. This problem can be avoided by use of alkaline CsCl centrifugation. Unfortunately, DNA strands are degraded to the ≈2-12 kb size range in alkaline CsCl gradients (our unpublished observation). The optimum procedure, in our experience, is to use at least two rounds of neutral CsCl gradient centrifugation with a denaturation step before each round.

Although very pure nascent single strands can be obtained, these strands are sensitive to breakage by shear or photolysis (BrdUrd-containing DNA is sensitive even to the small amount of UV emitted by ordinary fluorescent lamps). To obtain accurate estimates of minimum nascent strand sizes, it is important that nascent strand breakage be minimized. For that reason, we have developed the alternative approach diagrammed in Fig. 4B. In this case, nascent "strands" are released in double-stranded form. Although double-stranded BrdUrd-containing DNA is not reactive with anti-BrdUrd antibodies, most of it is sufficiently dense to be separable from normal DNA in a CsCl gradient. Regions that are GC-rich (such as the ribosomal DNA repeats in most eukaryotic organisms) and therefore don't incorporate much BrdUrd may not be sufficiently separated from nonreplicating DNA. Fortunately, most eukaryotic DNA is sufficiently AT-rich that satisfactory purification can be achieved by two rounds of CsCl gradient centrifugation (this problem does not arise with single-stranded DNA because of the larger density shifts achieved by BrdUrd substitution in single-stranded DNA). Double-stranded nascent DNA can be fractionated according to size by agarose gel electrophoresis. Pulsed-field electrophoretic techniques permit excellent resolution of molecules up to hundreds of kbs. Accurate fractionation of single-stranded DNA in this size range is more difficult. Alkaline sucrose gradients do not permit the level of resolution achievable by gel electrophoresis, but pulsed-field techniques have not yet been developed for single-stranded DNA, so effective resolution in gels is limited to 20 kb or less.

<u>Application of the alkaline and neutral nascent strand size analysis techniques to SV40 DNA.</u> To test whether the alkaline and neutral NSSA techniques are capable of correctly determining the directions of replication and origin location in the known SV40 replicon, we applied both techniques to SV40-infected CV-1 cells. In both cases, we isolated total DNA, host cell plus viral, just as we would do with uninfected cells. The cells were pulse-labeled with BrdUrd and their nascent strands were purified as described in the legends to Figs. 5 and 6. Due to the abundance of SV40 DNA in infected cells, the final purified nascent strands, fractionated according to size in alkaline or neutral agarose gels, could easily be detected by conventional hybridization probes, as shown in Figs. 5 and 6.

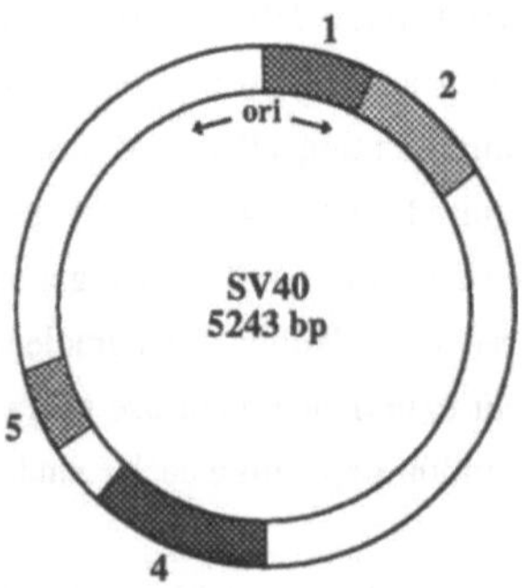

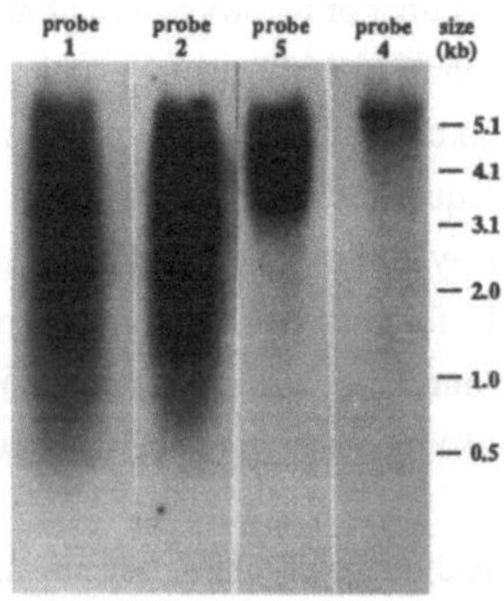

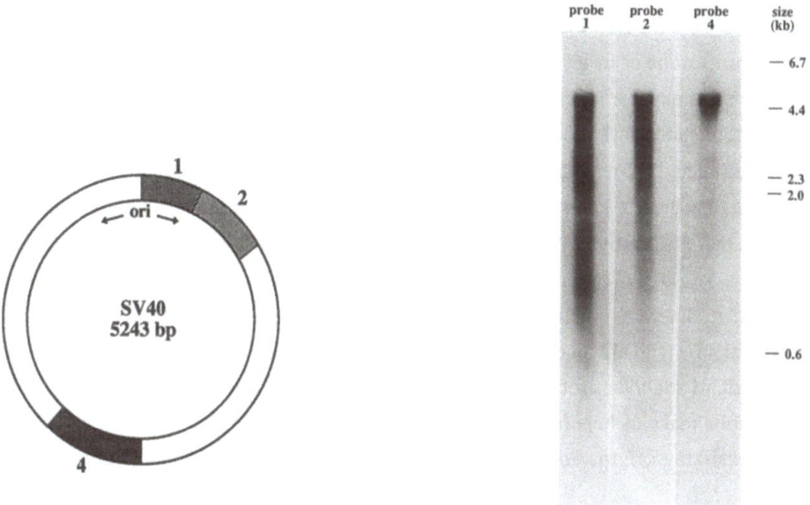

Fig. 6. Use of single-strand-nuclease-mediated nascent strand isolation to map directions of SV40 replication fork movement. SV40 infected CV-1 cells were pulse-labeled as in Fig. 5. Total cellular DNA was isolated by SDS lysis, proteinase K treatment, gentle phenol and chloroform extraction, and ethanol precipitation. The DNA was dissolved at 100 µg/ml in TE buffer (10 mM Tris, 1 mM EDTA, pH 8.0) and then incubated with 50-100 µg/ml P1 nuclease (Pharmacia) for 60 min at 37° C. The digestion efficiency can vary, and initial trials should be performed with small amounts of DNA to determine optimal conditions (25-40% of ^{3}H-DNA released to the heavy-light band in a CsCl gradient). Under these conditions, supercoiled non-replicating SV40 DNA is linearized due to a torsional-stress-dependent nuclease-sensitive site located in the T-antigen gene intron (Iacono-Connors and Kowalski, 1986). However, the replicated portions of replicating molecules are not torsionally stressed, and these portions are cut at replication forks, as diagrammed in Fig. 4. Nuclease digestion was stopped by addition of solid CsCl, and the heavy-light BrdUrd-containing nascent DNA was isolated by two rounds of isopycnic centrifugation in CsCl. The final fractions of dense DNA were pooled, concentrated by ethanol precipitation, fractionated in a neutral agarose gel, blotted to a nylon membrane, and sequentially hybridized to the SV40 probes shown in the diagram at the left of the figure.

Fig. 5. Use of alkaline nascent strand isolation to map directions of SV40 replication fork movement. CV-1 cells were infected with SV40, and 36 hours later they were labeled with BrdUrd (15 µg/ml) and ^{3}H-dCyd (1 µCi/ml) for 10 minutes. Subsequent procedures were carried out in dim yellow light. Pulse labeling was stopped by adding 2-3 volumes of ice cold PBS. Nuclei isolated from the labeled cells were lysed in an alkaline buffer (0.25 M NaOH, 0.5 M NaCl, 100 mM EDTA, 0.2% Sarkosyl) at 50° C for 30 min, and the lysate was gently layered onto an alkaline sucrose gradient (5% to 20% sucrose in the same buffer). After centrifugation at 17,000 rpm for 16 hrs at 21° C, the fractions containing DNA strands of 0 to 40 kb were collected, pooled, neutralized, and centrifuged to equilibrium in a neutral CsCl density gradient. The dense, ^{3}H-labelled DNA fractions were pooled, denatured, and centrifuged to equilibrium a second time. The final fractions of dense DNA were pooled, concentrated by ethanol precipitation, fractionated in an alkaline agarose gel, blotted to a nylon membrane, and sequentially hybridized to the SV40 probes shown in the diagram at the left of the figure.

It is evident that both the alkaline approach (Fig. 5) and the neutral approach (Fig. 6) yield nearly complete purification of intact SV40 nascent strands. Each of the probes used detects nascent strands of the sizes expected on the basis of the known location of the SV40 origin and known directions of replication fork movement. Clearly, if these experiments had been carried out on unknown material and the results in Figs. 5 and 6 had been obtained, it would have been possible to conclude that probes 1 and 2 are located close to an origin, that probe 4 is located far from an origin, and that forks move from the vicinity of probe 1 through probe 5 towards probe 4.

Similarly clear results were obtained by Vassilev and Johnson (1989) when they applied nascent strand size analysis (with alkaline denaturation, alkaline sucrose gradient size fractionation, and use of anti-BrdUrd antibodies for nascent strand purification) to the SV40 model system. These promising results make us hopeful that NSSA will prove useful in mapping directions of replication and localizing origins and termini in unknown regions of mammalian chromosomal DNA.

Summary. Techniques for measuring the direction(s) of DNA replication fork movement have permitted us to locate yeast replication origins and to compare directions of replication fork movement with directions of transcription in yeast. The nascent strand size analysis technique seems likely to permit similar measurements in mammalian cells.

LITERATURE REFERENCES

Brewer BJ (1988) When polymerases collide: replication and the transcriptional organization of the E. coli chromosome. Cell 53:679-686

Brewer BJ, Fangman WL (1987) The localization of replication origins on ARS plasmids in S. cerevisiae. Cell 51:463-471

Brewer BJ, Fangman WL (1988) A replication fork barrier at the 3' end of yeast ribosomal RNA genes. Cell 55:637-643

Burhans WC, Vassilev LT, Caddle MS, Heintz NH, DePamphilis ML (1990) Identification of an origin of bidirectional DNA replication in mammalian chromosomes. Cell 62:955-965

Cusick ME, DePamphilis ML, Wassarman PM (1984) Dispersive segregation of nucleosomes during replication of simian virus 40 chromosomes. J Mol Biol 178:249-271

Feldman JB, Hicks JB, Broach JR (1984) Identification of sites required for repression of a silent mating type locus in yeast. J Mol Biol 178:815-834

Handeli S, Klar A, Meuth M, Cedar H (1989) Mapping Replication Units in Animal Cells. Cell 57:909-920

Hanley PJB, Carrigan CM, Rowe DB, Wake RG (1987) Breakdown and quantitation of the forked termination of replication intermediate of *Bacillus subtilis*. J Mol Biol 196:721-727

Herman TM, DePamphilis ML, Wassarman PM (1979) Structure of chromatin at deoxyribonucleic acid replication forks: Okazaki fragments released from replicating SV40 chromosomes by single-strand specific endonucleases are not in nucleosomes. Biochemistry 18:4563-4571

Huberman JA, Zhu J, Davis LR, Newlon CS (1988) Close association of a DNA replication origin and an *ARS* element on chromosome III of the yeast, *Saccharomyces cerevisiae*. Nucleic Acids Res 16:6373-6384

Iacono-Connors L, Kowalski D (1986) Altered DNA conformation in the gene regulatory region of torsionally-stressed SV40 DNA. Nucl Acids Res 14:8949-8962

James CD, Leffak M (1986) Polarity of DNA replication through the avian alpha-globin locus. Mol Cell Biol 6:976-984

Leffak M, James CD (1989) Opposite replication polarity of the germ line c-*myc* gene in HeLa cells compared with that of two Burkitt lymphoma cell lines. Mol Cell Biol 9:586-593

Leu T-H, Hamlin JL (1989) High-resolution mapping of replication fork movement through the amplified dihydrofolate reductase domain in CHO cells by in-gel renaturation analysis. Mol Cell Biol 9:523-531

Linskens MHK, Huberman JA (1988) Organization of replication of ribosomal DNA in *Saccharomyces cerevisiae*. Mol Cell Biol 8:4927-4935

Nawotka KA, Huberman JA (1988) Two-dimensional gel electrophoretic method for mapping DNA replicons. Mol Cell Biol 8:1408-1413

Newlon CS, Green RP, Hardeman KJ, Kim KE, Lipchitz LR, Palzkill TG, Synn S, Woody ST (1986) Structure and organization of yeast chromosome III. In Yeast Cell Biology, UCLA Symposia on Molecular and Cellular Biology, New Series, Vol. 33, Hicks J, ed. Alan R. Liss, Inc. New York, pp. 211-223

Roufa DJ, Marchioni MA (1982) Nucleosome segregation at a defined mammalian chromosomal site. Proc Natl Acad Sci USA 79:3143-3147

Seidman MM, Levine AJ, Weintraub H (1979) The asymmetric segregation of parental nucleosomes during chromosome replication. Cell 18:439-449

St John TP, Davis RW (1981) The organization and transcription of the galactose gene cluster of *Saccharomyces*. J Mol Biol 152:285-315

Trempe JP, Lindstrom YI, Leffak M (1988) Opposite replication polarities of transcribed and nontranscribed histone H5 genes. Mol Cell Biol 8:1657-1663

Tsubota Y, Waqar MA, Davis LR, Spotila L, Huberman JA (1982) Both parental deoxyribonucleic acid strands at each replication fork of replicating simian virus 40 chromosomes are cut by a single-strand-specific endonuclease. Biochemistry 21:2713-2718

Vassilev L, Russev G (1988) Purification of nascent DNA chains by immunoprecipitation with anti-BrdU antibodies. Nucl Acids Res 16:10397

Vassilev L, Johnson EM (1989) Mapping initiation sites of DNA replication *in vivo* using polymerase chain reaction amplification of nascent strand segments. Nucl Acids Res 17:7693-7705

Vassilev L, Johnson EM (1990) An initiation zone of chromosomal DNA replication located upstream of the c-*myc* gene in proliferating HeLa cells. Mol Cell Biol 10:4899-4904

Vaughn JP, Dijkwel PA, Hamlin JL (1990) Replication initiates in a broad zone in the amplified CHO dihydrofolate reductase domain. Cell 61:1075-1087

Yoshikawa A, Isono K (1990) Chromosome III of *Saccharomyces cerevisiae:* An ordered clone bank, a detailed restriction map, and analysis of transcripts suggest the presence of 160 genes. Yeast 6:383-401

The Curious Case of the Novel 2-D Gel Artifact

Bonita J. Brewer
Department of Genetics
University of Washington
Seattle, WA 98195

The rDNA locus in the yeast *Saccharomyces cerevisiae* consists of approximately 200 tandem copies of a 9.08 kb repeating unit located on the right arm of chromosome XII (reviewed in Warner, 1989). Within each repeat is a 5S rRNA gene and the 18S, 5.8S and 26S rRNA genes transcribed as a 35S RNA precursor (Figure 1). The 5S gene is transcribed by RNA polymerase III and the 35S RNA is transcribed by RNA polymerase I. The orientation of the repeats on the chromosome results in PolIII moving toward the telomere and PolI moving toward the centromere.

Replication of this 1800 kb stretch of chromosome XII is initiated at origins of replication spaced at intervals of approximately 5 repeat units (Linskens and Huberman, 1988; Brewer and Fangman, 1988). These origins map to the ARS element previously located in the non-transcribed spacer between the 5S and 35S promoters (Kouprina and Larionov, 1983; Saffer and Miller, 1986). The two replication forks that are established at the origin proceed unequally: the fork moving toward the telomere is arrested at a site, the Replication Fork Barrier (RFB), roughly coincident with the 3' end of the adjacent 35S transcription unit while the fork moving toward the centromere is unimpeded and completes the replication of several repeats before encountering the fork arrested at the RFB in the adjacent replicon (Linskens and Huberman, 1988; Brewer and Fangman, 1988). This basic organization of replication results in the largely unidirectional replication of the rDNA locus.

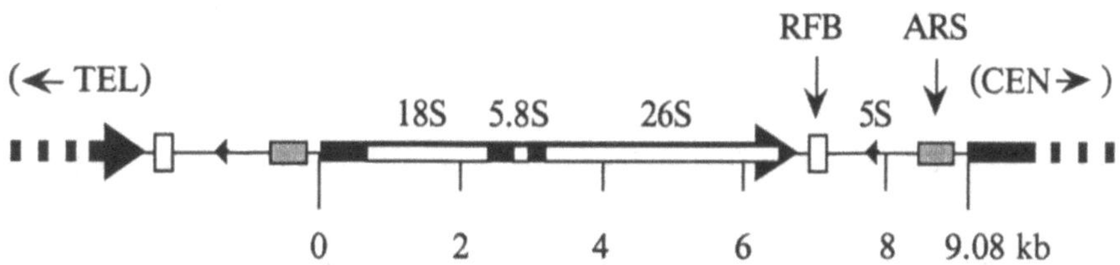

Figure 1. Physical Map of the Yeast rDNA Repeat. The horizontal line represents a section of the rDNA locus on chromosome XII; the horizontal arrows indicate the two major transcripts. The larger PolI transcript contains the sequences for the three rRNAs indicated as open boxes. The locations of the ARS and the RFB are indicated as boxed regions in the non-transcribed spacer regions. The orientation of the repeats is indicated relative to the centromere and the right telomere.

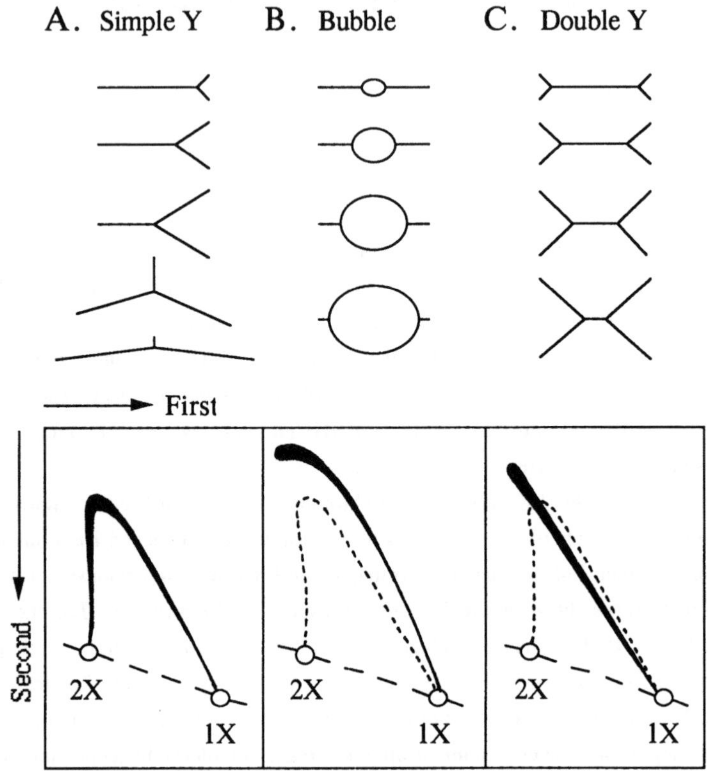

Figure 2. Cartoon of a Hypothetical Restriction Fragment that is Replicated by Three Different Mechanisms. (A) A single fork, (B) two forks diverging from an origin located within the restriction fragment, and (C) two forks converging toward a replication terminus. Patterns drawn are those expected for a large (8 kb) restriction fragment. See Brewer and Fangman (1987) for details.

Details of the replication of the rDNA were deduced by analyzing replication intermediates on two-dimensional (2-D) gels (Brewer and Fangman, 1988; Linskens and Huberman, 1988). In addition to the typical linear restriction fragments, nuclear DNA from an actively growing, asynchronous culture contains branched restriction fragments that are the intermediates in replication. A linear restriction fragment appears on a 2-D gel as a prominent spot on the diagonal of linear molecules. The intermediates in replication are retarded in both dimensions of electrophoresis relative to the spot of linears. In the first dimension of electrophoresis they are predominantly retarded by their increased mass. In the second dimension replication intermediates are retarded, relative to a linear molecule of equal mass, in characteristic ways that are the consequence of the length, number and arrangement of branches (Figure 2; Brewer

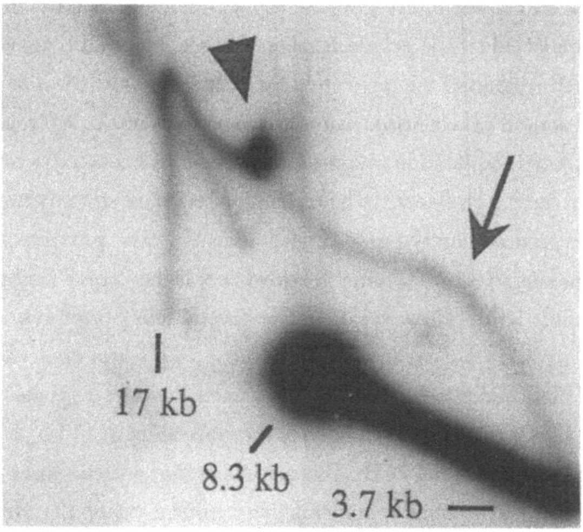

Figure 3. Replication of the 8.3 kb EcoRV Fragment from rDNA. Nuclear DNA isolated from an asynchronous population of yeast cells was cleaved with EcoRV and subjected to 2-D gel electrophoresis as described (Brewer and Fangman, 1987). The first dimension of electrophoresis was run from left to right; the second dimension of electrophoresis was run from top to bottom. To detect the 8.3 kb EcoRV fragment of rDNA a 950 bp fragment from the region containing the RFB was used as probe (see Figure 4). The 8.3 kb linear molecules are labeled; size markers are indicated along the arc of linears. Replication intermediates corresponding to the arrested RFB fragments are indicated by an arrow head. The unusual branched forms are indicated by the arrow.

and Fangman, 1987). For example, molecules with a single replication fork (simple Y intermediates) migrate as an arc of molecules that begins at the spot of linears (1X) and returns to the diagonal of linears at a mass of twice the original restriction fragment size (2X). Replication intermediates with an origin (bubbles) or a terminus (double-Ys) also arise from the linear spot. However, as these molecules double in mass their shape does not reapproach linearity and the intermediates do not return to the diagonal of linears. Regardless of the form of the replication intermediates, they occupy a position in the first dimension gel between the non-replicating linear fragment (1X) and one roughly twice its size (2X), and in the second dimension they will necessarily be continuous with the spot of 1X linears.

"The STATEMENT of the CASE"

In the course of our studies on the replication of the rDNA locus a particular combination of restriction enzyme and probe generated a novel and perplexing 2-D gel pattern. Yeast nuclear DNA from an exponentially growing asynchronous population of cells was cleaved with EcoRV and subjected to 2-D gel electrophoresis. The blot of this gel was then hybridized with a probe from the region of the replication fork barrier (Figure 3). The linear DNA detected with this combination of restriction enzyme and probe is an 8.3 kb fragment (see Figure 4 for restriction map). All replication intermediates were thus expected to occupy the region of the gel to the left of the 8.3 kb linears. The replication intermediates migrating to the left of the 8.3 linear spot are predominantly simple and double Ys. The prominent spot of branched molecules at approximately 11 kb (Figure 3, arrowhead) is the EcoRV fragment with a replication fork stalled at the RFB. However, there are several unexpected results on this 2-D gel. None of the intermediates appear to arise from the 8.3 kb linear spot. Instead, there is an unexpected arc of branched intermediates that intersects with the diagonal of linears at a mass of approximately 3.7 kb and extends in mass to the equivalent of 11 kb, at the position of the molecules with forks stalled at the RFB. This autoradiogram presents three puzzles: (1) From what rDNA fragments do the small replication intermediates arise? (2) How, in the course of replicating a fragment with the apparent mass of 3.7 kb (1X), can it achieve a mass that is equivalent to a fragment of >11 kb (approximately 3X)? (3) And, why are there no intermediates arising from the 8.3 kb fragment?

If branched intermediates arise from the position of 3.7 kb then there should be some linear 3.7 kb fragment that gives rise to them. Previous restriction maps of yeast rDNA show only a single EcoRV site (at position 0.35 kb). Restriction mapping of the rDNA in our strain revealed a second EcoRV site (at position 8.64 kb; see Figure 4B). These two sites result in fragments of 8.3 kb and 0.79 kb. In the 2-D gels the smear of hybridization along the linear diagonal below 8.3 kb may have obscured a 3.7 kb fragment if it were a minor species. To look specifically at the possibility of heterogeneity in the position of EcoRV sites in the rDNA repeats, DNA was analyzed at a lower exposure on a one dimensional gel. Nuclear DNA from the asynchronous culture was cut with different restriction enzymes and hybridized with the same RFB probe (Figure 4A). There is no specific restriction fragment of 3.7 kb released with EcoRV, or with any of the other enzymes, indicating that the replication intermediates are not arising from a subset of repeats with different restriction maps. However, in each lane, except for the AvaI digest, a continuous smear and somewhat discrete but fuzzy band of hybridization is observed below the main band. Its size varies with the enzyme used (1.9 kb EcoRV; 2.1 kb SmaI; 2.4 kb BglII; 3.1 kb XbaI). If we assume that these smaller fragments have a restriction site at one end and a chromosomal break (induced either *in vivo* or *in vitro*) at the other end, then the site of the break can be mapped. Using data from the digests shown in Figure 4 the chromosomal break maps to the position of the RFB. In the EcoRV digest a subfragment of approximately 1.85 kb, but not one of 3.7 kb, is liberated by cleavage at the EcoRV site because of a preexisting break at the RFB in a subset of repeats.

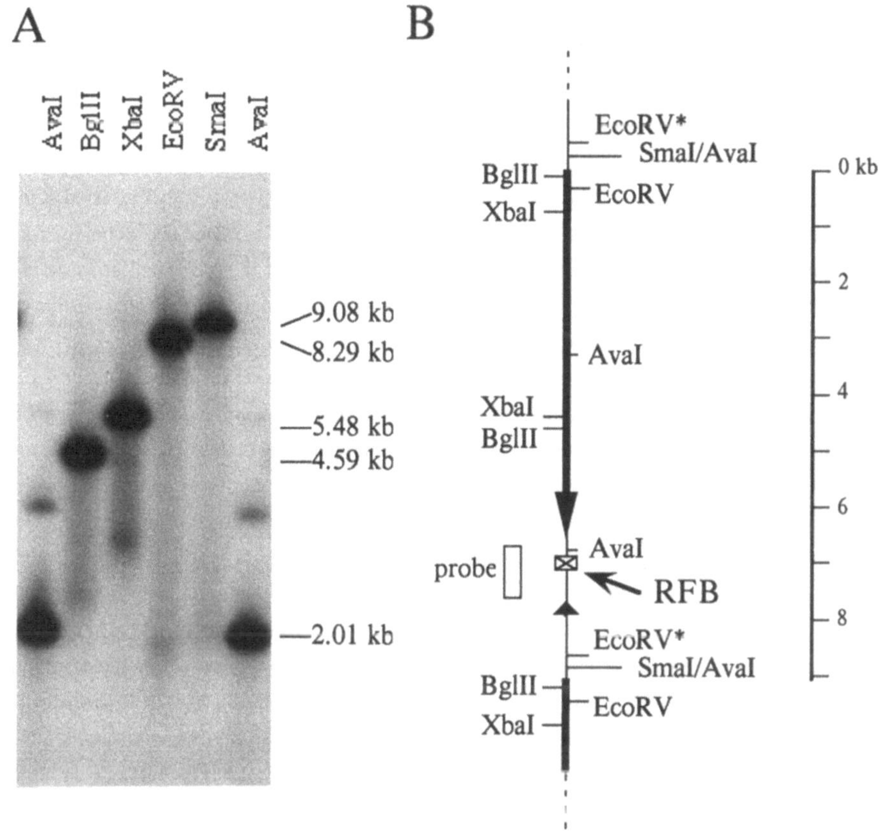

Figure 4. Genomic Restriction Analysis of the rDNA. Total nuclear DNA digested with the enzymes indicated was separated on a 0.4% gel at 1 V/cm for 18 hr. rDNA fragments were detected by hybridization to the 950 bp probe indicated by the open box in panel B. All restriction fragments confirmed previously identified enzyme sites with the exception of a second EcoRV* site within the ARS region at position 8.64.

"A CASE of IDENTITY"

Since there is no specific fragment of 3.7 kb in the EcoRV digest, the replication intermediates that appear to arise from the 3.7 kb intercept of the linear diagonal in Figure 3 remain unexplained. However the idea of preexisting (or induced) breaks proved to be instrumental in resolving the confusing 2-D pattern. As can be seen from both Figures 3 and 4, the majority of the linear rDNA fragments are the expected size of 8.3 kb; however, because of the normal fragmentation that occurs during the isolation and purification of the chromosomal DNA, there is a continuous smear of smaller fragments. The average size of chromosomal fragments recovered from cells is 50 to 100 kb (data not shown). Thus on average, every 5th to 11th repeat has suffered a break due to shear. Sites that have persistent branches--such as the RFB--may be more prone to breakage. With this exception, the breaks are likely to occur at random through the repeats.

Replication intermediates may also be broken by random shear. The most prominent replicating form among the 8.3 kb rDNA intermediates is a molecule with a fork arrested at the RFB--two replicated arms of 1.85 kb and a 6.45 kb unreplicated 35 S gene region (arrowhead in Figure 3). As already discussed, a break at the RFB in one of the replicated arms would generate a 1.85 kb fragment. However, if a break were to occur in the unreplicated portion of these replication intermediates, then Y forms of variable size will be produced. A series of fragmented intermediates are illustrated in Figure 5A, where a break is proposed to occur at 0.5 or 1.0 kb intervals along the unreplicated arm of the EcoRV fragment. Probing with the RFB portion of the molecule would detect this series of simple Y intermediates. However, these are not intermediates in the replication of a specific restriction fragment of a common length--the fork does not vary in its position, instead one end of the fragment varies in its position. The smallest branched form detected would be a 1.85 kb fragment that is nearly 100% replicated. Due to isomerization of the fragment during electrophoresis it can adopt a form in which the two replicated arms become aligned giving the illusion of a 3.7 kb fragment with a very small side branch consisting of the unreplicated, broken arm. The next intermediate illustrated in Figure 5A is a 2.35 kb fragment that has two replicated arms of 1.85 kb (79% replicated). Continuing in this fashion, the next intermediate would be a 2.85 kb fragment that is 65% replicated. As the position of the break moves away from the RFB, the apparent per cent replication continues to decrease, such that a molecule in which the break is 1.85 kb away from the RFB would appear as a 3.7 kb fragment that is 50% replicated. With breaks more distant from the RFB, the extent of replication of the fragment is reduced still further to less than 50%.

To predict the migration of each of the fragments of variable length and extent of replication it is useful to consider each fragment separately. For example, consider first the migration behavior of the complete arc of simple Ys for a hypothetical restriction fragment of 1.85 kb (Figure 5B). Along this arc the location of a specific branched structure can be estimated given its apparent per cent of replication. The idealized migration patterns of simple Y replication intermediates of various sizes are illustrated in Figure 5B by overlapping lines beginning at the right for a 1.85 kb fragment. The results for the specific branched intermediates that result from breakage of the RFB arrested intermediates are indicated by the filled circles along each simple Y arc. When these circles are connected--as they certainly would be if fragmentation occurred randomly and not just at 0.5 kb intervals--then the branched forms generate the observed 2-D gel pattern, beginning at the apparent size of 3.7 kb.

To determine if this explanation is correct, the blot was stripped and reprobed with a fragment from the opposite end of the EcoRV fragment. The expectation is that the smaller

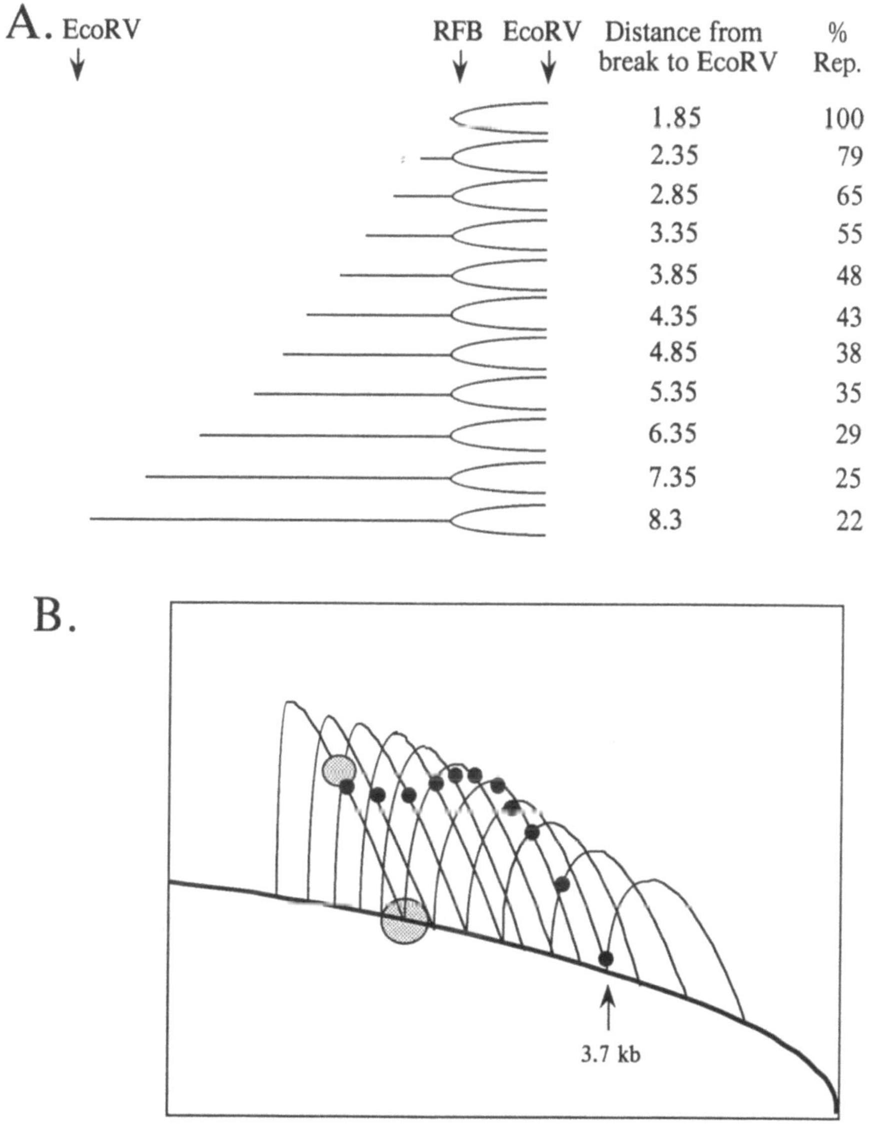

Figure 5. Broken Replication Intermediates. A. Replication intermediates with a single fork stalled at the RFB are illustrated with a break occurring at 0.5 or 1.0 kb intervals from the stalled fork. The distance from the break to the EcoRV site is indicated in kb. The extent of replication of each fragment is determined by dividing 1.85 kb by the length of the fragment. B. A hypothetical 2-D gel with simple Y intermediates of various sizes indicated by the overlapping arcs. The position of the 8.3 kb linear and the full length stalled RFB fragment are indicated by shaded circles. The position of intermediates in A are indicated along their respective simple Y arcs.

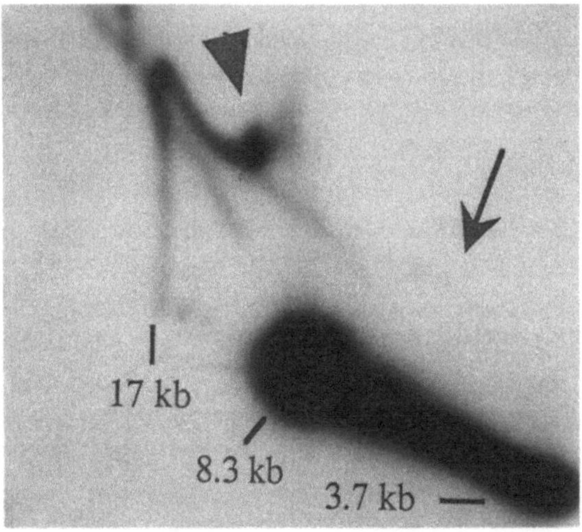

Figure 6. Rehybridization of the 2-D Gel. The 2-D gel shown in Figure 3 was stripped and rehybridized with a probe from the 18S portion of the rDNA repeat. The prominent arc of branched molecules with a mass less than 8.3 kb is absent (compare with Figure 3.)

members of the branched set would not be detected with this probe. Only those fragmented molecules that have a break within the region of the probe would be detected. The result, shown in Figure 6, confirms this prediction.

"The FINAL PROBLEM"

The answer to the third question--why are there no intermediates continuous with the 8.3 kb linear spot?--is still unclear. Rarity of a particular class of intermediates may simply signify that they are shorter lived or more labile. Alternately, their absence may result from a novel form of replication initiation. If, for example, initiation of replication at the rDNA origin were to involve the formation of a short D-loop, then one parental strand would be converted to duplex before the complementary parental strand (Figure 7). This asymmetry would result in a bubble with one single strand and one double strand. Since the new EcoRV site identified in our strain lies within the ARS region, only one of the daughter strands in newly initiated molecules may be cut (Figure 7). Cleavage would result in a molecule that is basically 9.08 kb with several short, duplex branches and a single stranded region covering the EcoRV site. The behavior of such molecules on 2-D gels is unknown. However, it is clear that if this explana-

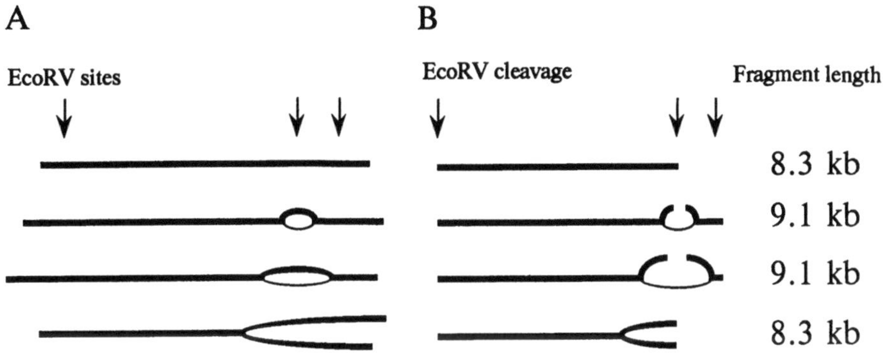

Figure 7. Hypothetical D-Loop at the rDNA Origin of Replication. A collection of chromosomal molecules at various stages of replication are shown. Double stranded DNA is indicated as a thick line; the single stranded D-loop is illustrated as a thin line. A. Molecules before cleavage by EcoRV. B. The same molecules after cleavage with EcoRV. The length of the molecule, without regard for the added mass contributed by the replicated portions, is indicated for each intermediate.

tion were true then one would not detect replication intermediates arising from the 8.3 kb spot. However, they should arise from the position of 9.08 kb unless they are especially labile, short lived or if their single stranded portions gave them an anomalous mobility. This model is currently being tested.

The replication of a conventional restriction fragment by a single replication fork generates a series of intermediate stages where the ends of the restriction fragment are invariant but the lengths of replicated and unreplicated branches undergo a reciprocal change--as the replicated arms increase in length the unreplicated arm decreases by the same amount. The mass of the fragment thereby increases from 1X to 2X as a function of the extent of replication. In the case described above random fragmentation can account adequately for both the appearance of intermediates arising from the 3.7 kb position along the linear arc and the apparent increase in mass of the intermediates from 1X (3.7 kb) to 3X (approximately 11 kb). The solution to this particularly puzzling 2-D gel pattern came from realizing that there wasn't a single 3.7 kb fragment that passes through stages of increased replication to become 11 kb. Instead the length of the replicated branches remains constant but one of the ends of the fragment is variable. It is perhaps surprising that an artifact can create such an artfully deceptive pattern on 2-D gels. This result reveals the importance of fully understanding the molecules in question when analyzing their replication intermediates with 2-D gels.

ACKNOWLEDGMENTS: I am indebted to Pieter Dijkwel and Michael Leffak for a most illuminating discussion at Les Arcs. It was Michael Leffak's suggestion that specific fragmentation might somehow be responsible for this bizarre pattern. The manuscript benefitted from the skillful editing of Walt Fangman. Subheading in the text were taken from the works of Sir Arthur Conan Doyle. I am grateful for the moral support of my friends and colleagues in the Genetics Department at the University of Washington and for the financial support from NIGMS.

REFERENCES:

Brewer, B. J. and Fangman, W. L. (1987) The localization of replication origins on ARS plasmids in *S. cerevisiae.* Cell 51, 463-471.

Brewer, B. J. and Fangman, W. L. (1988) A replication fork barrier at the 3' end of yeast ribosomal RNA genes. Cell 55, 637-643.

Kouprina, N. Y. and Larionov, V. L. (1983) The study of a rDNA replicator in *Saccharomyces.* Curr. Genet. 7, 433-438.

Linskens, M. and Huberman, J. A. (1988) Organization of replication of ribosomal DNA in *Saccharomycer cerevisiae.* Mol. Cell. Biol. 8, 4927-4935.

Saffer, L. D. and Miller, Jr., O. L. (1986) Electron miceroscopic study of *Saccharomyces cerevisiae* rDNA chromatin replication. Mol. Cell. Biol. 6, 1148-1157.

Warner, J. (1989) Synthesis of Ribosomes in *Saccharomyces cerevisiae.* Microbiol. Rev. 53, 256-271.

Structural and Functional Characteristics of Autonomously Replicating Mammalian Origin-Enriched Sequences (ORS)

M. Zannis-Hadjopoulos, C.E. Pearson, D. Bell, D. Mah, M. McAlear and G.B. Price
McGill Cancer Centre, McGill University, 3655 Drummond Street, Montréal, Québec H3G 1Y6, Canada

Cloned origin-rich sequences (ors) isolated by strand extrusion from mammalian (monkey and human) cells are capable of transient autonomous semiconservative replication in transfected HeLa cells. Sequence analyses reveal no primary consensus sequence features that can be associated with origin function, but several common structural features have been noted which include A/T-rich stretches of DNA and inverted repeats that can potentially give rise to cruciform structures. Some similarities to the yeast ARS consensus, the consensus for attachment to nuclear scaffold and for other regulatory elements have also been noted. The data suggest that origin activation is most probably dependent on the presence of structural determinants rather than on primary sequence.

Anti-cruciform DNA monoclonal antibodies, when introduced in replicating cells, enhance overall DNA synthesis up to 6-fold and amplify the relative copy number of low-copy elements up to 11-fold, seemingly by allowing multiple initiations at the same origin.

An *in vitro* system for ors DNA replication has been established that uses HeLa cell extracts and is dependent upon the presence of an ors template. *In vitro* replication initiates in the ors, is semi-conservative, bidirectional, aphidicolin-sensitive and is independent of SV40 T-antigen. Studies involving the anti-cruciform monoclonal antibody 2D3 *in vitro* confirm and extend our previous observations suggesting cruciform involvement in early initiation events.

The antibody has also been used to affinity-purify cruciform-containing genomic DNA, which was then cloned and analyzed for transient autonomous replication in HeLa cells. Four of nine randomly picked clones demonstrate semiconservative autonomous replication; this is in contrast to twelve independent clones that contain random genomic DNA fragments of equivalent size, which do not replicate in similar assays. Libraries of origin-containing DNA, enriched by affinity-purification with anti-cruciform antibody, from monkey, normal human and malignant human cells have been constructed.

INTRODUCTION

We are interested in the structural and functional characterization of mammalian origins of DNA replication. Using the strand extrusion method (Zannis-Hadjopoulos et al., 1981), we isolated from replicating mammalian (monkey and human) cells (Kaufmann et al. 1985; McAlear and Zannis-Hadjopoulos, submitted) chromosomal nascent origin-rich sequences (ors) that are activated at the onset of S phase, and cloned them in pBR322. Each of these ors should contain an origin of DNA replication at or near its center (Zannis-Hadjopoulos et al., 1983). We have completed detailed sequence (Rao et al., 1990; Landry and Zannis-Hadjopoulos, 1991), biochemical (Zannis-Hadjopoulos et al., 1984; 1985) and functional analyses on twenty-five monkey and five human ors, and found that they are capable of initiating autonomous semiconservative DNA replication upon transfection into HeLa cells (Frappier and Zannis-Hadjopoulos, 1987; Landry and Zannis-Hadjopoulos, 1991; McAlear and Zannis-Hadjopoulos, submitted). Autonomous replication was assayed by transient episomal replication in transfected HeLa cells using the Dpn I resistance assay and the bromodeoxyuridine (BrdUrd) substitution assay.

The primary nucleotide sequences of twenty five monkey and five human ors were obtained and subjected to computer-assisted analyses (Rao et al., 1990; Landry and Zannis-Hadjopoulos, 1991). Extensive pair-wise comparisons of the ors showed no significant sequence homologies, except for long asymmetrically distributed A/T-rich regions and inverted repeats (IRs). Five published consensus sequences [the B1 intron (Galson and Housman, 1988); the ß-globin control element, CACCC (Dierks et al., 1983); the yeast (S. cerivisiae) ARS consensus (Palzkill and Newlon, 1988); and the consensus for scaffold attachment regions, SAR-T and SAR-A (Gasser and Laemmli, 1986)] were also found to be statistically over-represented in at least twelve monkey and all of the human ors that we have analysed. Based upon the combined results of our analyses, we have arbitrarily classified our mammalian ors according to their association with highly or moderately repetitive and unique sequences (Landry and Zannis-Hadjopoulos, 1991).

Our results indicate that activation is dependent upon the presence of certain structural determinants. One of these structural determinants includes inverted repeats with the potential to form cruciforms. We have demonstrated that anti-cruciform antibody (Frappier et al., 1987) can enhance the replication of mammalian DNA up to 6-fold and increase the relative copy number of individual sequences up to 11-fold (Zannis-Hadjopoulos et al., 1988). Recently, we have developed a cell-free system which allows the specific initiation and replication of ors DNA templates (submitted), and further demonstrates the association of IRs and their potential to form cruciforms with ors replication function (in preparation).

Results and Discussion

In Vitro Replication of Ors

Using extracts from HeLa S3 cells, adapted for suspension culture, and reaction conditions similar to those described by Decker et al. (1986) for the in vitro replication of SV40 virus DNA, but excluding SV40 T antigen, we have developed a cell free system for ors replication (Frappier, 1988; Pearson et al., submitted). Four monkey ors plasmids (ors 3, 8, 9 and 12) that were previously shown to replicate autonomously when transfected in mammalian cells (Frappier and Zannis-Hadjopoulos, 1987) were used as templates in this system (Fig.1). Although a small amount of radioactive nucleotide precursor incorporation is sometimes observed in the pBR322 vector DNA, the incorporation into the four ors plasmids is consistently higher by several fold. This incorporation is sensitive to aphidicolin (Fig. 1A, lanes +A), suggesting the involvement of DNA polymerases α and/or δ, which are specifically inhibited by the drug. In contrast, the more limited and inconsistent incorporation that is observed in the vector DNA alone is not affected by the use of aphidicolin (not shown), and is attributed to repair synthesis (see below). Recovered products of the in vitro reaction include relaxed circular (form II) and linear (form III) DNAs, replicative intermediates (RI) that are migrating slower than form II, and a ladder of bands migrating between forms II and III and between forms I and III, that are indicative of topoisomeric molecules (Fig. 1A,B). Addition of high salt nuclear extracts to the reactions led to the formation

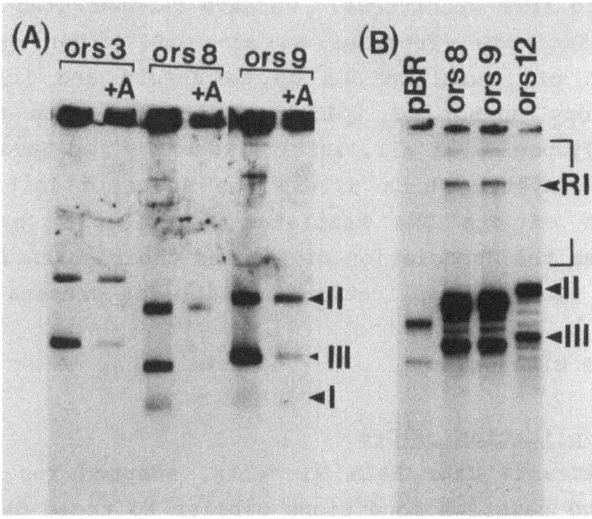

Figure 1. In vitro replication of ors plasmid DNAs. Ors 3, 8, 9 12 and pBR322 plasmid DNAs were incubated in reaction mixtures containing HeLa cytosolic and high-salt nuclear extracts (A) or cytosolic extracts alone (B) in the presence (+A) or absence of 30uM aphidicolin. Reactions conditions were as described in Decker et al. (1986) but without SV40 T antigen. After incubation at 30°C for 60 min, the DNAs were purified, concentrated, subjected to electrophoresis in a 0.6% agarose gel and the gel was then dried and exposed for autoradiography. Replicative intermediate (RI), linear (III), relaxed circular (II) and super-coiled (I) ^{32}P-labeled DNA products are indicated. The replication products of ors 8 DNA templates in (A) (without aphidicolin) and (B) represent a nucleotide incorporation rate of 5 and 40 pmoles/hr, respectively, based on total acid-insoluble radioactivity.

of completely supercoiled (form I) plasmid DNA (Fig. 1A), as has also been observed in the SV40 in vitro replication system (Stillman and Gluzman, 1985). In the absence of nuclear extracts, the incorporation of precursor nucleotide into ors plasmid DNA increased approximately 8-fold and no form I DNA products were usually recovered, although the presence of ladder bands on either side of form III DNA indicated the presence of a series of topoisomeric molecules with supercoils (Fig. 1B). Omission of exogenously added template DNA from the reactions yielded no products (not shown).

In order to further demonstrate the validity of the cell free system as a mammalian ors replication system, we measured the incorporation of radioactive precursor nucleotide as a function of time into ors plasmid DNA versus the vector DNA alone. As can be seen in Fig. 2 the incorporation into ors templates shows a

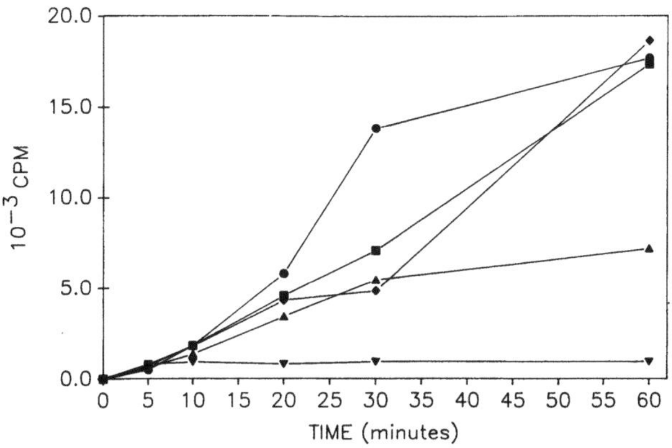

Figure 2. Incorporation of α-^{32}P-dCTP into ors 3, 8, 9, 12 and pBR322 as a function of time. In vitro reactions were carried out as in Fig. 1 using both cytoplasmic and nuclear extracts. The DNA products were linearized with PstI, run on 1% agarose and full length (form III) molecules were excised, soaked in Aquasol for 1h and counted for radioactivity in a scintillation counter, as described by Guo et al., (1989). ◆, ors 3; ■, ors 8; ▲, ors 9; ●, ors 12; ▼, pBR322.

steady increase over a period of 60 min after an initial lag of 5-10 min, whereas that for the vector alone is near basal levels.

Finally, the replication of ors in the cell free system is semiconservative, as shown by in vitro incorporation of BrdUTP and separation of substituted (heavy) and unsubstituted (light) DNA strands on neutral and alkaline CsCl density gradients (Fig 3). Figs. 3A and 3B show the resultant density distribution on alkaline gradient of ors 8 and pBR322 DNAs replicated in the absence and presence of BrdUTP and then either digested by AluI (Fig.3A), which generates multiple fragments, or PstI (Fig. 3B), which generates full length linear plasmid DNAs; in both enzyme treatments, ors 8 replicated in the presence of BrdUTP is recovered as a clear distinct peak at the density expected of fully substituted (H) single strands, whereas that replicated in

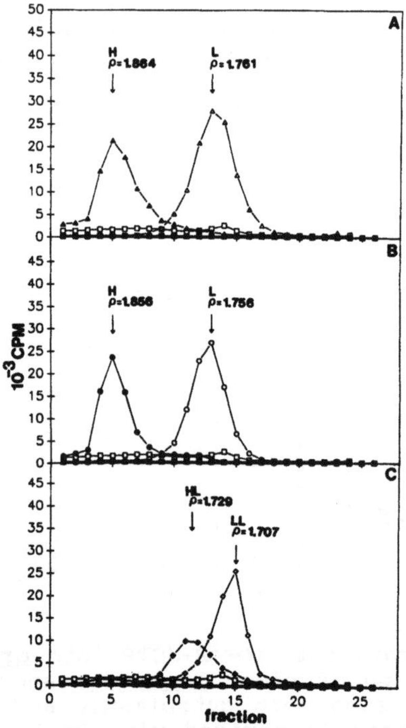

Figure 3. Isopycnic centrifugation in alkaline and neutral gradients of in vitro replication products of ors 8 and pBR322 Reactions were performed as described in Fig.1 using both cytoplasmic and nuclear extracts, except that 5-BrdUTP (final concentration of 100 uM) was used in place of dTTP and was added fresh to reaction mixes (■, pBR322;▲●◆, ors 8). For production of unsubstituted (LL) replication products reactions were performed in the presence of dTTP, as usual (□ , pBR322;△○◇ ors 8). After two ethanol precipitations, DNAs were run on G-50 Sephadex (Nick columns, Pharmacia), then re-precipitated. One third of the reaction (83 ng) were digested with AluI (A) or PstI (B and C), loaded onto alkaline (A and B) or neutral (C) CsCl gradients and centrifuged in a VTi80 rotor for 14 h at 65 Krpm, followed by 2 h at 70 Krpm. Approximately 30 fractions were collected from the bottom and counted for radioactivity. Densities were calculated from refractive indices. The positions of substituted (H) and unsubstituted DNA are indicated (arrows).

the absence of BrdUTP is recovered at the density expected of unsubstituted (L) single strands. The fact that both, AluI and PstI, digests lead to a similar density shift of the BrdUTP-substituted strands is suggestive of semiconservative replication and not of repair. If the latter were true, the amount of BrdUTP incorporation would not be sufficient to cause a density shift, particularly in the short fragments (Smith et al., 1981) generated by Alu I digestion. Fig. 3C shows the resultant density distribution on neutral gradient of in vitro replicated ors 8 and indicates that the products banded at the expected densities for semiconservative ors 8 replication in the absence (unsubstituted, LL) and presence (singly substituted, HL) of BrdUTP, suggesting one round of ors 8 replication in vitro. Similar distribution profiles, as those depicted in Fig.3 for ors 8, were also obtained for ors 3, 9 and 12 replicated in vitro in the presence or absence of BrdUTP (not shown). In contrast, there is no indication that the vector pBR322 replicated under the same in vitro conditions (Fig. 3A-C), suggesting that the low incorporation of precursor observed sometimes when this plasmid is used as template may be due to repair synthesis.

Since our results suggest that the in vitro replication system is semiconservative and ors specific, we evaluated the effect of the anti-cruciform DNA monoclonal antibody 2D3 on in vitro ors replication. Fig. 4 shows that the presence of 2D3 anti-cruci-

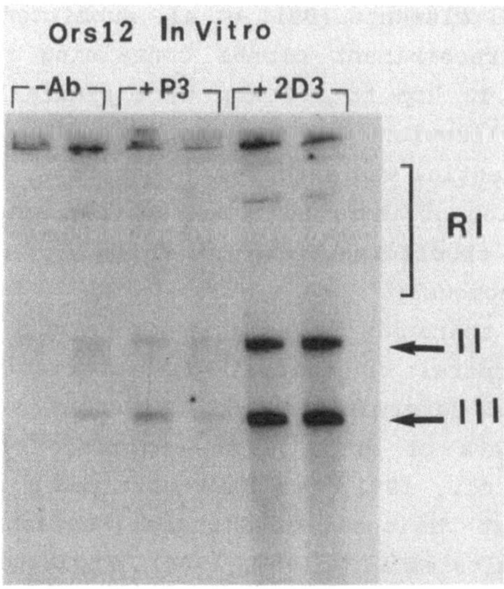

Figure 4. Effect of anti-cruciform DNA monoclonal antibody 2D3 on in vitro replication of ors 12. Reactions were carried out and processed as in Fig. 1B (without nuclear extracts) except that 1 ul of concentrated 2D3 monoclonal antibody was included in the mixture.

form antibody during the *in vitro* replication of *ors* 12 increases the relative copy number of *ors* 12 produced in a set period of time, presumably by the same mechanism as elucidated previously (Zannis-Hadjopoulos, et al., 1988).

Anti-Cruciform DNA Affinity Purification of Active Mammalian Origins:

We have used the anti-cruciform antibody to affinity purify cruciform-containing DNA from minimally sheared, high molecular weight monkey (CV-1) and human (WI38, SW48, and genital fibroblasts) DNA, in order to obtain libraries of sequences containing inverted repeats capable of extruding into cruciforms and investigate their involvement in DNA replication. A library of approximately 2,700 independent clones from log-phase growth CV-1 DNA has been obtained and several randomly selected clones were analysed for autonomous replicating potential. Of nine independent clones ranging in size from 0.15 kbp to 3.2 kbp, four were shown to replicate autonomously in transfected HeLa cells, by the BrdUrd semiconservative replication assay for extrachromosomal elements (Bell et al. submitted). In contrast, twelve pBR322 recombinant clones containing random CV-1 DNA fragments of 0.10 kbp to 3.0 kbp were similarly analysed for autonomous replication in transfected HeLa cells, and no replication potential was detected. Therefore, there appears to be an association of inverted repeats, that have the potential to extrude into cruciforms, with the ability of DNA sequences to replicate autonomously.

Conceivably, there may be variations of structure and sequence related to temporal or spatial characteristics; this would necessitate the determination of consensus sequences within different subsets of *ors*. By the extrusion method (Zannis-Hadjopoulos et al., 1981), we have obtained a set of *ors* that are activated at the onset of S phase (Kaufmann et al., 1985; Zannis-Hadjopoulos et al., 1984; 1985). The alternate method of anti-cruciform DNA affinity-isolation of genomic DNA with active (extruded) cruciforms provides a specifically enriched and larger population of autonomously replicating sequences (Bell et al., submitted), comprising active replication origins from all stages of the S phase. We anticipate that the wealth of mammalian autonomously replicating sequences now available to us and the convenience of a cell-free replication system will facilitate the analysis of the involvement of various DNA sequences and structures, and the definition of factors and mechanisms governing the replication of DNA in mammalian cells.

Acknowledgements

This work was supported by grants from the Medical Research Council of Canada (MA-7965) and the Natural Sciences and Engineering Research Council of Canada; DB and MM were supported by awards from the Cancer Research Society, Inc.

References

Decker RS, Yamaguchi M, Possenti R, DePamphilis ML (1986) Initiation of simian virus 40 DNA replication in vitro: aphidicolin causes accumulation of early-replicating intermediates and allows determination of the initial direction of DNA synthesis. Mol. Cell. Biol. 6:3815-3825

Dierks P, Van Ooyen A, Cochran MD, Dobkin C, Reiser J, Weissmann C (1983) Three regions upstream from the cap site are required for efficient and accurate transcription of the rabbit ß-globin gene in mouse 3T6 cells. Cell 32:695-706

Frappier L (1988) Structural and functional analysis of mammalian origins of DNA replication. Ph.D. Thesis, McGill University

Frappier L, Zannis-Hadjopoulos M (1987) Autonomous replication of plasmids bearing monkey DNA origin-enriched sequences. Proc. Natl. Acad. Sci. USA 84:6668-6672

Galson DL, Housman DE (1988) Detection of two tissue-specific DNA binding proteins with affinity for sites in the mouse ß-globin intervening sequence 2. Mol. Cell. Biol. 8:381-392

Gasser SM, Laemmli UK (1986) Cohabitation of scaffold binding regions with upstream/enhancer elements of three developmentally regulated genes of D.melanogaster. Cell 46:521-530

Guo Z-S, Gutierrez C, Heine U, Sogo JM, DePamphilis ML (1989) Origin auxiliary sequences can facilitate initiation of simian virus 40 DNA replication in vitro as they do in vivo. Mol. Cell. Biol. 9:3593-3602

Kaufmann G, Zannis-Hadjopoulos M, Martin RG (1985) Cloning of nascent monkey DNA synthesized early in the cell cycle. Mol. Cell. Biol. 5:721-727

Landry S, Zannis-Hadjopouos M (1991) Classes of autonomously replicating sequences are found among early-replicating monkey DNA. Biochim. Biophys. Acta in press

Palzkill TG, Newlon CS (1988) A yeast replication origin consists of multiple copies of a small conserved sequence. Cell 53:441-450

Rao BS, Zannis-Hadjopoulos M, Price GB, Reitman M, Martin RG (1990) Sequence similarities among monkey ori-enriched (ors)

fragments. Gene 87:233-242
Smith CA, Cooper PK, Hanawalt PC (1981) Measurement of repair replication by equilibrium sedimentation. In: DNA repair, a lab manual of research procedures, vol.I, part B, EC Friedman PC Hanawalt (eds), Marcel Dekker, Inc., New York and Basel
Stillman BW, Gluzman Y (1985) Replication and supercoiling of simian virus 40 DNA in cell extracts from human cells. Mol. Cell. Biol. 5:2051-2060
Zannis-Hadjopoulos M, Persico M, Martin RG (1981) The remarkable instability of replication loops provide a general method for the isolation of origins of DNA replication. Cell 27:155-163
Zannis-Hadjopoulos M, Chepelinsky AB, Martin RG (1983) Mapping of the 3'-end positions of simian virus 40 nascent strands. J. Mol. Biol. 165:599-607
Zannis-Hadjopoulos M, Kaufmann G, Martin RG (1984) Mammalian DNA enriched for replication origins is enriched for snap-back sequences. J. Mol. Biol. 179:577-586
Zannis-Hadjopoulos M, Kaufmann G, Wang S.-S, Lechner RL, Karawya E, Hesse J, Martin RG (1985) Properties of some monkey DNA sequences obtained by a procedure that enriches for DNA replication origins. Mol. Cell. Biol. 5:1621-1629
Zannis-Hadjopoulos M, Frappier L, Khoury M, Price GB (1988) Effect of anti-cruciform DNA monoclonal antibodies on DNA replication. The EMBO J. 7:1837-1844

Mutations Affecting Replication Origin Function in Yeast

Stephen E. Kearsey and David Kipling[1]
Department of Zoology
University of Oxford
South Parks Road,
Oxford
OX1 3PS
UK

ABSTRACT

A number of mutations that improve the replication competence of plasmids containing defective origins of replication have been isolated in *Saccharomyces cerevisiae*. Some of these are *cis*-acting, and generate origins *de novo* by mutation of prokaryotic vector DNA. Other mutations are *trans*-acting, and complementation analysis has identified a number of yeast genes whose products may have a role in chromosome replication. One gene thus identified, *RAR5*, encodes a 175 kDa product, and appears to behave as a repressor of replication origin function. The *RAR5* gene is identical to *DST2*, encoding a strand transferase activity, and *KEM1*, identified by mutations that enhance the karyogamy defect of a *kar1* mutation.

INTRODUCTION

Early investigations of eukaryotic chromosome replication revealed that, in contrast to bacterial chromosome replication, synthesis is initiated from a large number of origins during the S phase of the cell cycle. Initiation must show the appropriate once-per-cell-cycle timing to ensure that the replication is coordinated with other activities of the cell cycle. In addition, the number of origins that are used in S phase can affect the timing of replication of particular chromosomal

[1] Present address: MRC Human Genetics Unit, Western General Hospital, Crewe Road, Edinburgh, EH4 2XU, UK.

regions, and can be a determinant of the overall duration of S phase.

As in prokaryotes, *cis*-acting DNA sequences appear to be important in governing origin location in chromosomes, though information from higher eukaryotes is still very incomplete. In yeasts, ARS elements have the genetic properties of replication origins (for reviews, see Newlon, 1988; Umek et al., 1989) and in *Saccharomyces cerevisiae* it is apparent that DNA replication initiates close to ARS elements both in plasmids (Brewer and Fangman, 1987) and in the chromosome (Huberman et al., 1988). Although the cell cycle dependent activation of replication in a yeast chromosome requires special DNA sequences in *cis*, other factors can influence origin function. One unexpected result from comparing ARS activity in plasmids and in the chromosome is that many fragments of chromosomal DNA that can function as origins of replication on plasmids, do not do so in the chromosome (Reynolds et al., 1989). A related point is made by consideration of the identical rRNA gene repeats; although each repeat contains an ARS, only a subset is used in any single S phase (Brewer and Fangman, 1988). Although the reason for this phenomenon is not understood it is likely that other chromosomal activities may interfere with DNA replication. For instance, the effect of transcription directed towards an ARS is to cause an apparent reduction in initiation efficiency (Snyder et al., 1988), and transcription may also interfere with the progress of a replication fork (Brewer, 1988). It has been known for some time that not only is the timing of origin activation subject to regulation, but also the distribution of origins in eukaryotic chromosomes may be altered, implying that some potential origins are used only in special circumstances. For instance, treatments that interfere with DNA replication can reduce the spacing between origins, and developmental changes in origin spacing have been documented (for reviews, see Hand, 1978; Diffley and Stillman, 1990).

One approach for understanding the factors that determine origin function is to study mutations that alter the efficiency of origin use. We, and others, have isolated mutations that alter the mitotic stability of plasmids in *S. cerevisiae*, and some of these mutations appear to affect specifically ARS function (Maine et al., 1984; Kearsey and Edwards, 1987; Thrash-Bingham and Fangman 1989; Kipling and Kearsey, 1990). This approach clearly offers the possibility of identifying ARS-binding proteins that may be responsible for replication initiation but, in addition, may be of value for identifying more subtle factors that determine the distribution of replication origins in chromosomes. We describe here the isolation of mutations that improve the replication competence of a defective replicon. In one series of experiments *cis*-acting mutations were detected that generate origins *de novo*, while another strategy has provided *trans*-acting mutations that appear to increase the efficiency of ARS function.

CIS-ACTING MUTATIONS THAT GENERATE REPLICATION ORIGINS

Mutational studies have identified an essential-AT rich consensus sequence as a key component of the ARS element, and flanking regions of DNA can contribute to the efficiency of ARS function (see Newlon, 1988). We have studied mutations that allow a plasmid containing a defective ARS to replicate to provide a fresh perspective on the DNA sequence requirements for replication initiation (Kipling and Kearsey, 1990). We used a series of Ars^- derivatives of bacteriophage M13mp93 containing the *URA3* gene, and a nonfunctional ARS element, derived from the *HO* ARS by point mutagenesis (Kearsey, 1984). DNA was prepared in a mutator (*mutD5*) strain of *E. coli*, and used to transform a *ura3* yeast strain, selecting for uracil prototrophy. Plasmid was recovered from transformants that were unstable for the Ura^+ phenotype, and its ability to replicate autonomously was checked by yeast transformation, and by Southern hybridization of the transformants. From 146 independent mutagenesis and transformation experiments we obtained 10 Ars^+ plasmids that were of the same size as the original Ars^- plasmid. Some transformants clearly contained Ars^+ plasmids, but these could not be recovered in *E. coli* ; conceivably the mutation providing ARS function interfered with the function of a gene required for bacteriophage growth. We did not detect any *trans*-acting mutations that permit replication of an Ars^- plasmid by this approach.

Sequencing of the *HO* ARS region of the 10 Ars^+ revertants revealed, surprisingly, that only one showed a sequence change in this element (which restored a sequence known to be Ars^+, Fig. 1). The location of the mutations in the other revertants was narrowed down by a combination of marker rescue and subcloning experiments, and the relevant regions were sequenced. This showed that all the revertants contained one of three types of mutation (designated A17,

B12 and D10); these all occur in the M13 vector sequence and are created by point mutation or minor sequence duplications, and not by insertion of foreign DNA (Fig. 1). The A17, B12 and D10 mutations were transferred to other M13 vector genomes by marker rescue and shown to confer ARS function in the absence of the inactive *HO* ARS element, which is therefore not required in *cis* for the ARS activity of the mutation. In addition, restriction fragments containing the A17 and B12 mutations were transferred to the Ars$^-$ *CEN4*-containing plasmid pSE296 to allow quantitation of the ARS function by plasmid stability assays; both ARSs show similar activities to the *HO* ARS in this context (Table 1).

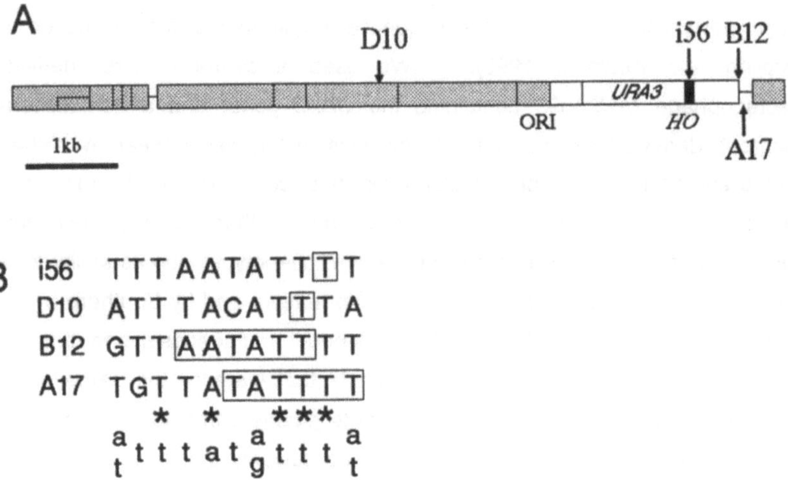

FIGURE 1. A. Location of mutations conferring autonomous replication in the bacteriophage M13 genome. Coding regions for M13 genes are shaded, and the location of the M13 origin of replication, and the *HO* ARS are indicated. Vertical lines indicate gene boundaries.

B. Comparison of the revertant ARSs with the ARS consensus sequence. Sequence changes caused by the mutations are boxed; for a full description of how the mutations relate to the wild-type sequence, see Kipling and Kearsey (1990). The ARS consensus of Broach et al., (1983) is shown in lower-case letters. Asterisks indicate nucleotide positions that are completely conserved in this comparison.

TABLE 1

MUTATIONS CONFERRING AUTONOMOUS REPLICATION

Ars+ mutation	Site of mutation	No. of isolates	Efficiency[1]
(*HO* ARS)	wild type		0.22 ± 0.01
i56	*HO* ARS	1	0.42 ± 0.02
A17	M13 (intergenic)	2	0.20 ± 0.01
B12	M13 (intergenic)	4	0.20 ± 0.01
D10	M13 (gene I)	2	ND

[1] The ARS elements were transferred to the Ars⁻ plasmid pSE296, and the mitotic stability was determined. The value given is the mitotic loss rate per cell per generation. ND = not determined.

The most striking feature of the sequences of the revertant ARSs is that they all provide a match to the ARS consensus sequence (Fig. 1). Thus although these experiments have generated a small number of novel ARSs, they serve to emphasize the essential role of this element and the rather simple sequence requirements for autonomous replication in yeast. The mutagenesis results of Newlon and colleagues stress similar conclusions. Efficient function of *ARS307* (C2G1) appears to depend on multiple matches to the ARS consensus (Palzkill and Newlon, 1988), although analysis of mutations in the essential region indicates that sequences deviating from the original Broach et al. (1983) consensus may contribute to replication origin activity (van Houten and Newlon, 1990). Our results also imply that many plasmids or fragments of yeast chromosomes that are incapable of autonomous replication may contain sequences that are only a single nucleotide change away from generating a detectable ARS element. It might have been possible to generate a greater variety of novel ARSs in these experiments had we not used a plasmid where much of the sequence was required for recovery of the phage in *E. coli*. Although the mutations described here clearly allow the plasmid to replicate autonomously, we cannot rule out the possibility that the original wild-type sequence may have been able to function very poorly as a replication origin and that this efficiency was merely increased by the mutation. Thus, one explanation of the close linkage of the A17 and B12 mutations is that this unmutated region might function as a very inefficient replication origin. The genetic assay for autonomous replication requires efficient plasmid replication over many generations; perhaps origins that, on average, function in less than 50% of S phases do not score as Ars^{+} in transformation experiments? It has been suggested that the lack of sequence requirements for replication in the *Xenopus* egg might in fact depend on origin-related sequences of the kind described here (Diffley and Stillman, 1990). In addition, we can imagine that the ARS-related sequences could mutate to provide efficient replication origin function in chromosome evolution and might even function under unusual physiological conditions or as a result of mutations in proteins involved in DNA replication or other aspects of chromosome function.

MUTATIONS IN THE *RAR5* GENE AFFECT PLASMID STABILITY

Centric plasmids that contain weak or defective ARS elements show low mitotic stabilities in dividing yeast cells, apparently because in a proportion of S phases the plasmid is not replicated, and thus a plasmid-free cell is produced in the next cell division (Hieter et al., 1985; Koshland et al., 1985). Using plasmids that are dependent on weak ARS elements for their replication, we have isolated a series of yeast mutants which maintain the plasmids at higher mitotic stabilities than wild-type strains, perhaps via a direct stimulation of ARS activity. These *trans*-acting mutations (designated *rar*) identify yeast genes that may have a role in plasmid replication. Using the procedures described previously (Kearsey and Edwards, 1987), we have identified a new gene, designated *RAR5*, using the plasmid pSE276-2 which is dependent on the *CEN3*-associated weak ARS for its replication (Table 2). Most recessive mutations isolated with the pSE276-2 plasmid specifically identify *RAR5*.

TABLE 2
COMPLEMENTATION ANALYSIS OF MUTATIONS ENHANCING PLASMID STABILITY

RAR complementation group	No. of alleles isolated using plasmid: pSE276-1	pSE276-2	Total
1	1	0	1
2	12	0	12
3 *(SPT13/GAL11)*[1]	2	0	2
4	6	1	7
5 (*DST2/KEM1*; STPβ)	1	6	7
other recessive	7	4	11
Total	29	11	40

[1] Kipling et al., 1991

A comparison of relative plasmid stabilities in wild-type and *rar5-1* strains is shown in Table 3. The *rar5*-associated stimulation of plasmid mitotic stability appears to be independent of the vector used, or other genetic elements (e.g. CEN elements) present on the minichromosome, although the magnitude of the stimulation varies between plasmids.

The *RAR5* gene was cloned by screening for complementation of the recessive plasmid-stabilization phenotype following transformation with a yeast cosmid library. This yielded a single cosmid, designated pF3, which can complement all seven *rar5* mutations, but has no effect on plasmid stability in other *rar* mutation backgrounds, or in a wild-type strain. The insert in the pF3 plasmid was shown to be allelic to the *rar5* mutation by directing integration of a *URA3*-marked plasmid. Sequencing of the region of pF3 required for *rar5* complementation revealed a large open reading frame, potentially encoding a 175 kDa protein, consisting of 1528 amino-acids.

TABLE 3

PLASMID LOSS RATES IN *RAR5-1* AND WILD-TYPE STRAINS

Plasmid	Genotype	Plasmid loss rate in rich medium	
		Wild-type (W303-1A)	*rar5-1* (R795-4A)
pDK296-HO	pBR322 *URA3 CEN4 SUP4-o HO* ARS	0.22±0.01	0.11±0.01
pDK296-A17	pBR322 *URA3 CEN4 SUP4-o* A17 ARS	0.20±0.01	0.09±0.01
pDK296-i56HO	pBR322 *URA3 CEN4 SUP4-o* 1062 *HO* ARS	0.42±0.02	0.29±0.01
pSE276-2	M13mp93 *URA3 CEN3 SUP4-o* 918 *HO* ARS	0.36±0.01	0.12±0.01

In order to determine whether the plasmid stabilization phenotype of the *rar5* mutant is due to loss of function, a disruption allele was constructed by inserting the *URA3* gene in a Sac I site situated 97 codons from the start of the *RAR5* reading frame. A fragment containing this gene disruption was integrated into a diploid strain, and the resulting diploid transformant (DKY120) was sporulated. This gene disruption is not lethal, since tetrads could give four viable spores, with 2:2 segregation of Ura^+:Ura^-. However a strain bearing this disruption does show a reduction in growth rate (doubling time of 130 minutes for the *rar5::URA3* strain DKY119, compared to 110 minutes for the parental W303-1A strain), implying that the function of the *RAR5* gene is important for normal mitotic growth. Disruption of the *RAR5* gene confers a plasmid-stabilization phenotype similar to the *rar5-1* allele; thus the loss rate of the pSE276-2 plasmid is 0.13 ± 0.02 in the *rar5::URA3* strain, compared to 0.12 ± 0.01 in the *rar5-1* strain. This result suggests that the original *rar5* alleles are null alleles and that loss of the *RAR5* gene product in some way promotes plasmid stabilization.

FUNCTION OF THE RAR5 PROTEIN

The similarity of the phenotypes of the original *rar5* mutations and the *rar5* disruption allele suggest loss of the RAR5 protein is responsible for the apparent stimulation of ARS function, and consequent elevation of plasmid stability, as if the RAR5 protein in some way acts to repress initiation of DNA replication. The apparent effect on ARS function may of course be indirect. For instance, as we discuss elsewhere, metabolic effects that cause the duration of S phase to be increased might have some effect on ARS function (Kearsey and Edwards, 1987). However, it is of interest that most Rar⁻ mutants detected using the pSE276–2 plasmid identify the *RAR5* gene, and this may reflect a more direct effect of the gene product in ARS function.

Sequencing the *RAR5* gene has provided important clues as to its function. The DNA sequence of the *RAR5* gene indicates that it is identical to *DST2*, encoding the DNA strand transferase β protein (STPβ; Dykstra et al., 1990a, 1990b), and *KEM1*, identified by mutations that enhance the karyogamy defect of a *kar1* mutation (Kim et al., 1990). STPβ was identified as a mitotic activity that can catalyze the transfer of a one strand of a double–stranded DNA molecule to a complementary single–stranded circle, in a reaction which is not dependent on a energy source such as ATP (Dykstra et al., 1990a). Although this property clearly implicates the protein in homologous recombination, depletion of this protein from the cell (caused by a *kem1* null allele) only causes a small reduction in the rate of mitotic recombination (Kim et al., 1990). In addition the slow growth phenotype of the *rar5* mutant implies that the protein has an important mitotic function, and this might not be expected of a protein uniquely involved in mitotic recombination. Although the strand transferase activity of STPβ reflects some aspect of its DNA–binding ability, it may be difficult to predict its range of physiological functions.

How might loss of STPβ affect plasmid replication? One trivial explanation would be that association of STPβ with the chromosomes in some way hinders access to replication origins, and thus removal of the protein can enhance ARS function. Alternatively, the strand transferase function may be important, but it is not immediately apparent why a reduction in this activity should enhance plasmid stability. Any reduction in mitotic recombination is unlikely to be relevant, as this process does not contribute significantly to the loss rate of small plasmids (Hieter et al., 1985; Koshland et al., 1985). One speculative possibility is that a strand transferase might inactivate potential origins by catalyzing the removal of D-loops in chromosomal DNA caused by the synthesis of short fragments that displace the other strand, structures that could otherwise serve as replication origins. A recent model of replication initiation proposed by Morrison et al. (1990) suggests that D-loop like structures, formed by DNA polymerase α, are an intermediate in replication initiation; perhaps the STPβ activity proofreads these structures and inactivates inappropriate origins?

In contrast to the presumed DNA-related role of the RAR5 gene protein which is provided by studies on STPβ, genetical analysis of the *KEM1* gene suggests that the gene product is implicated in microtubule function (Kim et al., 1990). The phenotypes of the *kar1* mutation itself suggest some involvement in spindle-pole body formation and microtubule regulation (Rose and Fink, 1987), and the enhancement of the *kar1* phenotype by the *kem1* mutation might reflect a normal contribution of the KEM1 protein to some aspect of microtubule function. In addition, the *kem1* mutation, even in the absence of the *kar1* mutation, shows certain defects in nuclear fusion, benomyl sensitivity, an increased rate of chromosome loss, and abnormalities in mitotic spindle formation and cell shape. Given the ability of STPβ to associate with DNA, the protein could in some way be involved in microtubule-chromosome connections; conceivably some alteration of centromere function could alter ARS function in the *rar5* mutant. Alternatively, the apparent effect on microtubule function in the *kem1* mutant might be an indirect consequence of loss of the DNA-related activity of the protein. Further physiological and biochemical studies should help to explain how the range of pleiotropic phenotypes of *rar5/dst2/kem1* mutations and the biochemical properties of the gene product are related.

ACKNOWLEDGEMENTS

This work was supported by the Cancer Research Campaign.

REFERENCES

Brewer, BJ (1988) Cell 53:679–686.
Brewer, BJ and Fangman, WL (1987) Cell 51:463–471.
Brewer, BJ and Fangman, WL (1988) Cell 55:637–643.
Broach, JR, Li, YY, Feldman, J, Jayaram, M, Abraham, J, Nasmyth, KA and Hicks, JB (1983) Cold Spring Harbor Symp. Quant. Biol. 47:1165–1173.
Diffley, JFX and Stillman, B (1990) Trends in Genetics 6:427–432.
Dykstra, CC, Hamatake, RK and Sugino, A (1990a) J. Biol. Chem. 265: 10968–10973.
Dykstra, CC, Kitada, K, Clark, AB, Hamatake, RK and Sugino, A (1990b) *DST2* sequence: EMBL Accession Number M36725.
Hand, R (1978) Cell 15:317–325.
Hieter, P, Mann, C, Snyder, M and Davies, RW (1985) Cell 40:381–392.
Huberman, JA, Zhu, LR, Davis, LR and Newlon, CS (1988) Nucleic Acids Res. 16:6373–6384.
Kearsey, S (1984) Cell 37:299–307.
Kearsey, SE and Edwards, J (1987) Mol. Gen. Genet. 210:509–517.
Kim, J, Ljungdahl, PO and Fink, GR (1990) Genetics 126:799–812.
Kipling, D and Kearsey, SE (1990) Mol. Cell. Biol. 10:265–272.
Kipling, D, Tambini, C and Kearsey, SE (1991) Nucleic Acids Res. (in press).
Koshland, D, Kent, JC and Hartwell, LH (1985) Cell 40:393–403.
Maine, GT, Sinha, P and Tye, B–K (1984) Genetics 106:365–385.
Morrison, A, Araki, H, Clark, AB, Hamatake, RK and Sugino, A (1990) Cell 62:1143–1151.
Newlon, CS (1988) Microbiol. Reviews 52:568–601.
Palzkill, TG and Newlon, CS (1988) Cell 53:441–450.
Reynolds, AE, McCarroll, RM, Newlon, CS, and Fangman, WL (1989) Mol. Cell. Biol. 9: 4488–4494
Rose MD, and Fink, GR (1987) Cell 48:1047–1060.
Snyder, M, Sapolsky, RJ and Davies, RW (1988) Mol. Cell. Biol. 8:2184–2194.
Thrash–Bingham, C and Fangman, WL (1989) Mol. Cell. Biol. 9:809–816.
Umek, RM, Linskens, MHK, Kowalski, D and Huberman, JA (1989) Biochem. Biophys. Acta 1007:1–14.
van Houten, JV and Newlon, CS (1990) Mol. Cell. Biol. 10:3917–3925.

Role of Transcriptional Elements in Activating Origins of Replication

Zong-Sheng Guo, Maria Wiekowski, Sadhan Majumder, Miriam Miranda and Melvin L. DePamphilis
Department of Cell and Developmental Biology
Roche Institute of Molecular Biology
Roche Research Center, Nutley, N.J. 07110, U.S.A.

Transcriptional elements found in promoters and enhancers function as components of several eukaryotic origins of replication (reviewed in DePamphilis, 1988; DePamphilis et. al., 1988). These origins consist of at least two primary components (Fig. 1). The core component (*ori*-core) is required for initiation of replication under all conditions, and generally is the place where DNA replication begins. Auxiliary components (*ori*-core) flank one or both sides of *ori*-core and consist of promoter or enhancer elements that facilitate the core's activity. Unlike the core component which is required for replication under all conditions, auxiliary sequences are dispensable in some cells (e.g. fertilized mouse eggs; Martinez-Salas et. al., 1988; 1989) and under some *in vitro* replication conditions (Guo et. al., 1989; Chen and Kelly, 1989; Gutierrez et. al., 1990).

We have undertaken two approaches directed at understanding the role of these *ori*-auxiliary sequences. The first is an analysis of the need for *ori*-auxiliary sequences in simian virus 40 (SV40) and polyomavirus (PyV) during replication *in vivo* and *in vitro*, and (2) the ability of mouse 1-cell embryos (fertilized eggs) to dispense with the need for enhancers to activate either *ori*-core or transcriptional promoters. Results from the first approach reveal that the activity of *ori*-auxiliary sequences depends upon binding specific proteins whose specific activity domains interact with the replication proteins binding to *ori*-core. Thus, the role of *ori*-auxiliary sequences in replication appear analogous to the role of promoter and enhancer elements in stimulating transcription. Results from the second approach reveal that the primary role of enhancers is to relieve repression resulting from chromatin modification that first appears upon formation of a 2-cell embryo. Together, these results demonstrate a dual role for enhancers in replication and transcription: relieving repression of *ori*-core and promoter components, and facilitating activation of replication and transcription complexes.

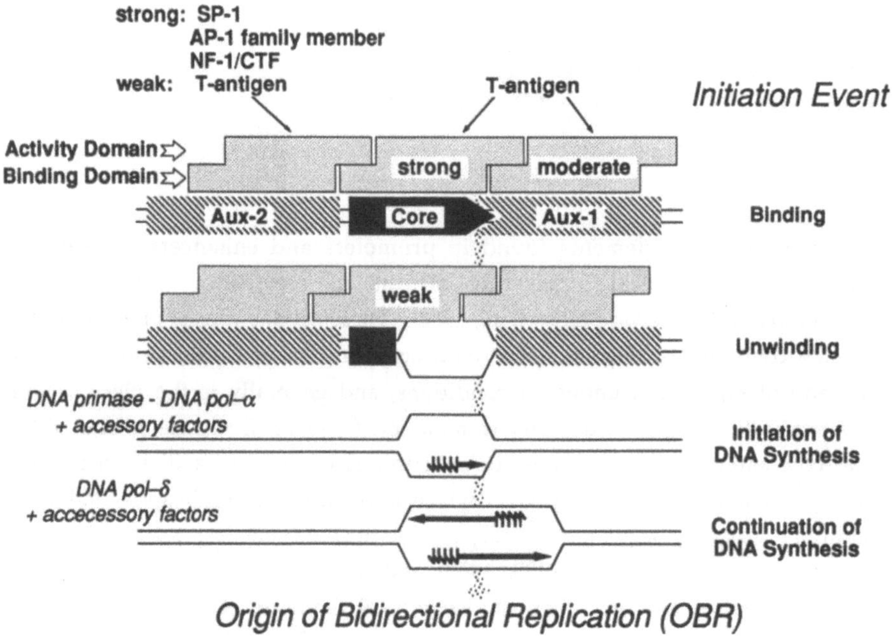

Figure 1. *Ori*-auxiliary activity of SV40 and PyV origins requires binding of specific transcription factors. Relative strength of T-antigen binding is indicated.

Ori-auxiliary activity requires binding of specific transcription factors

We have recently succeeded in reproducing the effects of *ori*-auxiliary sequences on DNA replicastion in SV40-infected monkey cells in an extract of these cells, and in identifying several parameters that determine the level of facilitation by *ori*-auxiliary sequences *in vitro* (Guo et. al., 1989). SV40 is a small circular DNA genome that replicates in the nucleus of its mammalian host as a "mini-chromosome"; with the exception of viral encoded large tumor antigen (T-ag), all of the factors needed to replicate and assemble this DNA into a chromosome are provided by the host cell. The role of T-ag is to bind to *ori*-core and initiate DNA unwinding. This, in turn, allows initiation of DNA synthesis by cellular enzymes. The complete SV40 origin of DNA replication (*ori*) consists of a required core sequence flanked by two auxiliary sequences that together increase the rate of DNA replication in monkey cells about 25-fold over core alone. *Aux*-1 consists of a strong T-ag binding site, and *aux*-2 consists of weak T-ag binding sites as well as strong SP-1 binding sites. Since *aux*-1 and *aux*-2 are not interchangeable and the origin of bidrectional DNA replication lies at one end of *ori*-core, *ori*-core has an inherent polarity (DePamphilis, 1988; DePamphilis et. al., 1988).

Although the replicationally active form of T-ag in these extracts had a strong affinity for *ori*-core, it had only a weak but specific affinity for *ori*-auxiliary sequences. Deletion of *ori*-auxiliary sequences reduced the affinity of *ori*-core for active T-ag by only 1.6-fold, consistent with the fact that saturating concentrations of T-ag in the cell extract did not reduce the stimulatory role of *ori*-auxiliary sequences in replication (Gutierrez et. al., 1990). This suggested that *ori*-auxiliary sequences promoted initiation of replication at some step following T-ag bidning to *ori*. Nevertheless, *ori*-auxiliary sequences clearly interacted with T-ag because, in the presence of a complete origin region, a molar excell of T-ag was required for *ori*-auxiliary sequences to facilitate replication (Guo et. al., 1989). In contrast, deletion of *ori*-auxiliary sequences reduced the efficiency of *ori* specific, T-ag dependent DNA unwinding in cell extracts at least 15-fold. Using only purified T-ag in the presence of topoisomerase I to unwind purified DNA, *oi*-auxiliary sequences strongly facilitated T-ag dependent DNA conformational changes consistent with melting the first 50 base pairs. Under these conditions, *ori*-auxiliary sequences had little effect on binding of T-ag to DNA. Therefore, one role of *ori*-auxiliary sequences in DNA replication is to facilitate T-ag dependent DNA unwinding after the T-ag preinitiation complex has bound to *ori*-core.

It has also been suggested that *ori*-auxiliary sequences function by excluding nucleosomes from the *ori*-core region that would otherwise interfere with T-ag activity (Chen and Kelly, 1989). We have examined the ability of subcellular DNA replication systems to assemble nucleosomes, and shown that 40% to 80% of the maximum number of nucleosomes possible are assembled *in vitro* in the presence of a nuclear extract (Gruss et. al., 1990). The efficiency of nucleosome assembly was independent of DNA replication; it required neither a functional origin nor T-ag. Since facilitation by *ori*-auxiliary sequences was independent of the number of nucleosomes assembled per DNA molecule, exclusion of nucleosomes alone cannot explain their function.

Our results reveal that SV40 *ori*-auxiliary sequences affect T-ag binding weakly, but DNA unwinding strongly. All of the cis-acting DNA sequence information required to initiate SV40 DNA replication (and its close relative, polyomavirus) is contained within *ori*-core, the flanking auxiliary sequences appear simply to increase the efficiency of initiation without altering its mechanism. We suggest that a primary function of SV40 *ori*-auxiliary sequences is to prevent unwound replication intermediates from dissociating before DNA primase-DNA polymerase–α can initiate DNA synthesis (Fig. 1). Strong T-ag binding to *ori*-core is eliminated as unwinding proceeds, because T-ag binds weakly and nonspecifically to single-stranded *ori*-DNA (Auborn et. al., 1988). Thus, T-ag helicase activity literally destroys the very DNA binding site required for it to initiate DNA replication. Under these conditions, continued DNA unwinding may be facilitated through weak but specific binding of T-ag to *ori*-auxiliary sequences. In this way, *ori*-auxiliary sequences would facilitate initiation of DNA synthesis by shifting the steady-state population of DNA molecules in favor of the transient unwound intermediates.

This model (Fig. 1) requires that the activity of *ori*-auxiliary sequences be mediated through specific proteins that bind to these sequences and interact with the T-ag complex that binds to *ori*-core. To investigate this hypothesis, unique synthetic oligonucleotide sequences containing specific transcription factor binding sites or DNA unwinding elements were substituted for the natural SV40 *ori*-auxiliary sequences that flank the SV40 *ori*-core component. Transfection of COS cells led to four major conclusions. First, *aux*-1, a strong T-ag binding site in SV40, functioned only in its natural orientation and only in its natural position relative to *ori*-core. Furthermore, "DNA unwinding sequences" described by Kowalski and coworkers (1989) that are readily unwound and serve as components of both yeast and E. coli origins of replication did not substitute for *aux*-1. Therefore, simply making the origin region easier to unwind is not sufficient for *aux*-1 activity; *aux*-1 must bind T-ag in a unique manner, presumably so that T-ag bound to *aux*-1 can interact with T-ag bound to *ori*-core.

Second, *aux*-2 activity requires binding of one or more transcription factors. Either AP-1 or SP-1 binding sites could substitute completely for the wild-type *aux*-2 sequence (20 to 30-fold stimulation of *ori*-core), whereas recreation of the natural T-ag binding sites in *aux*-2 alone stimulated only 2 to 3-fold, the same level observed *in vitro*. This stimulation required a specific protein activity domain as well as a DNA binding domain. No activity was observed in the absence of an appropriate binding protein, and the amount of activity observed depended on the specific activity domain of the binding protein. Replacement of *aux*-2 with tandem oligomers of the yeast GAL4 protein DNA binding sites failed to stimulate *ori*-core when GAL4 protein or chimeric proteins composed of the GAL4 protein DNA binding domain and either HSV-VP16 or human c-jun activity domains were provided as cotransfected expression vectors. These expression vectors did stimulate transcription in parallel assays. A similar analysis of PyV *ori*-core replication following transfection of FOP mouse cells revealed that tandem AP-1 sites could substitute completely for the wild-type PyV enhancer, but that GAL4, GAL4:VP16 or GAL4:c-jun provided only partial activity, and the GAL4 DNA binding domain alone was inactive. Thus, the ability of *ori*-auxiliary sequences to function depended upon binding proteins with a unique activity domain. Presumably, the activity domain must interact specifically with the core initiation complex (i.e. T-ag; Fig. 1, Gutierrez et. al., 1990), just as they do when activating promoters. The AP-1 bidnign site binds an as yet unidentified member of the AP-1 family of transcription factors.

Enhancers activate ori-core and promoter components by preventing repression from chromatin structure

In contrast to non-mammalian systems, promoters of transcription and origins of replication, along with their specific activation factors, were always required to initiate transcription and replication in an appropriate cell (Martinez-Salas et. al., 1988; 1989). The same principal cis-acting elements that are required to initiate DNA replication and transcription in the adult mouse (i.e. origins of replication, promoters and enhancers) are also required as early as the 2-cell stage in development where zygotic gene expression first begins. Specific embryo responsive enhancers have recently been identified (unpublished data). However, prior to this stage, enhancers are dispensable either for activating origins of replication or promoters of transcription. Mouse 1-cell embryos replicate but do not transcribe their DNA. Nevertheless, PyV DNA injected into embryos arrested morphologically at the 1-cell stage will undergo T-ag dependent replication and will express genes, because a biological clock exists that initiates zygotic gene expression at a certain time after fertilization, regardless of whether or not the 1-cell embryo has progressed morphologically to the 2-cell stage. Moreover, the levels of origin and promoter activity in arrested 1-cell embryos in the absence of an enhancer are equivalent to or greater than their levels in 2-cell embryos when linked to an enhancer. Therefore, if we could understand why enhancers are dispensable when DNA is injected into 1-cell embryos, we would understand the role of enhancers in activating origins and promoters.

Embryonic protein synthesis was characterized at various times after fertilization of mouse eggs to detect the onset of α–amanitin sensitive protein synthesis and to ascertain the effects of blocking DNA replication. These data were compared with the ability of the same embryos to express genes from injected plasmids. The results demonstrated that gene expression from injected plasmids was turned on and off at the same times after fertilization as zygotic gene expression, and this regulation was not altered by blocking DNA replication to prevent 1-cell embryos from forming 2-cell cmbyos. A biological clock appears to be triggered by fertilization that initiates zygotic gene expression about 24 hours after fertilization, regardless of whether or not 1-cell embryos have developed morphologically to the 2-cell stage. This means that DNA injected into arrested 1-cell embryos is exposed to the same set of transcription factors, and presumably at the same concentrations, as DNA injected into 2-cell embryos. Furthermore, we found that the same individual promoter elements required for transcription in 2-cell embryos to adult tissues are also required in 1-cell embryos. Thus, the role of an enhancer can be functionally distinguished from that of a promoter. Competition experiments to determine the relative levels of transcription factors revealed that factors binding to promoter elements (e.g. SP-1) are present at similar levels in 1-cell and 2-cell embryos. Therefore, the ability to dispense with an enhancer in 1-cell embryos does not result from the presence of saturating levels of initiation factors, suggesting that one primary function of an enhancer appears to be to relieve repression.

The changes that occur with formation of 2-cell embryos affected a diverse group of regulatory sequences that bear little sequence homology and interact with a variety of different initiation factors. This suggested that the apparent repression is mediated via a general change in nuclear or chromatin structure rather than changes in specific transcription and replication factors. Recent experiments suggest that the primary role of enhancers is to relieve a general represssion mechanism that involves modification of chromatin structure induced by the first mitosis.

Repression did not require formation of a zygotic nucleus. Unfertilized eggs were parthogenetically activated to produce 2-cell embryos with haploid nuclei derived exclusively from the female pronucleus. The same procedure carried out in the presence of cytochalasin to prevent loss of the second polar body yielded 2-cell embryos with maternally derived diploid nuclei. Extrusion of the female pronculeus from fertilized eggs produced 2-cell embryos with haploid nuclei derived exclusively from the male pronucleus. Each of these artificially derived 2-cell embryos responded to injected promoters in the same way that natural 2-cell embryos did; embryo responsive enhancers strongly stimulated promoter activity. Thus, repression did not depend on formation of a zygotic nucleus or doubling the number of chromosomes. Instead, repression appeared to depend on completing the first mitosis.

This hypothesis was supported by the unique effect of the first mitosis on replication and transcription of DNA injected into 1-cell embryos. As discussed above, when these embryos remained at the 1-cell stage, they efficiently replicated and transcribed DNA with no involvement of enhancers. However, when these embryos developed to the 2-cell stage or further, origin and promoter activity was strongly repressed. DNA stability was not affected. Moreover, this repression was not overcome either by enhancers or by transacting factors such as "E1a-like" protein in the case of the adenovirus EIIa promoter and cadmium in the case of the mettallothionein promoter. This repression was unique to the first mitosis, because the same DNA injected into 2-cell embryos passed through the first mitosis such that promoters can no longer respond to stimulatory factors. One obvious consideration would be a change in chromatin structure on the injected plasmids.

Nucleosome assembly from histones H3, H4, H2a and H2b can repress promoter (Grunstein, 1990) and origin (Chen and Kelly, 1989) activity by impeding binding of initiation factors. The repressed state of chromatin is dramatically affected by two modifications: (1) acetylation of lysine residues within the core histone amino-terminal tails, and (2) altered histone H1 interactions with nucleosomes. Hyperacetylated histones allow increased accessibility to transcription factors and other proteins, and are associated with chromatin unfolding and transcriptionally active genes. H1 or H1 variants can repress transcription by forming a more tightly organized higher-order chromatin structure. Actively transcribed chromatin is generally hyperacetylated and deficient in histone H1 (Kamakaka and Thomas, 1990; Taxi and Bird, 1990). Since DNA injected into embryos was converted into chromatin (Martinez-Salas et. al., 1989), it would be subject to these effects. Since all five major histones are synthesized in mouse oocytes and early embryos, repression likely results from modification of chromatin structure concomitant with the first mitosis.

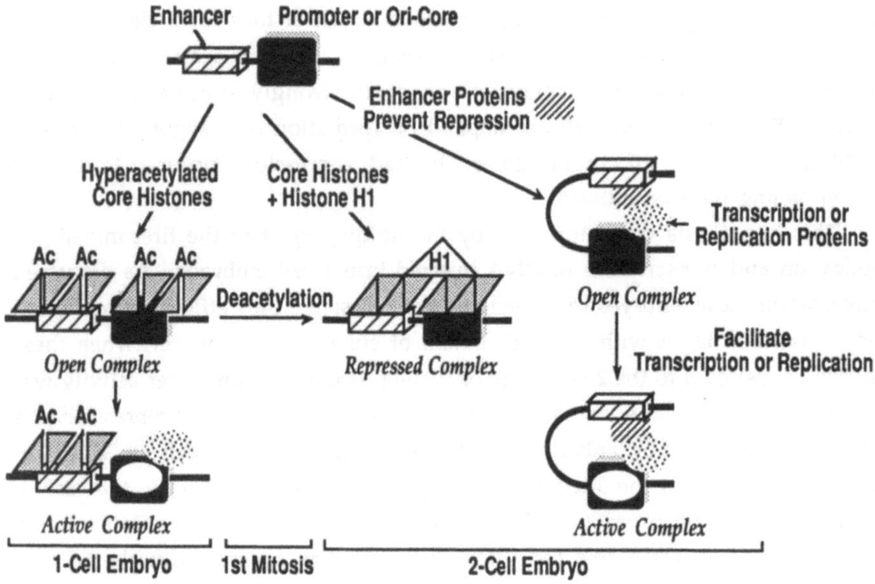

Figure 2. The need for enhancers to activate *ori*-core and promoter components first appears with formation of a 2-cell mouse embryo as a result of chromatin modification during the first mitosis.

If this modification inolves conversion of hyperacetylated histones present in 1-cell embryos to the normal steady-state level of acetylated histones in 2-cell embryos, then incubation of 2-cell embryos in butyrate would relieve the observed repression of promoter activity. Butyrate produces hyperacetylated histones by blocking histone deacetylase and thus prevents the normal turnover of histone acetyl groups. Butyrate treatment of 1-cell embryos did not stimulate the activity of injected promoters, but butyrate treatment of normal or parthenogenetic 2-cell embryos did. Butyrate stimulated promoter activity to the level observed in arrested 1-cell embryos, but relatively little stimulation was observed when the promoter was coupled to an enhancer. These results suggest that newly assembled chromatin may be hyperacetylated in pronuclei and rapidly deacetylated following the first mitosis.

The potential involvement of histone H1 was examined by coinjecting purified H1 with the plasmid expression vector. Preliminary results inidcate that HeLa H1 had little effect in 1-cell embryos, consistent with previous observations that H1 does not associate with plasmid chromosomes assembled from hyperacetylated histones (Reeves et. al., 1985). However, H1 repressed promoters in 2-cell embryos unless the promoter was linked to an embryo responsive enhancer, in which case, repression by H1 was alleviated. This result was consistent with H1 involvement in the need for enhancers to relieve repression of promoter activity in 2-cell embryos. Confirmation of these conclusions awaits a detailed analysis of the histone composition in embryos.

Conclusions

Enhancers fulfill two roles in replication and in transcription. First, they prevent repression of origins and promoters by allowing formation of an "open complex" that is accessible to replication or transcription factors (Fig. 2). This role accounts for 25 to 100-fold stimulation *in vivo*, and is the effect observed when DNA is injected or transfected into cells. Repression appears to result from modification of chromatin structure that occurs concurrently with formation of a 2-cell embryo. Second, enhancers facilitate formation of an "active complex", which accounts for a 2 to 3-fold stimulation *in vivo* or *in vitro*. This is the effect observed with SV40 DNA replication *in vitro* where T-ag is in great excess and presumably blocks binding of other transcription factors, and *in vivo* when synthetic T-ag DNA binding sites are substituted for the normal *aux*-2 sequence. Together, the dual role of enhancers can account for all of the effects of enhancers *in vivo* and in subcellular systems that support replication or transcription.

REFERENCES

Auborn, K. J., R. B. Markowitz, E. Wang, Y. T. Yu, and C. Prives (1988). SV40 T-antigen binds specifically to double-stranded DNA but do not to single-stranded DNA or DNA/RNA hybrids containing the SV40 regulatory sequences. J. Virol. **62**: 2204-2208.

Chen, L. and T. J. Kelly (1989). Transcriptional Activator nuclear factor I stimulates the replication of SV40 minichromosomes *in vivo* and *in vitro*. Cell **59**: 541-551.

DePamphilis, M. L. (1988). Transcriptional elements as components of origins of eukaryotic D A replication. Cell **52**: 635-638.

DePamphilis, M. L., E. Martinez-Salas,, D. Y. Cupo, E. A. Hendrickson, C. E. Fritze, W. R. Folk, and U. Heine (1988). Initiation of polyomavirus and SV40 DNA replication and the requirements for DNA replication during mammalian development", pp. 165-175, *In* Eukaryotic DNA Replication (B. Stillman and T. Kelly, eds.0, Cancer Cells, vol. 6, Cold Spring Harbor Laboratory, N.Y.

Grunstein, M. (1990). Nucleosomes: regulators of transcription. Trends Genet. 6; 395-400.

Gruss, C., C. Gutierrez, W. C. Burhans, M. L. DePamphilis Th. Koller, and J. M. Sogo (1990). Nucleosome assembly in mammalian cells extracts before and after DNA replication. EMBO J. **9**: 2911-2922.

Guo, Z.-S., C. Gutierrez, U. Heine, J. M. Sogo, and M. L. DePamphilis (1989). Origin auxiliary sequences can facilitate initiation of simian virus 40 DNA replication *in vitro* as they do *in vivo*. Mol. Cell. Biol. **9**: 3593-3602.

Gutierrez, C., Z.-S. Guo, J. Roberts, and M. L. DePamphilis (1990). Simian virus origin auxiliary sequences weakly facilitate binding of T-antigen, but strongly facilitate initiation of DNA unwinding. Mol. Cell Biol. **10**: 1719-1728.

Kamakaka, R. T. and J. O. Thomas (1990). Chromatin structure of transcriptionally competent and repressed genes. EMBO J. **9**: 3997-4006.

Kowalski, D. and M. J. Eddy (1989). The DNA unwinding element: a novel, cis-acting component that facilitates opening fo the E. coli replication origin. EMBO J. **8**: 4335-4344.

Martinez-Salas, E., D. Y. Cupo and M. L. DePamphilis (1988). The need for enhancers is acquired upon formation of a diploid nucleus during early mouse development. Genes and Development **2**; 1115-1126.

Martinez-Salas, E., E. Linney, J. Hassell, and M. L. DePamphilis (1989). The need for enhancers in gene expresssion first appears during mouse development with formation of a zygotic nucleus. Genes and Development **3**: 1493-1506.

Reeves, R., Gorman, C. M., and B. Howard (1985). Minichromosome assembly of non-integrated plasmid DNA transfected into mammalian cells. Nucl. Acids Res. **13**: 3599-3615.

Tazi, J. and A. Bird (1990). Alternative chromatin structure at CpG islands. Cell **60**: 909-920.

Telomere Replication in *Saccharomyces cerevisiae*

V.A. Zakian, S.-S. Wang, and R. Wellinger
Fred Hutchinson Cancer Research Center
1124 Columbia St.
Seattle, WA 98104
USA

Conventional DNA polymerases cannot start DNA synthesis *de novo* and can synthesize DNA only in the 5' to 3' direction. It was pointed out by Watson (1972) that these properties make it impossible for conventional DNA polymerases to replicate fully a linear DNA molecule: in the absence of special structures or mechanisms, replication by conventional DNA polymerases results in gaps at the 5' ends of newly replicated strands and hence in loss of sequence information with each round of DNA replication. Telomeres are thought to circumvent this dilemma by providing an alternative method for replication of the very ends of eukaryotic chromosomes.

Typically all telomeres in a cell bear a few hundred to many kilobase pairs of a simple sequence DNA. For example, telomeres in the macronucleus of *Tetrahymena* bear ~300 bps of 5'-C_4A_2-3', macronuclear telomeres of hypotrichous ciliates like *Oxytricha* and *Euplotes* bear ~30 bps of 5'-C_4A_4-3' DNA, and in *Saccharomyces cerevisiae*, chromosomal telomeres carry ~350 bps of a more heterogeneous sequence 5'-$C_{2\text{-}3}A(CA)_{1\text{-}6}$-3', commonly abbreviated $C_{1\text{-}3}A$ (reviewed in Zakian, 1989). In addition, in many organisms, there are interstitial tracts of the telomeric sequence (eg. Walmsley et al, 1984).

The enzyme telomerase was first identified in the ciliated protozoan *Tetrahymena* (Greider and Blackburn, 1985) then in other ciliates (Shippen-Lentz and Blackburn, 1989; Zahler et al, 1988) and in HeLa cells (Morin, 1989). *In vitro* telomerase can extend the 3' end of the G-rich strands of telomeric DNA in the absence of a DNA template. In ciliates and humans, telomerase is an RNA-dependent DNA polymerase. In *Tetrahymena* (Greider and Blackburn, 1989) and *Euplotes* (Shippen-Lentz and Blackburn,

1990), a small RNA that co-purifies with telomerase has been isolated and its gene cloned and sequenced. In both cases, the RNA contains about one and a half copies of the C-strand of the telomeric repeat. In _Tetrahymena_, this RNA has been shown to serve as a template for synthesis of telomeric DNA _in vivo_ (Yu et al, 1990). Since telomerase synthesizes telomeric DNA from an RNA template, it represents a specialized type of reverse transcriptase (Lundblad and Blackburn, 1990).

A number of lines of evidence suggest that the processes that replicate the ends of eukaryotic chromosomes are conserved among evolutionarily distantly related organisms. First, although telomeric sequences vary from organism to organism, most telomeric sequences can be represented by a loose consensus, 5'-$C_{1-8}T_{0-1}A_{0-4}$-3' (Zakian, 1989). Secondly, telomeric DNA from _Tetrahymena_, _Oxytricha_ and humans can serve as substrates for formation of new telomeres in _Saccharomyces_ even though the sequence of yeast telomeres differs from that of these other organisms (reviewed in Zakian, 1989). However, in each case the foreign telomeric sequence is modified by the terminal addition of yeast $C_{1-3}A$ repeats (Pluta et al, 1984; Shampay et al, 1984). Likewise, telomerase is a promiscuous enzyme at least _in vitro_ in that it can use different G-rich telomeric DNAs as a substrate for elongation (Greider and Blackburn, 1987; Morin, 1989). Although telomerase has not yet been detected in _Saccharomyces_, a reasonable model to explain all of these observations is that in yeast, as in _Tetrahymena_, telomere replication is accomplished by the action of a telomerase. Like the _Tetrahymena_ enzyme, the putative yeast telomerase would be able to use foreign telomeric sequences as templates, but during the elongation reaction it would always add yeast telomeric repeats. In support of this interpretation the _EST1_ gene, known to be required for maintenance of the $C_{1-3}A$ tracts at the ends of yeast chromosomes (Lundblad and Szostak, 1989), encodes a protein with sequence similarity to reverse transcriptase (Lundblad and Blackburn, 1990).

It has also been proposed that recombination between two stretches of telomere sequence could produce structures that serve as substrates for complete replication by conventional

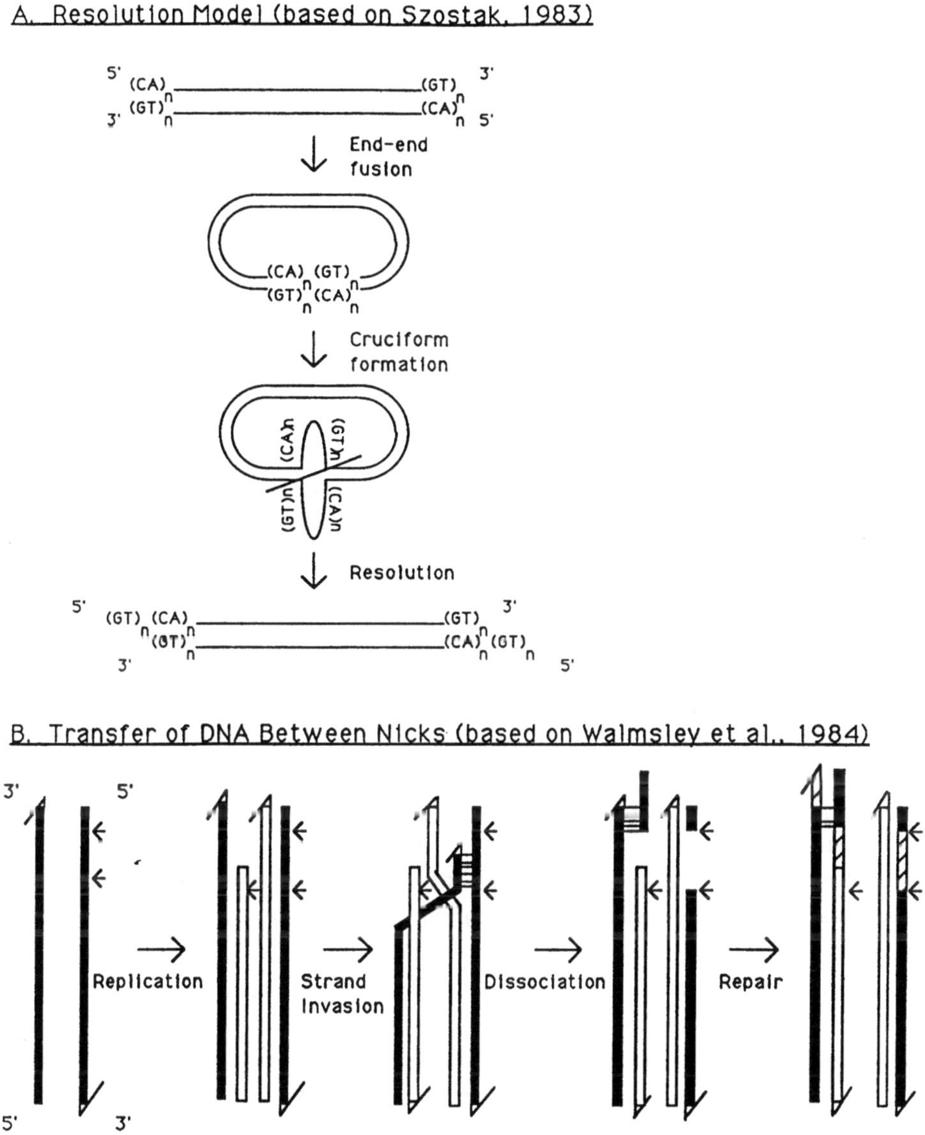

Figure 1: Two models for recombination-assisted replication of telomeres.
A. In the resolution model, telomeres fuse to form an inverted repeat of telomeric sequences. Branch migration within the inverted repeat leads to formation of a cruciform that can be resolved by enzymes that cleave Holliday junctions. The result is transfer of telomeric repeats from one end of the plasmid to

the other to produce a molecule with 5' extensions at each end. This intermediate can now be filled in by conventional replication enzymes, but the transferred repeats are inverted compared to the original repeats. **B.** In this model, the nicks in the C-rich strand (represented by small horizontal arrows) facilitate recombination. Parental strands are represented by solid and newly replicated strands by open lines. After replication and removal of the terminal RNA primer, the 3' single stranded end of the parental strand invades another telomere. When the invading strand dissociates, it can remove the portion of the C-rich donor strand found between nicks. Both the donor and recipient molecules are left with gaps that can be repaired by conventional replication enzymes.

replication proteins (reviewed in Zakian, 1989). These recombination events could occur between telomeres on the same or on different chromosomes or between terminal and interstitial tracts of telomeric DNA. Two different models for recombination facilitated telomere replication are presented in figure 1. In the resolution model (Szostak, 1983; fig.1A), telomeres fuse prior to replication to produce an inverted repeat of telomeric sequences. Branch migration within the inverted repeats can produce a cruciform which can be resolved into functional telomeres by activities that cleave Holliday junctions. Consistent with this model, head-to-head inverted repeats of C_4A_2 DNA can be processed into functional telomeres in yeast (Szostak, 1983; Murray et al, 1988). An alternative model (fig. 1b) proposes that the 3' G-rich single-strand tail left after removal of the most distal RNA primer on the newly synthesized strand can invade another telomeric region and base pair with the complementary C-rich strand (Walmsley et al, 1984). In many organisms, including Saccharomyces and Tetrahymena, nicks or gaps have been detected in the C-rich strand (reviewed in Zakian,1989). The model proposes that these nicks promote telomere-telomere recombination by allowing transfer of the repeats between nicks when the invading G-strand dissociates. Since in Tetrahymena the nicks occur every 12 to 18 nucleotides (Blackburn and Gall, 1978), this model predicts that only short stretches of telomeric DNA should be transferred.

In most situations, recombination between the very ends of chromosomes or linear plasmids is impossible to detect because it involves the exchange of essentially identical sequences located at all telomeres and results in no scorable phenotype. Previous work from our lab provided the first evidence that recombination occurs between the telomeric repeats at the very ends of DNA molecules (Pluta and Zakian, 1989). In these experiments, linear plasmids bearing different foreign telomeric sequences at each end were introduced by transformation into yeast. One end contained a short (28 to 108 bps) stretch of cloned C_4A_4 repeats. The other end was a terminal restriction fragment from Tetrahymena macronuclear ribosomal DNA that contains ~300 bps of C_4A_2 DNA. C_4A_4 and C_4A_2 repeats were known to serve as substrates for the addition of $C_{1-3}A$ repeats (Pluta et al, 1984; Shampay et al, 1984). As expected from these earlier studies, after propagation of the linear plasmids in yeast, ~300 bps of $C_{1-3}A$ DNA were added to each end of the input plasmids. In addition, 90% of the C_4A_4 ends acquired C_4A_2 repeats immediately internal to the added $C_{1-3}A$ DNA and ~20% of the C_4A_2 termini acquired C_4A_4 repeats internal to the added $C_{1-3}A$ DNA (Figure 2).

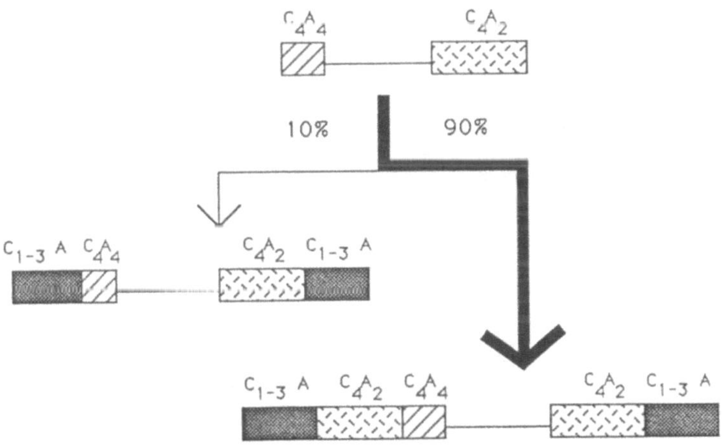

Figure 2: Recombination occurs during telomere formation in yeast (Pluta and Zakian, 1989). Linear plasmids bearing 28 to 108 bps of cloned C_4A_4 DNA at one end and ~300 bps of C_4A_2 DNA at the other were introduced by transformation into yeast. As expected, all linear plasmids had ~300 bps of $C_{1-3}A$ added to each end. In addition, the C_4A_4 and C_4A_2 ends had recombined in 90% of the plasmids.

To elucidate the mechanism of telomere-telomere recombination, the products of recombination were cloned and sequenced (Wang and Zakian, 1990). If telomere fusion occurs during the process of replication, recombination could result from the resolution reaction (fig 1a). However, this possibility requires that the transferred repeats be in the opposite orientation from the starting repeats, an expectation excluded by DNA sequence analysis. The sequencing data demonstrate that each recombination event involved the transfer of long tracts of telomeric DNA. This result eliminates the possibility that recombination is the result of the transfer of the short stretches of repeats between nicks (Fig.1b). Rather the sequencing studies support a model in which telomere-telomere recombination occurs by a gene conversion mechanism requiring very little homology (Wang and Zakian, 1990; figure 3).

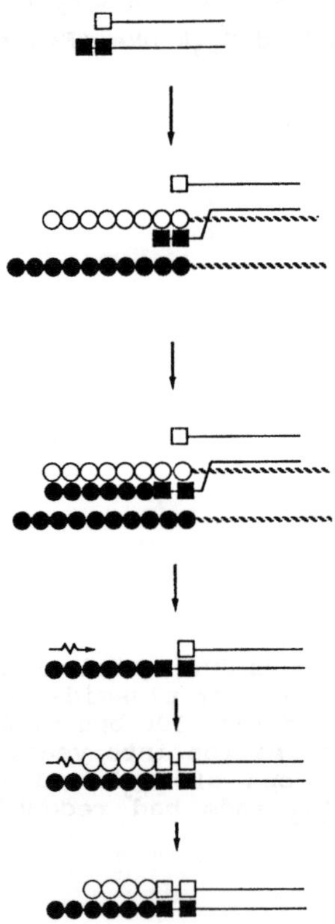

Figure 3: Telomere recombination occurs by gene conversion in yeast (based on data in Wang and Zakian, 1990 and model in Zakian et al., 1990). One end of a linear DNA molecule after replication is shown. Open symbols represent the C-rich strand of a telomere; closed circles represent the G-rich strand. After removal of the terminal RNA primer, a short 3' OH G-strand overhang (represented by squares) is left at the end of the molecule. This strand invades another telomere (represented by circles) near the junction of telomeric with non-telomeric DNA. The invading strand is extended using the "circle" telomere as template. After dissociation of the elongated strand, the newly replicated molecule is left with an extended G-rich strand derived from the "circle" telomere. This molecule can be repaired using standard replication enzymes. After replication, the DNA molecule still has a gap at the 5' end of the newly replicated strand, but there is a net gain of telomeric DNA.

The demonstration of telomere-telomere recombination revives interest in the role of recombination in telomere biology. Telomere recombination might play an ancillary or even exclusive role in telomere maintenance in organisms or cell types in which telomerase can not be detected. For example telomerase is postulated to be absent from human somatic tissues (de Lange et al, 1990; Hastie et al, 1990). Telomere-telomere recombination might play an additional role in organisms like Saccharomyces that have heterogeneous telomeric repeats rather than the precise repeats found in ciliates and humans. At this time, we have no evidence that the telomere-telomere recombination we detect in Saccharomyces is related to telomere replication. However, the fact that telomere-telomere recombination was detected in most transformants (Pluta and Zakian, 1989; figure 2) suggests that its occurrence is required for formation of new telomeres on foreign substrates. By extension, it might also be the mechanism by which yeast $C_{1-3}A$ repeats are added to foreign termini.

As one approach to ascertain the role of telomere-telomere recombination, we have determined the structure of a 7.5 kb linear acentric plasmid called YLpFAT10 at the time in the cell cycle when the ends of chromosomal DNA molecules are known to replicate (McCarroll and Fangman, 1988). For these experiments, we use a two-dimensional (2D) agarose gel electrophoresis method that facilitates detection of non-linear DNA molecules (Bell and Byers, 1979; Brewer and Fangman, 1987). Different models of telomere replication predict different structures for YLpFAT10 during telomere replication. For example, the resolution model predicts that telomeres fuse prior to replication to produce a covalently closed circular DNA molecule. Alternatively, if telomeres recombine during replication, some YLpFAT10 molecules might be held together by Holliday junctions. Circular DNA molecules or molecules containing Holliday junctions can be resolved easily from linear DNA molecules on 2D gels. The product of replication by telomerase or intermediates produced by the gene conversion model of telomere replication depicted in figure 3, are expected to have TG_{1-3} extensions on one or both ends. Duplex DNA molecules with single-strand tails at each end normally migrate as linear molecules in the 2D gel system. However, oligonucleotides with the sequence of the G-rich strand of many telomeric DNAs can base pair with each other by non-Watson Crick interactions (eg., Henderson et al, 1987; Sundquist and Klug, 1989; Sen and Gilbert, 1988; Williamson et al, 1989). Therefore, if a plasmid molecule acquires 3' single-strand TG_{1-3} tails, these tails might interact to form a non-linear molecule.

Yeast cells containing YLpFAT10 were synchronized by blocking cells late in the G1 phase of the cell cycle. When the cells are released from the block, they progress through the cell cycle in a synchronous manner (Zakian et al, 1979; Brewer and Fangman, 1987). DNA was prepared from samples at various points in the cell cycle and subjected to 2D gel electrophoresis and Southern hybridization. (The data on replication of YLpFAT10 will be published elsewhere; Wellinger, Wolf, and Zakian, in preparation). Late in the S phase YLpFAT10 molecules with the structure expected for replication intermediates are detected. The structure of these intermediates demonstrates that replication begins at, and proceeds bi-directionally from, the single centrally placed replication origin on the plasmid. The replication origin on YLpFAT10 derives from an endogenous circular plasmid of yeast called 2μm DNA, a plasmid that normally replicates early in S phase (Zakian et al, 1979). Thus, the linear plasmid YLpFAT10 replicates late in the S phase, the same time when the ends of chromosomal DNA molecules replicate (McCarroll and Fangman, 1988), rather than at the time when circular plasmids containing the 2 μm origin are replicated.

In addition to conventional replication intermediates, another form of the plasmid is detected late in S phase. This form has a mobility indistinguishable from that of a nicked circle with a mass equal to that of YLpFAT10. Like conventional replication intermediates, this form is about five times more abundant in late S phase than in G1, early S, or post S phase cells. The retarded form of the plasmid has been detected in four different strain backgrounds including the tel1 and tel2 strains both of which have short telomeres compared to wild type cells (Lustig and Petes, 1986). Plasmids similar to YLpFAT10 but containing an origin of replication, ARS1, from chromosome IV display the same retarded structure.

A number of experiments suggest that circularization of the plasmid is a result of interaction of TG_{1-3} single strand tails. First, plasmid and chromosomal termini from the late S phase cells hybridize without prior denaturation of the DNA to a $C_{1-3}A$ but not a TG_{1-3} probe. Secondly, recombinant DNA plasmids were constructed which after restriction digestion terminate in ~350 bps of $C_{1-3}A$ with the telomeric repeats in the same orientation as on linear plasmids maintained in yeast. When these plasmids are treated with T7 gene 6 exonuclease to produce 3' TG_{1-3} single strands at both ends, a fraction of the molecules migrates in 2D gels as a nicked circle with the same mass as the starting linear plasmid. However, when the plasmid is treated with T4 DNA polymerase to produce 5' $C_{1-3}A$ strands at both ends, all of the DNA migrates to the position of untreated linear plasmid molecules in 2D gels.

These results demonstrate that the ends of chromosomes and plasmids acquire TG_{1-3} extensions late in the S phase. As indicated above, molecules with TG_{1-3} extensions are the expected product of both recombination mediated gene conversion and telomerase mediated replication (fig. 3). Currently, this project is being pursued in two ways. First is to determine if the G-strand extensions are produced by telomerase or by recombination. To this end, we will determine if G-strand extensions occur in cells with null mutations in the EST1 gene, a gene hypothesized to encode a protein subunit of a putative yeast telomerase (Lundblad and Blackburn, 1990). However, whether or not the G-strand extensions are acquired by telomerase, molecules with long 3' single strand extensions are likely to be excellent substrates for recombination since 3' single strand tails are known to promote strand transfer in vivo and in vitro (Smith, 1988). Thus, the occurrence of these intermediates provides an explanation for our earlier data demonstrating that telomere-telomere recombination is frequent. The second goal is to determine if the G-strand interactions that allow circle formation in vitro also occur in vivo. To this end, in vivo cross linking methods are being developed. If plasmid circularization is detected in vivo, it will provide the first hint that the novel, non B-form DNA structures that telomeric sequences can adopt in vitro have a role in normal telomere biology.

References:

Bell, L., and Byers, B. (1979). Occurrence of crossed strand-exchange forms in yeast DNA during meiosis. Proc. Natl. Acad. Sci. USA 76:3445-3449.

Blackburn, E. H., and Gall, J. G. (1978). A tandemly repeated sequence at the termini of the extrachromosomal ribosomal RNA genes in Tetrahymena. J. Molec. Biol. 120:33-53.

Brewer, B. J., and Fangman, W. L. (1987). The localization of replication origins on ARS plasmids in Saccharomyces cerevisiae. Cell 51, 463-471.

de Lange, T., Shiue, L., Myers, R. M., Cox, D. R., Naylor, S. L., Killery, A. M., and Varmus, H. E. (1990). Structure and variability of human chromosome ends. Molec. Cell. Biol. 10:518-527.

Greider, C. W., and Blackburn, E. H. (1985). Identification of a specific telomere terminal transferase activity in Tetrahymena extracts. Cell 43: 405-413.

Greider, C. W., and Blackburn, E. H. (1987). The telomere terminal transferase of Tetrahymena is a ribonucleoprotein enzyme with two kinds of primer specificity. Cell 51:887-898.

Greider, C. W., and Blackburn, E. H. (1989). A telomeric sequence in the RNA of Tetrahymena telomerase required for telomere repeat synthesis. Nature 337:331-337.

Hastie, N. D., Dempster, M., Dunlop, M. G., Thompson, A. M., Green, D. K., and Allshire, R. C. (1990). Telomere reduction in human colorectal carcinoma and with ageing. Nature 346:866-868.

Henderson, E., Hardin, C. C., Walk, S. K., Tinoco, J., I, and Blackburn, E. H. (1987). Telomeric DNA oligonucleotides form novel intramolecular structures containing guanine-guanine base pairs. Cell 51:899-908.

Lundblad, V., and Blackburn, E. H. (1990). RNA-dependent polymerase motifs in EST1: tentative identification of a protein component of an essential yeast telomerase. Cell 60:529-530.

Lundblad, V., and Szostak, J. W. (1989). A mutant with a defect in telomere elongation leads to senescence in yeast. Cell 57:633-643.

Lustig, A. J., and Petes, T. D. (1986). Identification of yeast mutants with altered telomere structure. Proc. Natl. Acad. Sci. USA 83:1398-1402.

McCarroll, R. M., and Fangman, W. L. (1988). Time of replication of yeast centromeres and telomeres. Cell 54:505-513.

Morin, G. (1989). The human telomere terminal transferase enzyme is a ribonucleoprotien that synthesizes TTAGGG repeats. Cell 59:521-529.

Murray, A. W., Claus, T. E., and Szostak, J. W. (1988). Characterization of two telomeric DNA processing reactions in Saccharomyces cerevisiae. Molec. Cell. Biol. 8:4642-4650.

Pluta, A. F., Dani, G. M., Spear, B. B., and Zakian, V. A. (1984). Elaboration of telomeres in yeast: recognition and modification of termini from Oxytricha macronuclear DNA. Proc. Natl. Acad. Sci. USA 81:1475-1479.

Pluta, A. F. and Zakian, V. A. (1989). Recombination occurs during telomere formation in yeast. Nature 337:429-433.

Sen, D., and Gilbert, W. (1988). Formation of parallel four-stranded complexes by guanine-rich motifs in DNA and its implications for meiosis. Nature 334:364-366.

Shampay, J., Szostak, J. W., and Blackburn, E. H. (1984). DNA sequences of telomeres maintained in yeast. Nature 310:154-157.

Shippen-Lentz, D., and Blackburn, E. H. (1989). Telomere terminal transferase activity from Euplotes crassus adds large numbers of TTTTGGGG repeats onto telomeric primers. Molec. Cell. Biol. 9:2761-2764.

Shippen-Lentz, D., and Blackburn, E. H. (1990). Functional evidence for an RNA template in telomerase. Science 247:546-552.

Smith, G. R. (1988). Homologous recombination in procaryotes. Microbiol. Rev. 52:1-28.

Sundquist, W. I., and Klug, A. (1989). Telomeric DNA dimerizes by formation of guanine tetrads between hairpin loops. Nature 342:825-829.

Szostak, J. W. (1983). Replication and resolution of telomeres in yeast. Cold Spring Harbor Symp. Quant. Biol. 47:1187-1194.

Szostak, J. W., and Blackburn, E. H. (1982). Cloning yeast telomeres on linear plasmid vectors. Cell 29:245-255.

Walmsley, R. W., Chan, C. S. M., Tye, B.-K., and Petes, T. D. (1984). Unusual DNA sequences associated with the ends of yeast chromosomes. Nature 310:157-160.

Wang, S.-S., and Zakian, V. A. (1990). Telomere-telomere recombination provides an express pathway for telomere acquisition. Nature 345:456-458.

Watson, J. D. (1972). Origin of concatemric DNA. Nature (New Biology) 239:197-201.

Williamson, J. R., Raghuraman, M. K., and Cech, T. R. (1989). Monovalent cation-induced structure of telomeric DNA: The

G-Quartet model. Cell 59:871-880.

Yu, G.-L., Bradley, J. D., Attardi, L. D., and Blackburn, E. H. (1990). In vivo alteration of telomere sequences and senescence caused by mutated Tetrahymena telomerase RNAs. Nature 344:126-132.

Zahler, A. M., and Prescott, D. M. (1988). Telomere terminal transferase activity in the hypotrichous ciliate Oxytricha nova and a model for replication of the ends of linear DNA molecules. Nucleic Acids Res. 16:6953-6985.

Zakian, V. A. (1989). Structure and function of telomeres. Annu. Rev. Genet. 23:579-604.

Zakian, V. A., Brewer, B. J., and Fangman, W. L. (1979). Replication of each copy of the yeast 2 micron DNA plasmid occurs during the S-phase. Cell 17:923-934.

Zakian, V. A., Runge, K., and Wang, S.-S. (1990). How does the end begin? Trends in Genetics 6:12-16.

The Role of DnaA-DnaA Box Interaction in Initiation of Chromosomal Replication in *Bacillus subtilis*

N. Ogasawara, S. Moriya and H. Yoshikawa
Departement of Genetics
Osaka University Medical School
2-2, Yamadaoka, Suita 565
Japan

Introduction

In bacteria, an autonomously replicating sequence (*ars*) was first isolated from the replication origin region of the *Escherichia coli* chromosome (Yasuda and Hirota, 1977) and was defined as *oriC* (origin of the chromosomal replication). Genetic and biochemical studies in *E. coli*, on the other hand, showed that DnaA protein acts at the first step in the initiation process by binding to *oriC*, and then opening the double strand to make ready for the formation of primosome (von Meyenburg and Hansen, 1987; Bramhill and Kornberg, 1988). Thus, *oriC* and DnaA protein are considered to be the replicator and initiator of the *E. coli* chromosome. Similar studies in *Bacillus subtilis* have failed to identify either the *ars* or a *dnaA* gene counterpart in this bacterium. Furthermore, a *B. subtilis* initiator gene *dnaB* was found to be similar to *E. coli dnaA* in function but has no structural similarity (Ogasawara et al., 1986). Therefore it was once considered that each bacterium has its unique machinery for the initiation of chromosomal replication and its regulation.

However, DNA sequence determination of the replication origin regions of *B. subtilis* (Ogasawara et al., 1985), *Pseudomonas putida* (Fujita et al., 1989), *Micrococcus luteus* (Fujita et al., 1990) and *Mycoplasma capricolum* (Fujita et al., in preparation) demonstrated that DnaA protein (trans-acting element) and its binding sequence (cis-acting element) are conserved in most of eubacteria (Ogasawara et al., 1990). The highly conserved 9 nucleotide of the DnaA protein binding site, TTATCCACA, is called DnaA-box (Ogasawara et al., 1985). Later we could isolate a temperature sensitive *dnaA* mutant and could demonstrate that the DnaA protein functioned as a limiting factor for initiating chromosomal replication in *B. subtilis* through binding to the DnaA-box (Moriya et al., 1990). Recently we could finally isolate an *ars* from the *B. subtilis* chromosome containing many repeats of DnaA-box. Replication origins of *Pseudomonas* reported by Yee and Smith (Yee and Smith, 1990) also contains DnaA-boxes as an essential element. We therefore propose that the interaction of DnaA protein with DnaA-box is the essential feature of the initiation of the chromosomal replication and its regulation in eubacteria.

In this article we first describe conservation and variation of the two elements in eubacteria. Second, the role of DnaA and DnaA-box in the initiation of chromosomal replication and its regulation in *B. subtilis* will be compared with those in *E. coli*.

Conservation of the gene organization in the replication origin region of the bacterial chromosome

We have determined some 15 kb sequence in the replication origin region of the *B. subtilis* chromosome (Fig. 1). Two characteristic features are noted in gene organization. First, there are three non-coding regions containing repeats of DnaA-box located close to *dnaA* gene. Second, these non-coding regions are arranged symmetrically in relation to the direction of transcription of the gene. Thus, all genes located at the left side of the central non-coding region are transcribed from right to left and genes located at the right side were transcribed in the opposite direction. These structural features suggest strongly that replication starts from the central non-coding region and proceeds bidirectionally in harmony with the direction of transcription.

The gene organization of the *B. subtilis* origin region is compared with that of *dnaA* gene regions from *E. coli* and *P. putida* (Fig. l). In addition to the 6 homologous genes, from *rnpA* to *gyrB*, reported previously (Ogasawara et al.,1985), two open reading frames, *50kd* and *60kd* which are highly homologous between *E. coli* and *P. putida*, were found to be homologous with *orf459* and *orf261* of *B. subtilis*, respectively. Central non-coding region is also conserved in the three

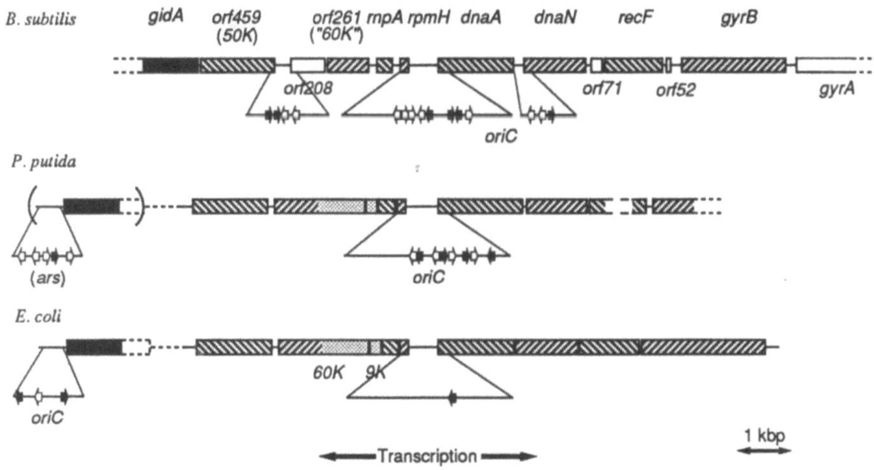

Fig.1 Gene organization in the replication origin region of the chromosome of *B. subtilus, P. putida* and *E. coli*. Data of *E. coli* were from the literature and a personal communication by F. Hansen. Sequence of *P. putida* was determined by us except for the second origin (Yee and Smith, 1990). Arrangements of *orf* are schematically shown by boxes. Genes conserved in the *dnaA* regions of three bacteria are shaded. Black boxes indicate the *gidA* gene. Non coding regions are magnified to indicate DnaA-boxes by small arrows: blacks for consensus TTATCCACA and opens for sequences one base different from the consensus.

bacteria except that the region in *E. coli* contains only one DnaA-box. Second and third non-coding regions present in *B. subtilis* were not found near the *dnaA* of the other two bacteria. However a second non-coding region linked to *gidA* gene is located about 40kb left of *dnaA* in the *E. coli* chromosome. This region contains 3 DnaA-boxes and acts as the *oriC* of the *E. coli* chromosome (Oka et al., 1980). A *Pseudomonas* counterpart of the second non-coding region was isolated as the second origin of the *P. aeruginosa* (Yee and Smith, 1990). The location of this region has not been determined yet. Comparison of the gene organization between *B. subtilis* and *E. coli* suggests strongly that a chromosomal segment of some 50kb from *oriC* to *tnaA* is inverted in the *E. coli* chromosome during evolution.

We assume that the gene organization in *B. subtilis* is a prototype of the replication origin region, because 1) the central non-coding region flanking *dnaA* is conserved in all 5 eubacteria so far examined and all except for *E. coli* contain many repeats of DnaA-box, 2) in the case of *E. coli*, harmony between the direction of transcription and replication is disturbed shortly after the initiation takes place, although the coincidence between the direction of transcription and replication has been shown to be a general feature of the *E. coli* chromosome (Brewer, 1990). This deviation from the general rule can be ascribed to the inversion mentioned above.

The gene organization of the same region is conserved also in other two gram positive bacteria, *Micrococcus luteus* (Fujita et al., 1990) and *Mycoplasma capricolum* (Fujita et al., in preparation). As in *B. subtilis*, the *dnaA* gene is flanked by two non-coding regions in these bacteria, indicating that this is the characteristic of gram positive bacteria. As described below *ars* isolated from the *B. subtilis* chromosome requires the presence of the two regions simultaneously in cis.

Isolation of autonomously replicating sequences (*ars*) from the *B. subtilis* chromosome

Shotgun cloning of *ars* fragments from the *B. subtilis* chromosome has not been successful. Therefore we have tried to isolate *ars* from various fragments which had been cloned from the origin region of the chromosome. In the course of this study we found that the introduction of extra DnaA-boxes being carried in plasmids inhibited the cell growth strongly. The inhibition was dependent on the sequence specific interaction of DnaA-box with DnaA protein, indicating that plasmids containing DnaA-boxes are incompatible with the host chromosome by competing with the limited amount of DnaA protein (Moriya et al., 1988). This is analogous with the negative regulation between the plasmids sharing a common initiation factor. The three non-coding regions containing DnaA-boxes can be divided into four regions differing in the degree of the incompatibility as shown schematically in Figure 2. The central non-coding region was divided into two regions. The *incA* region contains 5 DnaA-boxes and yet has almost no incompatibility even when the region was carried in a high-copy plasmid (50 copies/cell). The *incB* containing 3

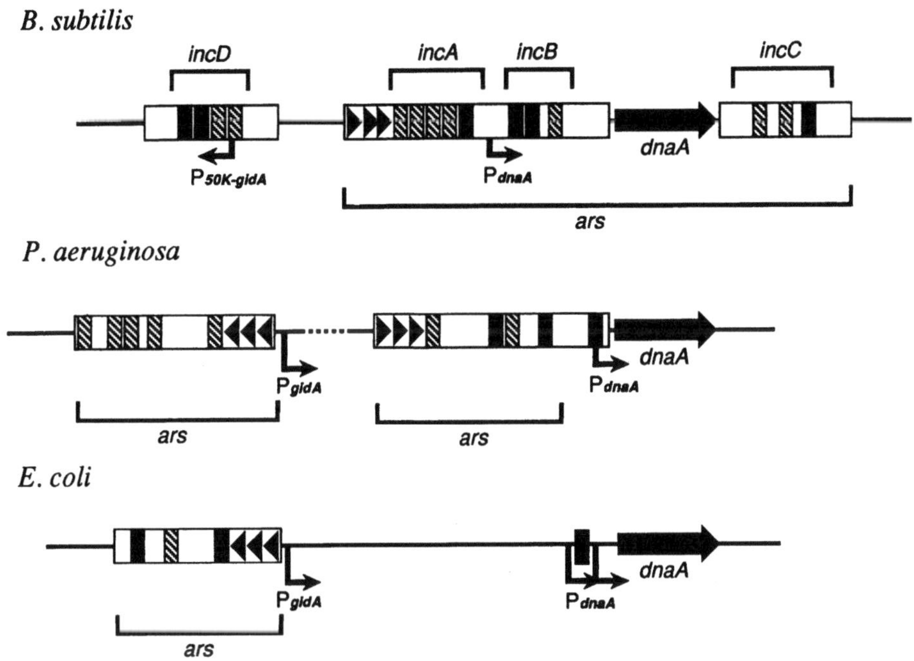

Fig.2 Functional elements for initiation and its regulation. DNA regions for initiation protein (DnaA), *ars* and incompatibility (*inc*) are schematically shown. Non-coding regions containing DnaA box are boxed. Consensus sequences are indicated by black and those with one base change are shaded. Black arrowheads are AT rich oligomers (15 in *B. subtilus* and 13 in *E. coli* and *P. putida*). Minimum regions essential for autonomous replication are indicated by *ars*.

DnaA-boxes was completely incompatible even in a low-copy plasmid (4-5 copies/cell). *IncC* and *IncD* regions containing 3 and 4 DnaA-boxes, respectively, are incompatible in the high-copy plasmid but not incompatible in the low-copy plsmid.

From these experiences, we decided to adopt the following strategies in isolating *ars* from the origin region. First, attempts were made to clone large fragments which would have been missed in shotgun experiments. Second, a plasmid which is stable and effective for selection even at one copy/cell was constructed as an *ars* searching vector. In collaboration with T. Atlung's group in Copenhagen, we have succeeded in isolating an *ars* using a derivative of pBR322 containing a chloramphenicol resistant gene from pC194 as an *ars* searching vector (Moriya et al., in preparation). As expected the first *ars* fragment we isolated was as large as 4.5kb and contained the intact *dnaA* gene with the two flanking non-coding DnaA-box regions (Fig. 3). Deletions and mutations of the starting fragment have revealed that the two DnaA-box regions should be present simultaneously in cis, but the *dnaA* gene in between was not essential for *ars* activity. Deletion of one of the three

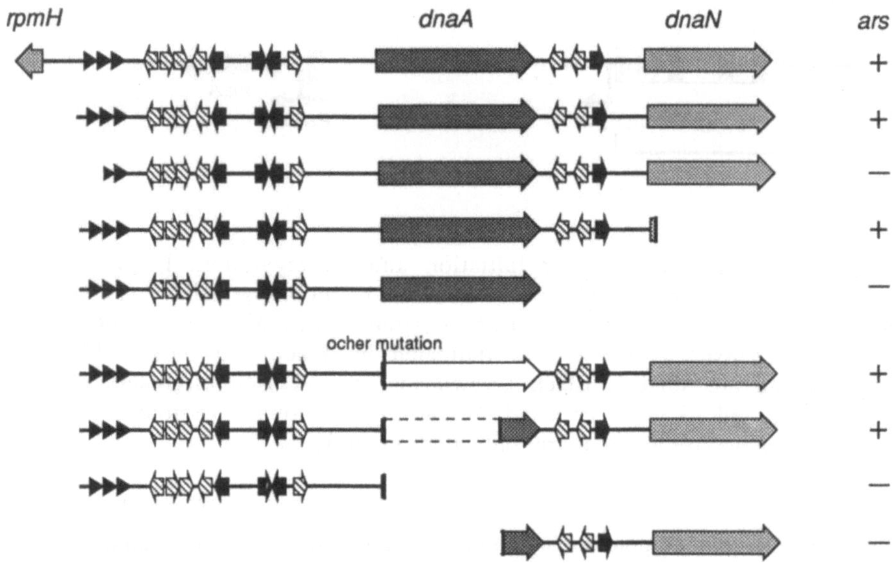

Fig.3 ars activity of various fragments from the *dnaA* region of the *B. subtilus* chromosome. DNA fragmnts assayed for ars activity are schematically shown. Black arrowheads; AT rich oligomers (15 in *B. subtilus* and 13 in *E. coli* and *P. putida*), small arrows; DnaA-boxes (consensus sequences are indicated by black and those with one base change are shaded), and large hatched arrows; *orf* (not drawn in the same scale as non-coding regions). Mutation or deleted is shown by open arrow or dotted line, respectively.

15mers rich in AT present in the 5'-portion of the DnaA-box region upstream of the *dnaA* destroyed the *ars* activity, suggesting that the opening of double strand is initiated in this region as in the case in *E. coli* (Bramhill and Kornberg, 1988). Although the *dnaA* region can be deleted artificially, the smallest native fragment which is active as *ars* is as large as 2.7kb. The failure in shotgun experiments can be explained because there are many sites for conventional restriction enzymes in the *dnaA* gene. The copy number of plasmids containing various *ars* fragments was estimated as low as 0.5-1.0/chromosome. Furthermore the plasmid tends to integrate into chromosome in stationary phase cells even in a *recE* mutant. Two DnaA-box regions are required simultaneously for the *ars* activity. None of the three DnaA-box regions are active as *ars* by themselves. The requirements for the two regions may be related to the incompatibility exerted by these DnaA-box regions. DnaA-box plasmids may be incompatible due not only to competition with DnaA protein, but also by direct interaction with DnaA-box region in the chromosomal *oriC* through interaction between DnaA proteins bound to DnaA-box regions. The two distantly located DnaA-box regions may construct a DNA conformation essential for the activation of *oriC* through interaction of the two DnaA-box regions mediated by bound DnaA proteins.

Function of DnaA and DnaA-box in Initiation in *B. subtilis* as compared with *E. coli*

We have studied the function of DnaA and DnaA-box in initiation of chromosomal replication in *B. subtilis* and compared with those in *E. coli*. The results are summarized in Table 1 including data not shown in this article.

Basic features are common between the two species. Thus, DnaA protein was shown to be essential for initiation of chromosomal replication in *B. subtilis* by isolation of a temperature sensitive mutant with a single amino acid change in the c-terminal conserved domain (Moriya et al., 1990). Direct and specific interaction between DnaA protein and DnaA-box are demonstrated clearly for both *B. subtilis*

Table 1. The function of DnaA and DnaA-box in initiation: comparison between *B. subtilis* and *E. coli.*

Functions	*B. subtilis*	*E. coli*
1) DnaA binds specifically to DnaA-box in vitro.	yes	yes
2) DnaA is essential for initiation.	yes	yes
3) DnaA-box regions act as *ars*.	yes	yes
Number of DnaA-box elements required.	3	1
4) Extra DnaA-box regions inhibit cell growth (incompatibility).	yes	no
5) Overproduction of DnaA on initiation.	inhibit	stimulate
6) Autogenous suppression of *dnaA* gene expression.	yes	yes
7) Gene expression of *dnaA* couples initiation of replication.	yes	no

(Fukuoka et al., 1990) and *E. coli* (Bramhill and Kornberg, 1988) proteins in vitro. However, there seems to be striking differences in the mechanism and regulation of the initiation between the two organisms. First, incompatibility by extra DnaA-boxes was observed only in *B. subtilis.* In consequence, less than one copy of the *oriC* plasmid can coexist with the chromosome of *B. subtilis*, while multiple copies of *E. coli oriC* plasmid can be maintained. Requirement of three DnaA-box regions ,of which two are closely located and the third is distantly located, for *ars* activity in *B. subtilis* may be related to the complex interaction among DnaA-box regions through binding with DnaA protein. Similar complex interaction among regulatory DNA elements was proposed to explain the copy number control of plasmids containing multiple repeats of binding sites for initiator protein (Nordström, 1990). Activation and inactivation of the *oriC* function may be regulated through 1) primary interaction between DnaA protein and DnaA-box regions and 2) secondary interaction among DnaA-box regions mediated by DnaA proteins bound to the DnaA-boxes.

In addition to the complex nature of the interaction between DnaA and DnaA-box regions, regulation of *dnaA* gene expression is also complex in *B. subtilis*. Autoregulation of the gene expression is so stringent that the expression seems to be coupled with the replication of its gene (Ogasawara et al., 1985). In consequence, DnaA protein may be synthesized only at the early phase of the DNA replication cycle and then gradually degraded with the half life of about 20 min at 37°C (Moriya et al., 1990). This suggests the presence of another regulatory factor which determines the timing of the initiation. The fact that overproduction of DnaA protein inhibits the growth in *B. subtilis* suggest the presence of an essential factor which is titrated out by interaction with the excess DnaA protein. Comparative studies on interaction of DnaA protein with DnaA-box between *B. subtilis* and *E. coli* will help understand the mechanism and regulation of the initiation of chromosomal replication in bacteria.

References

Bramhill D, Kornberg A (1988) A model for initiation at origins of DNA replication. Cell 54:915-918

Brewer BJ (1990) Replication and transcriptional organization of the *Escherichia coli* chromosome. in The Bacterial Chromosome (Drlica K, Riley M eds) pp61-83 Amer Soc Microbiol Washington DC

Fujita MQ, Yoshikawa H, Ogasawara N (1989) Structure of the *dnaA* region of *Pseudomonas putida*. Mol Gen Genet 215:381-387

Fujita MQ, Yoshikawa H, Ogasawara N (1990) Structure of the *dnaA* region of *Micrococcus luteus*: Conservation and variation among eubacteria. Gene 93:73-78

Moriya S, Fukuoka T, Ogasawara N, Yoshikawa H (1988) Regulation of initiation of the chromosomal replication by DnaA-boxes in the origin region of the *Bacillus subtilis* chromosome. EMBO J 7:2911-2917

Moriya S, Kato K, Yoshikawa H, Ogasawara N (1990) Isolation of a *dnaA* mutant defective in initiation of replication in *Bacillus subtilis*. EMBO J 9:2805-2910

Nordström K (1990) Control of plasmid replication-how do DNA iterons set the replication frequency? Cell 63:1121-1124

Ogasawara N, Fujita MQ, Moriya S, Fukuoka T, Hirano M, Yoshikawa H (1990) Comparative anatomy of *oriC* of eubacteria. in The Bacterial Chromosome (Drlica K, Riley M eds) pp287-298 Amer Soc Microbiol Washington DC.

Ogasawara N, Moriya S, Mazza PG, Yoshikawa H (1986) Nucleotide sequence and organization of *dnaB* gene and neighboring genes on the *Bacillus subtilis* chromosome. Nucleic Acids Res 14:9989-9999

Ogasawara N, Moriya S, von Meyenburg K, Hansen F, Yoshikawa H (1985) Conservation of genes and their organization in the chromosomal replication origin of *Bacillus subtilis* and *Escherichia coli*. EMBO J 4:3345-3350

Oka A, Sugimoto K,Takanami M, Hirota Y (1980) Replication origin of the *Escherichia coli* K-12 chromosome. J Mol Biol 176:443-458

von Meyenburg K, Hansen FG (1987) Regulation of chromosome replication. in E.coli and S. typhimurium: Cellular and Molecular Biology (Neidhardt FC, Ingraham JL, Low KB, Magasanik, B, Schaechter M, Umbarger HE eds) pp1555-1577 Am Soc Microbiol Washington DC

Yasuda S, Hirota Y (1977) Cloning and mapping of the replication origin of *Escherichia coli*. Proc Natl Acad Sci USA 74:5458-5462

Yee YW, Smith DW (1990) *Pseudomonas* chromosomal replication origins: A bacterial class distinct from *Escherichia coli* type origins. Proc Natl Acad Sci USA 87:1278-1282

DnaA Protein Mediated Transcription Termination in the GUA Operon of *Escherichia coli*

Christoph Schaefer[1], Andreas Holz and Walter Messer
Max-Planck-Institut für molekulare Genetik,
Abt. Trautner,
Ihnestrasse 63,
D-1000 Berlin 33,
Germany

SUMMARY

The *guaB* gene is the first gene in GMP biosynthesis. It harbours a potential DnaA protein binding site (*dnaA* box, 5'-TTATCCACA) in its coding region. This box is in the active orientation for DnaA-mediated transcription termination. The location of the *guaB* promoter was defined using a promoter search plasmid with *galK* as a monitoring gene. Transcription originating at the promoter and entering the gene was quantified under varying levels of DnaA protein. The formation of the DnaA protein/*dnaA* box complex in the coding region of *guaB* resulted in a decrease in *guaB* expression due to transcription termination. DnaA protein, therefore, represents a regulatory element in *guaB* expression.

---

[1] present address: Institute of Brain Research, August-Forel-Straße 1, CH-8029 Zürich, Switzerland

INTRODUCTION

The DnaA protein of *E. coli* plays a key role in the initiation of chromosomal DNA replication. DnaA protein binds specifically to a 9 bp consensus sequence, the *dnaA* box (5'-(T/C)(T/C)(A/T/C)T(A/C)C(A/G)(A/C/T)(A/C)) (Fuller et al., 1984; Schaefer and Messer, 1991).

Binding of DnaA to 4 *dnaA* boxes present in the minimal replication origin, *oriC*, leads to replisome assembly and initiation of DNA replication (Sekimizu et al., 1988; Messer et al., 1988). Promoters with an adjacent *dnaA* box are repressed by DnaA protein. This was demonstrated for the *dnaA* promoter, resulting in autoregulation (Atlung et al., 1985; Braun et al., 1985; Kücherer et al., 1986; Wang and Kaguni, 1987), and for the *mioC* (Lother et al., 1985; Rokeach et al., 1986; Stuitje et al., 1986; Loebner-Olesen et al., 1987; Schauzu et al., 1987), *uvrB* (van den Berg et al., 1985), *rpoH* (Wang and Kaguni, 1989), and *drpA* (Zhou and Syvanen, 1990) promoter. Recently we could demonstrate a third function for DnaA protein. The complex of DnaA protein with a *dnaA* box in one orientation was found to act as a transcription terminator (Schaefer and Messer, 1988; Messer et al., 1988; Gielow et al., 1988; Schaefer and Messer, 1989; Schaefer and Messer, 1991). We expect, by extrapolation from known sequences, approximately 1600 *dnaA* boxes in the *E. coli* chromosome. Many genes contain such boxes in an active orientation for transcription termination within their coding regions.

The first gene in the *gua* operon, the *guaB* gene, encodes the inosine monophosphate (IMP) dehydrogenase. IMP dehydrogenase catalyses the NAD^+ dependent conversion of IMP to XMP (xanthosine monophosphate), the first reaction in GMP (guanosine monophosphate) biosynthesis. *guaB* is the operator-proximal gene in the *guaBA* operon. The operon is subject to dual control by adenine and guanine nucleotides. Guanine nucleotides repress and adenine nucleotides induce expression of *guaBA* (Mekra and Drabble, 1981). The *guaB* gene harbors a *dnaA* box in its coding region (5'-TTATCCACA) at position 514-522 (Tiedeman and Smith, 1985) in the orientation which is active in transcription termination (Schaefer and Messer, 1989). Binding of DnaA protein to this internal box might mediate transcription termination and thus modulate the expression of IMP dehydrogenase.

MATERIALS AND METHODS

Bacterial strains and plasmids.

Host strain for the plasmids was *E. coli* EC558 (*ara*-14, *argE*3, Δ(*gpt-proA*)62, *galK*2, *his*-4, *lacY*1, *leuB*6, *mtl*-1, *rpsL*31, *supE*44, *thi*-1, *tsx*-33, *xyl*-5, *dnaA46*) (Polaczek and Ciesla, 1984). The plasmid pHK27 (Schaefer and Messer, 1988) which carries the *dnaA* gene under the control of the *tac* promoter, was used for overexpression of DnaA protein. Minimal medium (with casamino acids) was as described (Schaefer and Messer, 1988).

***In vitro* construction of *galK* fusions.**

Restriction enzymes were purchased from Boehringer Mannheim, FRG. We used standard techniques for plasmid preparation and cloning (Maniatis et al., 1982). Three plasmids were constructed each with a different part of the *gua* operon fused to the *galK* gene in the promoter search plasmid pGAL21. pGAL21 is a derivative of pGAL21-lac (Schaefer and Messer, 1991) harboring the *galK* gene, preceded by stop codons in all three reading frames in order to avoid translational fusions, the *lacIq* gene and the pBR322 replicon.

The plasmid pGUA1 harbors an EcoRI-NspHI fragment of about 150 bp (Fig.1) with a part of the regulatory region of *guaB* fused to the EcoRI-SphI fragment of pGAL21. Plasmid pGUA2 contains a 250 bp DpnI-PvuII fragment, including part of the regulatory region and part of the reading frame of the *guaB* gene (without the *dnaA* box), inserted into the SmaI site of pGAL21. pGUAB2 is an insertion of a 430 bp DpnI-DpnI fragment into the SmaI site of pGAL21, containing the same sequences as pGUA2 and additional *guaB* sequences with the *dnaA* box (Fig.1).

Galactokinase and β-lactamase assays.

The assay procedure for galactokinase (McKenney et al., 1981) and the modifications were described (Schaefer and Messer, 1988). To determine the relative copy number, the activity of the β-lactamase (*bla*) gene encoded by the plasmids was measured (Sargent, 1968). The GalK units were corrected for Bla units according to the equation described in (Schaefer and Messer, 1989).

RESULTS

Promoter analysis in the *guaBA* operon.

Previous results, based on S1 mappings of the *guaB* promoter region by Tiedeman and Smith (1985), suggested a promoter at position 12-43. During our analysis we observed promoter activity in a different region. Therefore we cloned two different fragments of the upstream non-coding region of the *guaBA* operon into the promoter search plasmid pGAL21, in front of the *galK* gene to monitor for transcription activity. pGUA1 contains the 149 bp EcoRI-NspHI fragment from the *guaB* regulatory region. This fragment should harbor the promoter predicted by Tiedeman and Smith (1985) (Fig. 1B). pGUA2 contains a 250 bp DpnI-PvuII fragment downstream of the fragment in pGUA1. Only the plasmid pGUA2 showed promoter activity, expressing 138 units GalK, an activity comparable to an induced *lac* promoter. Plasmid pGUA1 showed only background GalK activity (3 units). From these results we conclude that the potential -35 region of the *guaB* promoter is located at position 164-169 and the -10 region at 189-194 (positions of nucleotides refer to Tiedemann and Smith, 1985).

The role of the *guaB dnaA* box in gene regulation.

In order to analyse transcription termination at the *dnaA* box in *guaB*, we integrated the *guaB* promoter and the start of the *guaB* coding region, including the potential *dnaA* box, in front of the *galK* gene into plasmid pGAL21, resulting in plasmid pGUAB2. Transcription activity distal to the *dnaA* box was monitored by *galK* expression. pGUA2 with a shorter fragment of the *guaB* gene without the *dnaA* box served as a control (Fig. 1A).

GalK activity was measured in the presence of three different levels of DnaA protein: Upon inactivation of DnaA protein at 42° C in the temperature sensitive mutant host EC558 (*dnaA*46), at wild type level, and at an elevated DnaA level, due to the presence of the DnaA overproducing plasmid pHK27 (Schaefer and Messer, 1988). GalK expression in plasmid pGUA2 (without the *dnaA* box) was unaffected by the different levels of DnaA protein. The GalK activity was about 140 units (Fig. 1A). Expression of *galK* in the plasmid pGUAB2 with the *dnaA* box in the coding region of *guaB* showed a dependence on the amount of DnaA protein in the cell. Upon inactivation of DnaA protein (-), *galK* expression increased from 80 to 140 units, the same rate of expression as found in the absence of a *dnaA* box in pGUA2. In the wild type situation (wt) and at an excess of DnaA protein (++) the rate of expression decreased to 60% and 30%, respectively. This suggests an increasing transcription termination by the DnaA

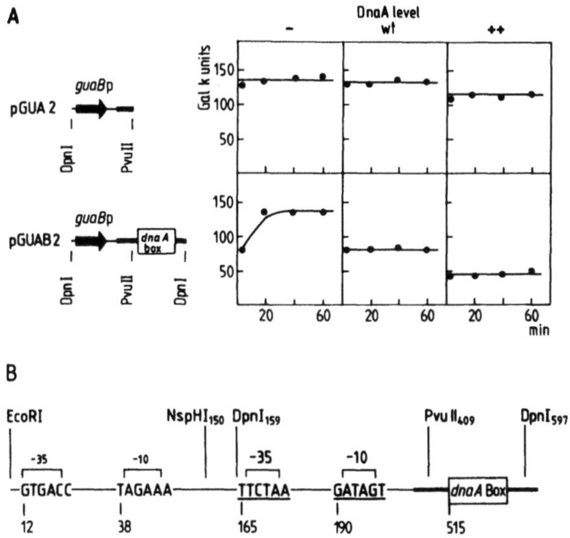

Figure 1: A. GalK activity from *guaB*p with (pGUAB2) and without (pGUA2) *dnaA* box at different DnaA concentrations.
DnaA level (-): EC558 (*dnaA*46) host cells were shifted at t=0 min from 30°C to 42°C.
(wt): EC558 at 30°C.
(+ +): EC558 with the DnaA overproducing plasmid pHK27 at 30°C, which was induced at t=0 with 1 mM IPTG.
B. Promoter region of the *guaBA* operon (not drawn to scale).
Subscript numbers indicate the positions of restriction sites in the sequence of Tiedeman and Smith (1985). Thick line: *guaB* coding region.

protein/*dnaA* box complex with increasing levels of DnaA protein, eventually resulting in termination of about 70% of the transcripts (Fig.1A).

The interaction of DnaA protein with the *dnaA* box in the coding region of *guaB* led to transcription termination, to a decrease of *guaB* expression and therefore represents a regulatory element of *guaB* expression.

DISCUSSION

We cloned different fragments of the *guaB* promoter region with and without a potential DnaA protein binding site (*dnaA* box) in front of the *galK* gene, to monitor transcription. The DnaA protein concentration in the cell was varied to measure its influence on the rate of *guaB* expression.

Transcription originating at the *guaB* promoter was not affected by different levels of DnaA protein on plasmid pGUA2 (without *dnaA* box). In contrast plasmid pGUAB2 (with *dnaA* box 5'-TTATCCACA) showed transcription termination (up to 60%) with increasing amounts of DnaA protein in the cell. This clearly indicates that the actively oriented *dnaA* box in the *guaB* gene binds DnaA protein and thereby modulates *guaB* expression.

The *guaB* gene is the first gene of an operon followed by the *guaA* gene. We defined a strong promoter at position 164-194. The position of the promoter described here is different from the position (12 - 43) suggested by Tiedeman and Smith (1985) based on S1 mapping studies. However, we have not excluded the possibility that the promoter found by Tiedeman and Smith was affected by our cloning procedure. The *gua* operon may be transcribed from two different promoters, which are sensitive to different controls.

Many genes contain *dnaA* boxes in transcribed regions where they can potentially function as regulatory elements of gene expression. The *dam* (Jonczyk et al., 1989) and the *ftsA* and *ftsQ* genes (Masters et al., 1989) are examples for which modulation of expression by DnaA protein has been demonstrated. We reported transcription termination at a *dnaA* box within the reading frame of the *dnaA* gene itself (Messer et al., 1988). However, in most host strains we found extensive DnaA independent transcription termination in a 200 bp segment around this *dnaA* box. This effect tends to mask an additional termination by DnaA protein within *dnaA* (unpublished results).

The *gua* operon is responsible for GMP synthesis and thereby affects protein biosynthesis (via GTP), DNA synthesis (via dGTP), RNA synthesis (via rGTP) and stringently regulated promoters (via ppGpp). The *gua* operon itself is regulated, in addition to other controls, by the replication protein DnaA. Apparently, DnaA protein may function as a global regulatory protein linking the expression of many genes to the replication cycle.

Acknowledgement

We thank John M. Smith for a gift of plasmids containing *guaB*.

References

Atlung, T., E. Clausen, and F.G. Hansen. 1985. Autoregulation of the *dnaA* gene of *Escherichia coli*. Mol. Gen. Genet. **200**:442-450.

Braun, R.E., K. O'Day, and A. Wright. 1985. Autoregulation of the DNA replication gene *dnaA* in *E. coli*. Cell **40**:159-169.

Fuller, R.S., B.E. Funnell, and A. Kornberg. 1984. The *dnaA* protein complex with the *E. coli* chromosomal origin (*oriC*) and other sites. Cell **38**:889-900.

Gielow, A., C. Kücherer, R. Kölling, and W. Messer. 1988. Transcription in the region of the replication origin, *oriC*, of *Escherichia coli*. Termination of *asnC* transcripts. Mol. Gen. Genet. **214**:474-481.

Jonczyk, P., R. Hines, and D.W. Smith. 1989. The *Escherichia coli dam* gene is expressed as a distal gene of a new operon. Mol. Gen. Genet. **217**:85-96.

Kücherer, C., H. Lother, R. Kölling, M.A. Schauzu, and W. Messer. 1986. Regulation of transcription of the chromosomal *dnaA* gene of *Escherichia coli*. Mol. Gen. Genet. **205**:115-121.

Loebner-Olesen, A., T. Atlung, and K.V. Rasmussen. 1987. Stability and replication control of *E. coli* minichromosomes. J. Bact. **169**:2835-2842.

Lother, H., R. Kölling, C. Kücherer, and M. Schauzu. 1985. Initiation of replication in *Escherichia coli* involves *dnaA* protein regulated transcription within the replication origin. EMBO J. **4**:555-560.

Maniatis, T., E.F. Fritsch, and J. Sambrook. 1982. Molecular Cloning. Cold Spring Harbor Laboratory, Cold Spring Harbor, NY.

Masters, M., T. Paterson, A.G. Popplewell, T. Owen-Hughes, J.H. Pringle, and K.J. Begg. 1989. The effect of DnaA protein levels and the rate of initiation at *oriC* on transcription originating in the *ftsQ* and *ftsA* genes: In vivo experiments. Mol. Gen. Genet. **216**:475-483.

McKenney, K., H. Shimatake, D. Court, U. Schmeissner, C. Brady, and M. Rosenberg. 1981. A system to study promoter and terminator signals recognized by *Escherichia coli* RNA polymerase, p.383-415. In J.G. Chirikijan, and T.S. Papas (eds.), Gene Amplification and Analysis. 2: Analysis of Nucleic Acids by Enzymatic Methods. North Holland Publishing Company, Amsterdam.

Mehra, R.K., and W.T. Drabble. 1981. Dual control of the *gua* operon of *Escherichia coli* K12 by adenine and guanine nucleotides. J. Gen. Microbiol. **123**:27-37.

Messer, W., W. Seufert, C. Schaefer, A. Gielow, H. Hartmann, and M. Wende. 1988. Functions of the DnaA protein of *Escherichia coli* in replication and transcription. Biochim. Biophys. Acta **951**:351-358.

Polaczek, P., and Z. Ciesla. 1984. Effect of altered efficiency of the RNA I and RNA II promoters on *in vivo* replication of colE1-like plasmids in *Escherichia coli*. Mol. Gen. Genet. **194**:227-231.

Rokeach, L.A., A.E. Chiaramello, D.E. Junker, K. Crain, A. Nourani, M. Jannatipour, and J.W. Zyskind. 1986. Effects of DnaA protein on replication and transcription events at the *Escherichia coli* origin of replication, *oriC*. UCLA Symp. Mol. Cell. Biol. **47**:415-427.

Sargent, M.G. 1968. Rapid fixed-time assay for penicillinase. J. Bact. **95**:1493-1494.

Schaefer, C., and W. Messer. 1988. Termination of the *E. coli asnC* transcript. The DnaA protein/*dnaA* box complex blocks transcribing RNA polymerase. Gene **73**:347-354.

Schaefer, C., and W. Messer. 1989. Directionality of DnaA protein / DNA interaction. Active orientation of the DnaA protein / *dnaA* box complex in transcription termination. EMBO J. **8**:1609-1613.

Schaefer, C., and W. Messer. 1991. DnaA protein/DNA interaction. Modulation of the recognition sequence. Mol. Gen. Genet.: in press.

Schauzu, M.A., C. Kücherer, R. Kölling, W. Messer, and H. Lother. 1987. Transcripts within the replication origin, *oriC*, of *Escherichia coli*. Nucl. Acids Res. **15**:2479-2497.

Sekimizu, K., D. Bramhill, and A. Kornberg. 1988. Sequential early stages in the *in vitro* initiation of replication at the origin of the *Escherichia coli* chromosome. J. Biol. Chem. **263**:7124-7130.

Stuitje, A.R., N. de Wind, J.C. van der Spek, T.H. Pors, and M. Meijer. 1986. Dissection of promoter sequences involved in transcriptional activation of the *Escherichia coli* replication origin. Nucl. Acids Res. **14**:2333-2344.

Tiedeman, A.A., and J.M. Smith. 1985. Nucleotide sequence of the *guaB* locus encoding IMP dehydrogenase of *Escherichia coli* K12. Nucl. Acids Res. **13**:1303-1316.

van den Berg, E.A., Geerse, R.H., Memelink, J., Bovenberg, R.A.L., Magnee, F.A., and P. van de Putte. 1985. Analysis of regulatory sequences upstream of the *E. coli uvrB* gene; involvement of the DnaA protein. Nucl. Acids Res. **13**:1829-1840.

Wang, Q., and J.M. Kaguni. 1987. Transcriptional repression of the *dnaA* gene of *Escherichia coli* by dnaA protein. Mol. Gen. Genet. **209**:518-525.

Wang, Q., and J.M. Kaguni. 1989. dnaA Protein regulates transcription of the *rpoH* gene of *Escherichia coli*. J. Biol. Chem. **264**:7338-7344.

Zhou, Z., and M. Syvanen. 1990. Identification and sequence of the *drpA* gene from *Escherichia coli*. Mol. Gen. Genet. **172**:281-286.

Screens for Proteins Binding to the ARS Consensus Sequence

James F. Theis and Carol S. Newlon
Department of Microbiology and Molecular Genetics
UMDNJ-New Jersey Medical School
185 South Orange Avenue
Newark, N.J. 07103
USA

A subset of autonomously replicating sequence (*ARS*) elements, identified on the basis of their ability to promote high efficiency transformation and extrachromosomal maintenance of plasmids in *Saccharomyces cerevisiae*, has recently been shown to function as *bona fide* chromosomal replication origins (Huberman *et al.* 1988, Dubey *et al.* 1991, Greenfeder and Newlon, ms. in prep.). Extensive analysis of the DNA sequences required for *ARS* function on plasmids has revealed the necessity for an eleven base pair (bp) consensus sequence and a variable number of bases 3' to the T-rich strand of the consensus sequence (reviewed by Newlon 1988, Campbell and Newlon 1991). Mutational analysis of the consensus sequence demonstrated a stringent requirement for most of the bases in the consensus sequence, a finding which suggests that this sequence may function as a protein binding site. *In vivo* footprinting at *ARS1* is consistent with the binding of a protein in the region of the consensus sequence (Lohr and Torchia 1988).

A major challenge in understanding the initiation of DNA replication is the identification of the protein(s) that interact with replication origins. The use of small *ARS*-containing fragments in DNA binding assays has led to the identification by several laboratories of an abundant protein called ABF1 (Diffley and Stillman 1988, Eisenberg *et al.* 1988, Sweder *et al.* 1988). This protein binds not to the consensus sequence but to a site that is present near the consensus sequence in many but not all *ARS* elements. The *ABF1* protein appears to function as an enhancer of *ARS* function (Walker *et al.* 1990). Two other *ARS* binding factors, ABFII and ABFIII, have been reported but not yet fully characterized (reviewed by Campbell and Newlon 1991). However, it is clear that neither of these factors binds to the consensus sequence.

In this manuscript we report on the use of two different approaches designed to identify proteins that bind to the *ARS* consensus sequence. The first of these, the south-western screen, differs from previous attempts to isolate *ARS* binding proteins in two respects. First, binding activities are not assayed in yeast cell extracts, but rather, bacterially-produced proteins are assayed after adsorption to filters. The second difference is in the probes used to assay binding. In previous studies small restriction fragments were used to assay binding. Such fragments potentially contain binding sites for many different proteins. Therefore, binding of a rare protein might be obscured by binding of an abundant one. To avoid the presence of multiple protein binding sites, we made use of a pair of complementary 23 base oligonucleotides carrying the *ARS* consensus sequence that were previously employed to create a semi-synthetic *ARS* element (Palzkill and Newlon 1988).

The second approach was based upon the observation of Brent and Ptashne (1984) that the binding of a protein between an upstream activation sequence (UAS, the *S. cerevisiae* enhancer analog) and a promoter reduces expression from that promoter. If the insertion of an *ARS* element repressed the synthesis of the reporter gene by virture of protein binding to the *ARS* consensus sequence, then it could be used as the basis of a genetic selection for mutations that reduce the affinity of the binding factor for the consensus sequence and thus allow expression of the reporter gene.

RESULTS

South-western screen

We screened a λgt11 library containing *S. cerevisiae* inserts provided by M. Snyder (Yale University) for phages that express, in *E. coli*, proteins capable of binding double-stranded oligonucleotides containing the *ARS* consensus sequence. The pair of complementary oligonucleotides previously used to construct a synthetic *ARS* element were kinased with γ-^{32}P-ATP, annealed, and ligated to form multimers. The λgt11 library was screened with these oligonucleotides using the method of Vinson *et al.* (1988) as modified by Kalionis and O'Farrell (pers. comm.). Phage were plated on *E. coli* strain Y1090, expression of *lacZ* fusion proteins was induced by overlaying with IPTG-impregnated nitrocellulose filters, and the expressed proteins were adsorbed to these filters (Young and Davis 1983). The proteins on the filters were then denatured in 6M guanidine hydrochloride and renatured by serial 1:1 dilutions. The probe was incubated with the filters, and unbound probe was washed off.

In the first round, 10^6 phage were screened. Positive phage were picked, diluted, and reisolated through two additional rounds of screening. Multiple isolates of two overlapping phage were identified (Figure 1). Both phage produced a *lacZ* fusion protein of 180 kD upon induction induction with IPTG. Hybridization of the phage inserts to blots of *S. cerevisiae* chromosomes separated on pulsed-field gels mapped the inserts to chromosome IV.

At this stage of our analysis, we learned of a clone, isolated from the same λgt11 library, encoding a protein that bound an A+T-rich element in centromeric DNA, CDEII, and also mapped to chromosome IV (Moscatelli, Gariboldi and Panzeri 1990). The centromere-binding activity had been shown by DNA sequence analysis to be the *PHO2/BAS2* gene product. We therefore tested our phage inserts for hybridization to the cloned *PHO2/BAS2* gene, and our phage clearly contain the *PHO2/BAS2* gene (Figure 1). This gene encodes a transcriptional regulator whose activity affects a number of metabolic pathways including phosphate utilization, histidine, adenine and tryptophan biosynthesis (Arndt, Styles and Fink 1987, Braus *et al.* 1989, Yoshida *et al.* 1989). The best studied case is its regulation of the *HIS4* gene. *PHO2/BAS2* is required for the basal transcription of *HIS4*, and the encoded protein binds to a site in the *HIS4* promoter that resembles $(TTAA)_n$ (Tice-Baldwin, Fink and Arndt 1989).

PHO2/BAS2 null mutants are viable, but require adenine and inorganic phosphate for growth (Arndt, Styles and Fink 1987).

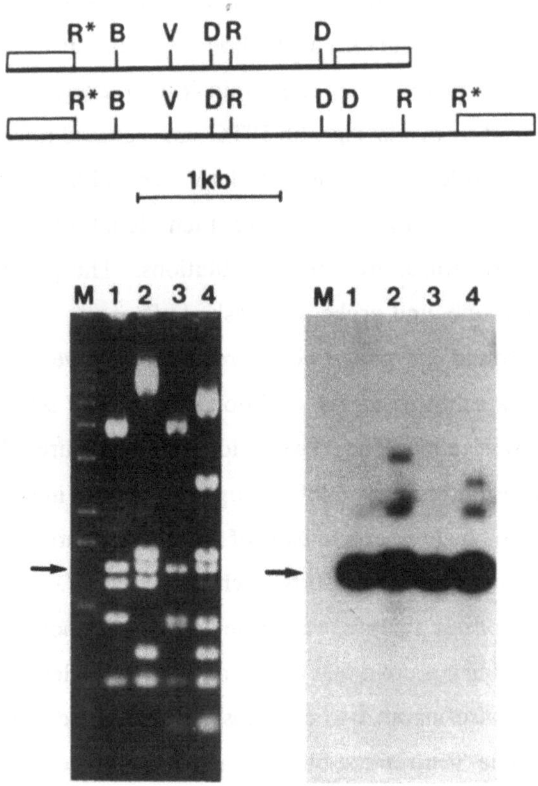

Figure 1. The *ARS*-consensus binding activity is the *PHO2/BAS2* gene product. The upper part of the figure shows the restriction maps of the two classes of λgt11 phages identified. Boxes represent the phage arms. The fusion protein is synthesized from the *lacZ* promoter in the left arm. B=*Bgl*II, D=*Dra*I, R=*Eco*RI, V=*Eco*RV. The lower part of the figure shows an ethidium bromide-stained agarose gel and the autoradiogram of that gel after it was blotted and probed with the 1.2 kb *Eco*RI fragment included in the larger phage. Lane M contains molecular weight markers (BRL 1 kb ladder). Lanes 1 and 3 contain digests of a plasmid carrying the insert from the larger phage. Lanes 2 and 4 contain digested DNA from the plasmid *BAS2*-YCp50 (Arndt, Styles and Fink 1987). DNA in lanes 1 and 2 was digested with *Eco*RI; DNA in lanes 3 and 4 was digested with *Eco*RI and *Bgl*II.

There is a precedent for multifunctional proteins that are transcription factors and also appear to be involved in some aspect of DNA replication. The *ABF1* and *MCM1* gene products fall into this category (Buchman *et al.* 1988, Diffley and Stillman 1989, Jarvis, Clark and Sprague 1989, Passmore, Elble and Tye 1989, Rhode *et al.* 1989, Shore *et al.* 1987). We therefore further tested the role of the *BAS2/PHO2* gene product in DNA replication. First, we used South-western blotting experiments to test the ability of the λgt11 fusion protein to bind two other pairs of complementary oligonucleotides. The first pair contained the *ARS* consensus sequence with a more A+T-rich flanking sequence (Palzkill and Newlon, 1988). The second pair contained a mutant consensus sequence that gives an Ars⁻ phenotype. The DNA binding activity of the *PHO2/BAS2* clones recognized both pairs of oligonucleotides with a higher affinity than the pair used for the initial screening, and the binding affinity of the oligonucleotides correlated with how well they match the binding site for *PHO2/BAS2*, $(TTAA)_n$. These observations suggest that *PHO2/BAS2* is unlikely to encode the replication initiator protein. However, it could still play a role in initiation since most *ARS* elements contain multiple imperfect matches to the consensus sequence, and weak binding to any one site might be overcome by protein-protein interactions conferring cooperative binding on iterated sites. To further test the possible role of the *PHO2/BAS2* gene product in DNA replication, we compared the stability of a plasmid minichromosome carrying *CEN4* and a derivative of *ARS307* in a *bas2-2* deletion strain (AY860, Arndt, Styles and Fink 1987) and a *BAS2* strain. The *ARS307* derivative, shown to bind the *PHO2/BAS2* gene product *in vitro*, was equally stable in both strains (mitotic stability 0.039 vs 0.037), suggesting that the *PHO2/BAS2* gene product plays no role in *ARS* function *in vivo*.

Thus the protein identified in the South-western screen appears to play no significant role in DNA replication. Rather, it was identified by virtue of the similarity of its binding site to the *ARS* consensus sequence. There are many possible explanations for the failure of the South-western screen to identify an *ARS* consensus binding activity that functions in the initiation of replication. The screen requires that an active fusion protein be present in the library and that the binding activity be reconstituted following denaturation in guanidine hydrochloride. In addition, the protein must be capable of binding alone, without protein or small molecule cofactors, and it must have a sufficiently high binding affinity for the *ARS* consensus oligonucleotide to produce a signal above background in the binding assay.

Brent-Ptashne screen

To test whether insertion of an *ARS* element between a UAS and the promoter it controls reduces expression from that promoter, we cloned one or more copies of semi-synthetic *ARS* element, C4A.155 (Palzkill and Newlon 1988), into a plasmid, Δ20B, obtained from R. Brent (Figure 2). This plasmid contains a *Xho*I site between the UAS_{Gal} and the *GAL1* promoter that drives the expression of a *GAL1-lacZ* fusion protein. The expression of the fusion protein from this construct can be induced by growth on galactose, and the β-galactosidase activity of the fusion protein can be monitered by standard colorimetric assays. This construct contains a 180 bp deletion of sequences that normally lie between UAS_{Gal} and the *GAL1* promoter (West, Young and Ptashne 1984), and it has been shown that insertions of up to 234 bp of DNA at the *Xho*I site of Δ20B have no effect on reporter gene expression (Brent and Ptashne 1984). The semi-synthetic *ARS* element consists of a 45 bp fragment of *ARS307* that contains the exact copy of the consensus sequence with most of the 3'-flanking sequences replaced by four copies of an oligonucleotide carrying the *ARS* consensus sequence (Palzkill and Newlon 1988). The presence of one or two copies of this semi-synthetic *ARS* element at the *Xho*I site of Δ20B repressed the expression of β-galactosidase activity 10- to 100-fold compared to plasmid JK101, a construct carrying the *E. coli lexA* binding site (Figure 2, Brent and Ptashne 1984).

This result suggests that either a protein binding to the semi-synthetic *ARS* element or some feature of the *ARS* element itself interferes with the expression

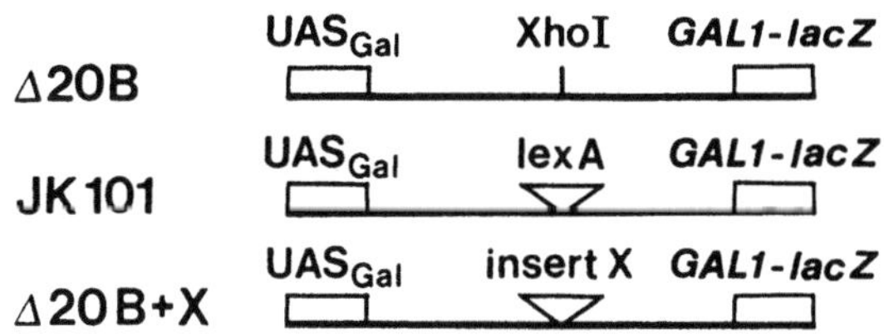

Figure 2. Structure of the reporter gene region of plasmids used for protein binding assay. Δ20B and JK101 were obtained from R. Brent. JK101 was derived from Δ20B by insertion of the *lexA* binding site from the ColE1 promoter (Brent and Ptashne 1984). *ARS307* derivatives or oligonucleotides were inserted into the *Xho*I site of Δ20B as indicated by "insert X" to create Δ20B+X plasmids as shown in the third line.

of the reporter gene. To test whether the repression was due to protein binding at the consensus sequence, we took advantage of *ARS307* constructs that had been shown to lack *ARS* function because the exact match to the consensus sequence either had been deleted or replaced by a sequence carrying a point mutation (Van Houten and Newlon 1990). Table 1 shows that the presence of a wild type derivative of *ARS307* in either orientation at the *Xho*I site of Δ20B causes the repression of β-galactosidase activity (Δ20B+WTA, Δ20B+WTB). However, nonfunctional *ARS307* derivatives carrying either a point mutation in the consensus sequence (Δ20B+T9A, Δ20B+T9B) or a deletion of the consensus sequence (Δ20B+0A, Δ20B+0B) were as effective as the wild type *ARS* element at repressing reporter gene activity. The repression by these *ARS307* inserts was seen even in a *bas2-2* deletion strain, demonstrating that binding of the *BAS2* gene product cannot responsible for the observed repression. Thus, it appears that the repression of reporter gene expression by *ARS307* and its derivatives must be caused either by a protein that binds to domain B of the *ARS* element or by some feature of its DNA sequence *per se*.

As a final test of whether this assay could be used to detect protein binding to the *ARS* consensus sequence, we cloned oligonucleotides carrying either an exact match to the consensus sequence (ATTTATGTTTT) or a two base-pair

Table 1

Plasmid	β-galactosidase activity (U/mg protein)[a]
JK101	130
Δ20B+WTA[b]	18
Δ20B+WTB[b]	5
Δ20B+T9A[c]	20
Δ20B+T9B[c]	6
Δ20B+0A[d]	23
Δ20B+0B[d]	5

[a]β-galactosidase assays were performed as described by Guarente (1983). Data shown are from strain AY860 (*MATα bas1-2 bas2-2 ura3-52*). Similar results were obtained from a wild type strain.
[b]Δ20B+WT carries the *ARS307 Bam*HI to *Sal*I fragment from pVH402 (Van Houten and Newlon 1990) inserted at the plasmid *Xho*I site with the aid of a *Xho*I linker at the *ARS307 Bam*HI site. A and B refer to the orientation of the *ARS* element. In orientation A, the consensus sequence is nearest the *gal1-lacZ* fusion.
[c]Δ20B+T9 plasmids are as Δ20B+WT except that the *ARS* element carries a point mutation that changed the T at position 9 of the consensus sequence to an A. This point mutation causes an Ars⁻ phenotype (Van Houten and Newlon 1990).
[d]Δ20B+0 plasmids are as Δ20B+WT above except that the *ARS* element was cloned from pVH400 in which the consensus sequence was deleted (Van Houten and Newlon 1990).

mismatch (ATTaATGTaTT) into the *Xho*I site of Δ20B. Neither of these oligonucleotides was capable of repressing reporter gene expression. We therefore conclude that this approach cannot be used to identify an *ARS* consensus sequence binding protein by virtue of its ability to repress the expression of a reporter gene. The plasmid used in these experiments is a high-copy-number, 2 μm-based construct. The repression of the reporter gene on this plasmid may require a relatively abundant protein that has a high affinity for the binding site. If the putative *ARS* consensus binding protein is low in abundance or has a relatively low affinity for its binding site, then it could easily escape detection in this assay.

These observations raise several questions with regard to *ARS* structure and function. First, does the *ARS* consensus sequence actually serve as a protein binding site? It is likely that it does. There are two recent reports of proteins that bind to the *ARS* consensus sequence and show stronger binding to single-stranded DNA than to double-stranded DNA (Kuno, Murakami and Kuno 1990, Hofmann and Gasser 1991). In the best studied case, the ability of the protein to bind mutant consensus sequences correlated well with the ability of the sequences to promote *ARS* activity (Hofmann and Gasser 1991). It is not yet clear whether these proteins function in DNA replication *in vivo*.

Second, why is the 3'-flanking region (domain B) of *ARS307* capable of repressing expression of the reporter gene in plasmid Δ20B? One possibility is that it contains high affinity binding sites for one or more as yet undiscovered proteins. Our observation that a derivative of *ARS307* with an almost completely synthetic domain B, C4A.155 (Palzkill and Newlon 1988), also represses expression of the reporter gene makes this possibility somewhat less likely. An alternative explanation is that some feature of the DNA structure formed by this region is responsible for the observed repression. Umek and Kowalski (1988, 1990) have demonstrated that in supercoiled plasmids the 3'-flanking region of *ARS* elements is sensitive to mung bean nuclease as a result of thermodynamically stable DNA unwinding. This local DNA unwinding could influnce the ability of proteins to bind in the region either directly or indirectly.

Acknowledgements

We thank L. Panzeri for communication of results prior to publication and

R. Brent for supplying plasmids. Support for this work was provided by NIH research grant GM35679 to CSN.

References

Arndt KT, Styles C, Fink GR (1987) Multiple global regulators control *HIS4* transcription in yeast. Science 237:874-880

Braus G, Mösch H-U, Vogel K, Hinnen A, Hütter R (1989) Interpathway regulation of the *TRP4* gene of yeast. EMBO J 8:939-945

Brent R, Ptashne M (1984) A bacterial repressor or a yeast transcriptional terminator can block upstream activation of a yeast gene. Nature 312:612-615

Buchman AR, Kimmerly WJ, Rine J, Kornberg RD (1988) Two DNA-binding factors recognize specific sequences at silencers, upstream activating sequences, autonomously replicating sequences, and telomeres in *S. cerevisiae*. Mol Cell Biol 8:210-225

Campbell JL, Newlon CS (1991) Chromosomal DNA replication. In J.R. Broach, E.W. Jones and J.R. Pringle (ed.) The Molecular and Cellular Biology of the Yeast Saccharomyces: Genome Dynamics, Protein Synthesis, and Energetics Cold Spring Harbor Laboratory Press, in press

Diffley JFX, Stillman B (1988) Purification of a yeast protein that binds to origins of DNA replication and a transcriptional silencer. Proc Nat Acad Sci USA 85:2120-2124

Diffley JFX, Stillman B (1989) Similarity between the transcriptional silencer proteins ABF1 and RAP1. Science 246:1034-1037

Dubey DD, Davis LR, Greenfeder SA, Ong LY, Zhu J, Broach JR, Newlon CS, Huberman JA (1991) The ARS elements associated with silencers of the yeast mating type locus, HML, do not function as chromosomal DNA replication origins. Submitted

Eisenberg S, Civalier C, Tye B-K (1988) Specific interaction between a *Saccharomyces cerevisiae* protein and a DNA element associated with certain autonomously replicating sequences. Proc Natl Acad Sci USA 85:743-746

Guarente L (1983) Yeast promoters and lacZ fusions designed to study expression of cloned genes in yeast. Meth Enzymol 101:181-191

Hofmann JFX, Gasser SM (1991) Identification and purification of a protein that binds the yeast *ARS* consensus sequence. Cell 65:1-10

Huberman JA, Zhu J, Davis LR, Newlon CS (1988) Close association of a DNA replication origin and an *ARS* element on chromosome III of the yeast, *Saccharomyces cerevisiae.* Nucl Acids Res 16:6373-6384

Jarvis EE, Clark KL, Sprague GF (1989) The yeast transcription activator PRTF, a homolog of the mammalian Serum Response Factor, is encoded by the *MCM1* gene. Genes Dev 3:936-945

Kuno K, Murakami S, Kuno S (1990) Single-strand-binding factor(s) which interact with *ARS1* of *Saccharomyces cerevisiae*. Gene 95:73-77

Lohr D, Torchia T (1988) Structure of the chromosomal copy of yeast *ARS1*. Biochemistry 27:3961-3965

Moscatelli F, Gariboldi M, Panzeri L (1990) The transcriptional factor *PHO2* binds *in vitro* to the centromeric region CDEII. Yeast 6:S60

Newlon CS (1988) Yeast chromosome replication and segregation. Microbiol Rev 52:568-601

Palzkill TG, Newlon CS (1988) A yeast replication origin consists of multiple copies of a small conserved sequence. Cell 53:441-450

Passmore S, Elble R, Tye B-K (1989) A protein involved in minichromosome maintenance in yeast binds a transcriptional enhancer conserved in eukaryotes. Genes Dev 3:921-935

Rhode PR, Sweder KF, Oegema KF, Campbell JL (1989) The gene encoding *ARS* binding factor I is essential for the viability of yeast. Genes Dev 3:1926-1939

Shore D, Stillman DJ, Brand AH, Nasmyth KA (1987) Identification of silencer binding proteins from yeast: possible roles in *SIR* control and DNA replication. EMBO J 6:461-467

Sweder KS, Rhode PR, Campbell JL (1988) Purification and characterization of proteins that bind to yeast *ARS*s. J Biol Chem 263:17270-17277

Tice-Baldwin K, Fink GR, Arndt KT (1989) BAS1 has a Myb motif and activates *HIS4* transcription only in combination with BAS2. Science 246:931-935

Umek RM, Kowalski D (1988) The ease of DNA unwinding as a determinant of initiation of yeast replication origins. Cell 52:559-567

Umek RM, Kowalski D (1990) Thermal energy suppresses mutational defects in DNA unwinding at a yeast replication origin. Proc Nat Acad Sci USA 87:2486-2490

Van Houten JV, Newlon CS (1990) Mutational analysis of the consensus sequence of a replication origin from yeast chromosome III. Mol Cell Biol 10:3917-3925

Vinson CR, LaMarco KL, Johnson PF, Landschulz WN, McKnight SL (1988) In situ detection of sequence-specific DNA-binding activity specified by a recombinant bacteriophage. Genes Dev 2:801-806

Walker SS, Francesconi SC, Eisenberg S (1990) A DNA replication enhancer in *Saccharomyces cerevisiae*. Proc Nat Acad Sci USA 87:4665-4669

West RW, Yocum RR, Ptashne M (1984) *Saccharomyces cerevisiae GAL1-GAL10* divergent promoter region: Location and function of the upstream activating sequence UAS_G. Mol Cell Biol 4:2467-2478

Yoshida K, Kuromitsu Z, Ogawa N, Oshima Y (1989) Mode of expression of the positive regulatory genes *PHO2* and *PHO4* of the phosphatase regulon in *Saccharomyces cerevisiae*. Mol Gen Genet 217:31-39

Young RA, Davis RW (1983) Yeast RNA polymerase genes: Isolation with antibody probes. Science 222:778-782.

Protein-DNA Interaction at Yeast Replication Origins: an ARS Consensus Binding Protein

J.F.X. Hofmann and S.M. Gasser
Swiss Institute for Experimental Cancer Research (ISREC)
Ch. des Boveresses
CH-1066 Epalinges/Lausanne
Switzerland

Eukaryotic DNA replication initiates at many sites on each chromosome during the S phase of the cell cycle. The identification of yeast chromosomal DNA sequences that act in cis to allow the extrachromosomal maintenance of plasmids (ARS or autonomously replicating sequences, Stinchcomb et al., 1976) in yeast led to the suggestion that they function as origins of replication. Proof that initiation of DNA synthesis occurs at or near ARS elements comes from two-dimensional gel electrophoresis systems (Brewer and Fangman, 1987; Huberman et al., 1987) that allow the separation of replication intermediates and the subsequent mapping of initiation sites for DNA replication.

The DNA sequence requirements of ARS elements have been reviewed recently (Newlon, 1988; Umek et al., 1989). In general, ARS elements have a significantly higher A+T content than the bulk chromosomal DNA and they all contain one or more copies of an 11 nucleotide consensus sequence: 5'-(A/T)TTTAT(A/G)TTT(A/T)-3'. The essential regions for ARS function include the consensus sequence and a variable number of nucleotides 3' of the T-rich strand of the consensus (Celniker et al., 1984; Kearsey, 1984; Palzkill and Newlon, 1988). Extensive mutational analysis of the C2G1 ARS addressed the functional contribution of each nucleotide in the ARS consensus (Van Houten and Newlon, 1990). Mutations that reduced or abolished ARS activity occurred at each position in the consensus sequence, demonstrating that each position of this sequence contributes to ARS function. Mutations that created alternative perfect matches to the consensus sequence had no major effect on ARS function. The region required for full ARS efficiency extends beyond the core consensus in both directions.

Recently attempts were made to revert cis acting ARS mutations by *in vitro* mutagenesis (Kipling and Kearsey, 1990). Surprisingly, most cis acting mutations that restored ARS activity were found to create improved close matches to the ARS consensus (or additional matches to the ARS consensus) in bacterial vector sequences. This suggests that the ARS consensus and perhaps near matches to it are essential for ARS function and that the sequence requirements of the *S. cerevisiae* DNA replication apparatus for initiation are not very complex. Furthermore, the specificity of effect that point mutations have on ARS function are compatible with the role of the ARS consensus as a recognition site for a protein(s) that is involved in the initiation of DNA synthesis.

In contrast to the studies of the core consensus, sequence comparisons and mutational analyses have been less revealing about the role of the 3' flanking sequence. It exhibits little primary sequence homology from one ARS to another. In one hypothesis it was suggested that the flanking sequence requires to contain a repeated sequence of near matches to the core consensus sequence (Palzkill and Newlon, 1988). However only the exact match at the core consensus sequence exhibits the sensitivity to point mutations. Another explanation of the ARS 3' flanking region has been proposed by Umek and Kowalski (1988). They suggested that DNA unwinding is the primary role of the ARS flanking sequence. Stable origin unwinding *in vitro* proved to be temperature dependent and certain large deletions of the ARS flanking sequence generate cold sensitive ARS mutations *in vivo* (Umek and Kowalski, 1990). This demonstrates that thermal energy contributes to origin unwinding *in vivo*. In addition, insertion mutants, in which the 3' flanking region is moved significantly downstream, are replication deficient (Palzkill et. al., 1986; Diffley and Stillman, 1988). We will suggest below that perhaps both the unwinding propensity and near ARS matches are important features of the 3' flanking region.

Considerable effort has been put into the identification of trans-acting factors that bind yeast ARS elements (reviewed by Newlon, 1988). An abundant nuclear protein called ABF-1 has been purified and cloned by several groups, based on its affinity for a site that is often found 3' of the ARS consensus (Diffley and Stillman, 1988; Eisenberg et al., 1988; Sweder et al., 1988). Deletion of the ABF-1 binding site at ARS1 slightly reduces the efficiency of the ARS element *in vivo*, but the site is clearly not essential for plasmid replication. Although neither ABF-1 nor other previously identified factors bind directly to the essential ARS consensus sequence, protein-DNA interaction at the yeast ARS consensus was proposed based on in vivo footprinting data at the chromosomal copy of ARS-1 (Lohr and Torchia, 1988).

In this paper we present the identification of a yeast protein that binds *in vitro* to the ARS consensus and some close matches of it in the 3' flanking region. The ARS consensus binding protein (ACBP) is unusual in that it specifically binds the T-rich single-strand of its recognition sites.

Identification of a yeast factor that binds to the histone H4 ARS consensus sequence and a 9/11 match to it in the 3' flanking domain

The ARS element downstream of the copy-I Histone H4 gene has been extensively characterized in order to define the sequences required for H4 ARS function (Bouton and Smith, 1986; Holmes and Smith, 1989). A minimal region of at least 66 bp permits autonomous replication of non-replicating plasmids in yeast. This minimal element contains an 11/11 bp ARS consensus sequence and several 9/11 bp close matches to the ARS core in the 3' flanking domain. ARS unwinding and origin function have similar sequence requirements in the 3' flanking region of the H4 ARS consensus (Umek and Kowalsky, 1988).

Recently it was shown that DNA binding activities from yeast extracts interact with single stranded sequences that contain the ARS consensus (Hofmann and Gasser, 1991; Kuno et al., 1990). This binding activity has been purified by DNA affinity chromatography using single stranded oligonucleotides that contain the T-rich strand of the histone H4 ARS consensus (Hofmann and Gasser, 1991). The ARS binding factor (ACBP) purifies as a 67 kD protein.

Since the 3' domain of the Histone H4 ARS is known to be important for its function, it was determined whether there are any binding sites for ACBP in this region. Several close matches to the ARS consensus are present in the 3' domain of the ARS core (Holmes and Smith, 1989). Binding sites were mapped by gel retardation assays using single-stranded restriction fragments of the H4 ARS. Purified ACBP was bound to these restriction fragments in the presence of single-stranded competitor DNA (Figure 1A). Binding was competed with an oligonucleotide containing the H4 ARS consensus (Hofmann and Gasser, 1991). A summary of the mapping data is shown in Figure 1B. In addition to the binding site at the ARS core consensus, a single binding site of ACBP is present at a T-rich region on the opposite strand in the 3' domain of the ARS core consensus in which there are two overlapping 9/11 bp matches to the consensus.

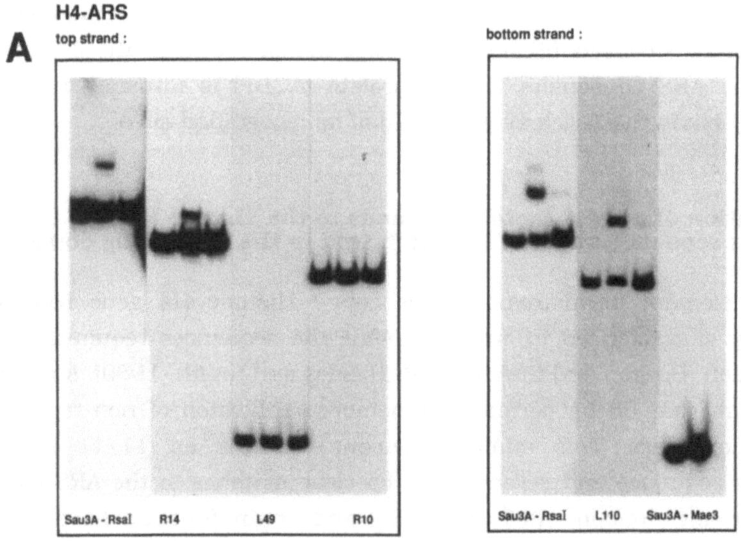

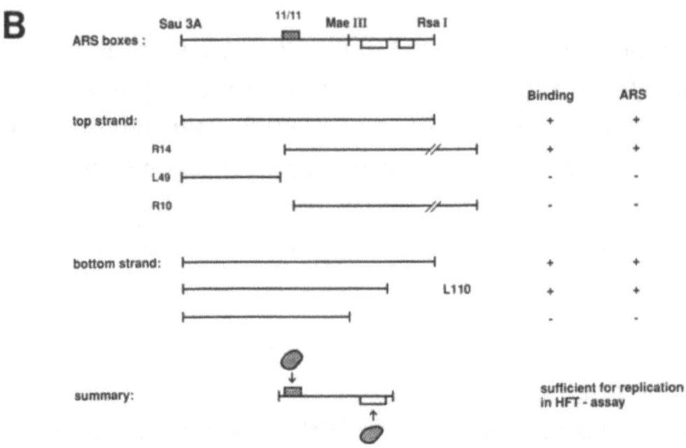

Figure 1. Binding of ARS Factor to the Histone H4 ARS Correlates with H4 ARS Activity.

A. Mapping of binding sites of the ARS binding factor on the two opposite single strands of the Histone H4 ARS by gel retardation. Top and bottom strand refer to the strands of the published sequence (Smith and Andrésson, 1983; Bouton and Smith, 1986), the top strand contains the T-rich strand of the ARS core consensus. The two ends (genomic restriction sites) of each fragment used are indicated at the bottom of each panel. For each individual strand three incubations were loaded on the band shift gel: first, end-labelled single-stranded DNA; next specific complex formation after incubation with purified ARS binding factor, and third competition for complex formation with 100 ng of cold T-rich strand of the H4 ARS box oligo. For the SauIIIA-MaeIII fragment the competition was omitted. Incubation and UV crosslinking of protein to DNA were done as described (Hofmann and Gasser, 1991).

B. Summary of binding sites of the ARS binding factor on the Histone H4 ARS. ARS boxes: boxes above or below the straight line indicate the presence of the T-rich strand of an ARS sequence in the top or bottom strand respectively. The stippled box represents an 11/11 bp ARS consensus sequence. The empty boxes represent 9/11 matches to the ARS consensus, with the wider box representing two overlapping 9/11 matches (Holmes and Smith, 1989). Top strand and bottom strand: Individual single-strands that show complex formation in the gel retardation are indicated by + in the panel "binding". If the corresponding double stranded fragment confers high frequency of transformation on a nonreplicating plasmid (Bouton and Smith, 1986), this is indicated by + in the panel "ARS". The summary shows the two binding sites that map to the T-rich strand of the indicated ARS consenses.

The borders of the two protein binding sites at the H4 ARS precisely correspond to the borders of the minimal fragment that confers autonomous replication on non-replicating plasmids (Holmes and Smith, 1989). Linker insertion, but not point mutations, in the 9/11 bp ARS consensus match to which ACBP binds increases the loss rate of the mutant H4 ARS on plasmids, whereas deletion of the near match element in the minimal ARS 3' flanking domain results in loss of high frequency of transformation (Holmes and Smith, 1989). It is not known if this is the result of deleting an ACBP binding site or an other, undefined DNA structure. Two other 9/11 matches to the ARS consensus in the 3' flank of the H4 ARS are not recognized by ACBP in vitro (indicated in Figure 1).

ACBP recognizes the ARS consensus at HMR-E and at ARS1

An important consideration at this point is whether ACBP binds other endogenous ARS elements and, if so, at which sites. To investigate this, ACBP binding sites were mapped on ARS1, and on the silencer sequence HMR-E which has been shown to contain ARS function (Brand et al., 1987). Figure 2 demonstrates the detection and mapping of binding sites on the two ARSs. Protein binding to the single strand that contains the T-rich sequence of the HMR-E ARS consensus can be detected. This binding activity can be competed by oligonucleotides containing the H4 ARS consensus (Figure 2A, see also Figure 3A). A 7 bp deletion in the ARS core that abolishes ARS function of the 138 bp AhaIII-AluI silencer fragment (Brand et al., 1987), but recreates a 9/11 bp close match to the consensus, reduces the binding efficiency of purified ARS binding factor by approximately 65% (ΔA, Figure 2A).

A

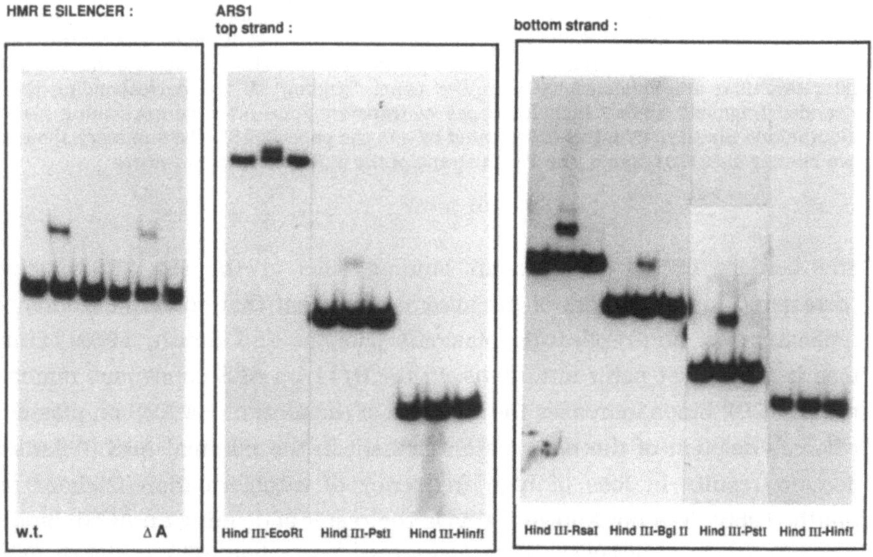

B

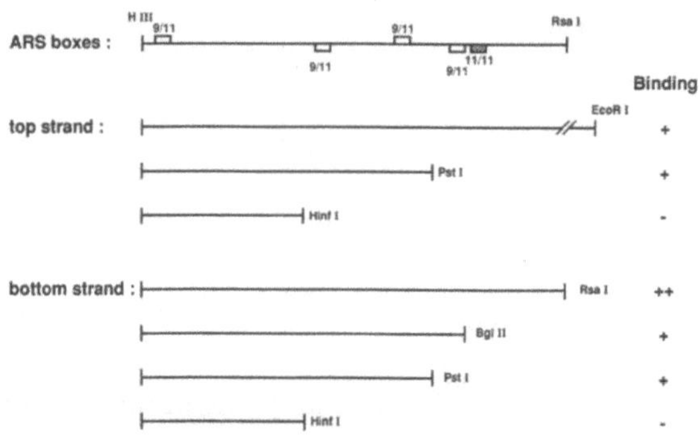

Figure 2. Binding of the ARS Binding Factor to ARS1 and to the ARS Consensus of the HMR-E silencer.

A. Gel retardation assay was done as described in Figure 1. Using the 138 bp AhaIII-AluI fragment of the HMR-E silencer, binding to HMR-E silencer was only tested on the strand that contains the T-rich sequence of the ARS consensus (Abraham et al., 1984). w.t. indicates wild type sequence, ΔA is the restriction fragment derived from pΔA that contains a 7 bp internal deletion in the ARS core consensus (Brand et al., 1987), leaving the following 9/11bp consensus: 5' TTTTAGGTATT 3'. Restriction sites indicated at the bottom are the flanking genomic sites of the restriction fragments used.

B. Summary of binding data on the two individual strands of ARS-1. The bottom strand contains the T-rich strand of the ARS core consensus. Symbols are as described in Figure 1. ++ stands for two shifted bands of different mobility in the gel retardation assay.

Similar to the arrangement at the H4 ARS, several binding sites for the same factor could be mapped on the two strands of ARS1 (Figure 2A and B). A single site is present on the top strand of the sequence depicted in Figure 2B. This binding site is found between the PstI and the HindIII site of ARS1, which contains the T-rich strand of the single 9/11 bp close match to the ARS consensus. Gel retardation experiments using the HindIII-RsaI fragment of the bottom strand show two shifted complexes with different mobility (Figure 2A). We interpret this to be due to the presence of two potential binding sites for ACBP: probably the T-rich strand of the ARS consensus (between the BglII and the RsaI site) and the T-rich strand of a 9/11 bp close match situated between the HinfI and the PstI site. Cleavage of the HindIII-RsaI fragment at the BglII site eliminated one complex, and cleavage at HinfI removed the second (Figure 2B).

These binding sites in ARS1 are found exclusively in domains that play dominant roles in ARS function, i.e. domain A, that contains the essential ARS consensus; and the proximal portion of domain B, 3' to the ARS consensus, which is also crucial for ARS function of ARS1 (Diffley and Stillman, 1988).

Single point mutations in the C2G1 ARS consensus affect ACBP binding

To test the specificity of the observed interactions of ACBP with single-stranded DNA, sequences containing point mutations in the 11bp ARS consensus were examined in gel retardation assays. By quantifying the ability of the wild type ARS consensus sequence to compete for the binding of mutant forms, the relative affinity of ACBP for wild type and mutant binding sites can be established. The semisynthetic C2G1 ARS was chosen for this study, since an exhaustive analysis of the effect of point mutations in the C2G1 ARS consensus on ARS activity has been published (VanHouten and Newlon, 1990).

Three of the four point mutations tested significantly reduce the affinity of the ARS binding factor for the ARS consensus (Figure 3). Two of these alter thymidine residues (positions 3 and 10) to C and G residues, respectively; both of these modifications also eliminate ARS activity at the C2G1 ARS. In one oligonucleotide the guanidine at position 7 is changed to a thymidine, thus enriching the consensus for T residues. Nonetheless this reduces the ability of this sequence to compete for binding to the ARS factor, suggesting that the binding is not due solely to the T-rich character. Each of these tested mutations renders the C2G1 ARS non-functional for plasmid replication in vivo (Van Houten and Newlon, 1990).

Figure 3. Single Point Mutations in the C2G1 ARS Consensus Affect ACBP Binding

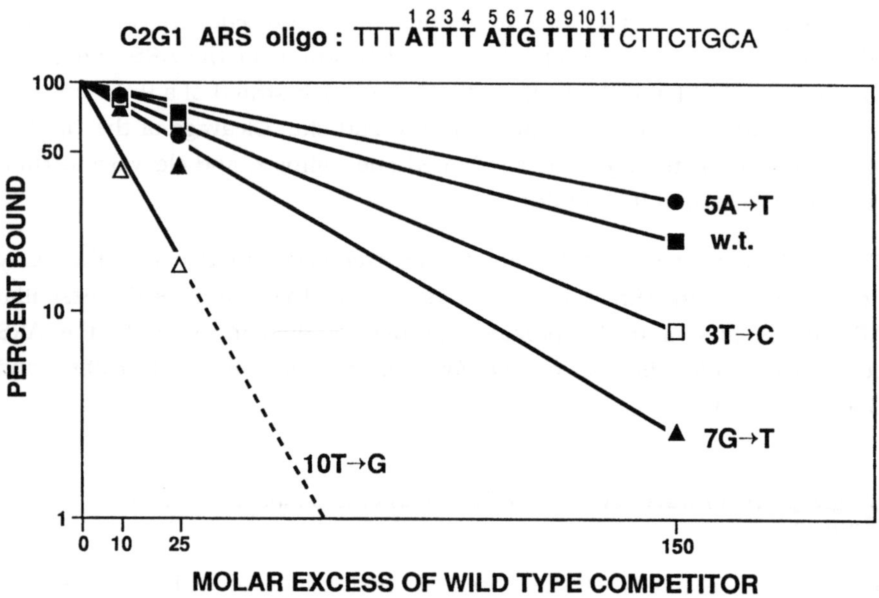

Gel retardation assays were done with a 22nt oligonucleotide that contains the ARS consensus of the C2G1-ARS and authentic flanking sequences of the semisynthetic ARS element that was used for functional analysis of point mutants therein (Van Houten and Newlon, 1990). Four mutated sequences with single nucleotide changes at the indicated positions were tested as well. The ARS consensus is shown in bold letters and the the individual positions of the consensus are indicated. The end-labelled single stranded oligonucleotides were incubated in 20 µl bandshift buffer for 20 min at room temperature in the presence of 2 µg single stranded competitor DNA and 100ng purified ACBP and subsequently loaded on a 5% native polyacrylamide gel. Binding reactions for the C2G1 ARS and the indicated point mutants were also done in the presence of increasing amounts of the wildtype semisynthetic C2G1 ARS oligonucleotide. Competition for complex formation was evaluated by densitometric scanning. The complex formation using the 10T to G mutation in the presence of 150 fold excess of wildtype ARS consensus was below detectable levels, as indicated by the dotted line.

Another point mutation that inactivates the C2G1 ARS (A to T in position 5) appears to enhance the affinity of ACBP for the consensus *in vitro* (Figure 3). If the *in vitro* interaction reflects *in vivo* affinities, this may indicate that increased binding to the single-strand consensus is also detrimental to ARS function. In conclusion, three point mutations which abolish activity also decrease binding, a further one increases it. The significant variation of binding affinities due to single nucleotide changes demonstrates that the ACBP-DNA interaction is sequence specific.

DISCUSSION

The data mapping ACBP binding sites are in good agreement with proposed requirements for ARS activity (Palzkill and Newlon, 1988). Based on extensive deletion analysis at the yeast genomic ARS C2G1, the authors proposed a requirement for one or several close matches to the ARS core consensus present 3' of the T-rich strand of the 11/11 bp ARS consensus and preferentially in inverted orientation. Such an organization corresponds to the ACBP binding sites mapped in this paper at the H4 ARS and ARS1.

What might be the role of the ARS binding factor in initiation of DNA replication ? The preference of ACBP for one single strand of the ARS consensus provides a potential mechanism for specifying the template for leading strand synthesis (Figure 4). A simple model would predict that bidirectional DNA replication requires the presence of two single-strand binding sites for the ARS binding factor on opposite strands with a 3' orientation to each other, in order to initiate two diverging replication forks. Due to the 5' to 3' directionality of the replicating enzymes, the positioning of two binding sites on opposite strands 5' to each other, would result in two converging replication forks which might finally collide and result in a non-functional origin. Experiments by Holmes and Smith (1989) are consistent with our proposed requirement for an ACBP-binding site on each strand. Inversion of the perfect ARS consensus at the H4 ARS, which placed the T-rich sequence on the opposite strand, resulted in an inactive ARS element. This result may either reflect the need for binding sites on opposite strands, or the disruption of the ACBP binding site by the inversion. By testing these constructs and additional constructs for bidirectional DNA replication on plasmids and for ACBP binding, the model in Figure 4 can be readily tested. A similar requirement for the correct orientation of the ARS consensus to its 3' flanking domain has been found at ARS1 (Diffley and Stillman, 1988).

Figure 4. A Model Proposing Early Steps in the Initiation of DNA Replication.

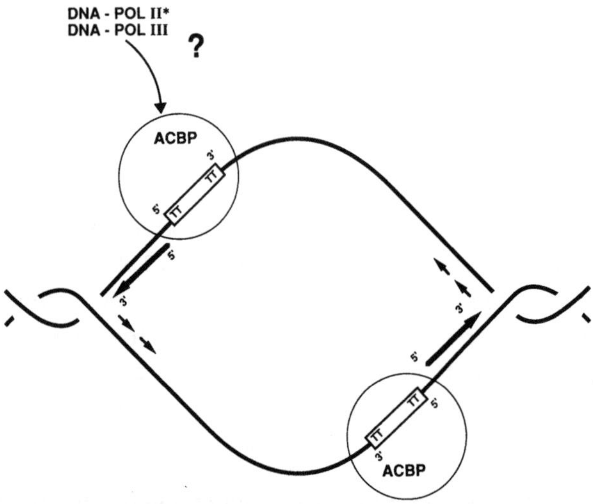

It is proposed that the ARS binding factor (represented by circles) binds the T-rich strand of the ARS consensus sequence (or a close match to the latter) in the unwound origin of DNA replication. Physical interaction of the bound ARS factor with components of the DNA replication machinery (probably a leading strand DNA polymerase) might then permit the initiation of DNA replication. The leading strand is represented by a thick continuous arrow pointing in 5' to 3' orientation. Bidirectional replication may require the existence of two binding sites on opposite strands with a 3' orientation to each other.

REFERENCES

Abraham, J, Nasmyth, K, Strathern, KA, Klar, AJS and Hicks, J B (1984) Regulation of mating type information in yeast. **J Mol Biol** 176, 307-331

Bouton, AH and Smith, M (1986) Fine structure analysis of the DNA sequence requirements for autonomous replication of Saccharomyces cerevisiae plasmids. **Mol Cell Biol** 6, 2354-2363

Brand, AH, Micklem, G and Nasmyth, K (1987) A yeast silencer contains sequences that can promote autonomous plasmid replication and transcriptional activation. **Cell** 51, 709-719

Brewer, BJ and Fangman, WL (1987) The localization of replication origins on ARS plasmids in S. cerevisiae. **Cell** 51, 463-471

Celniker, SE, Sweder, K, Srienc, F, Bailey, JE and Campbell, JL (1984) Deletion mutations affecting autonomously replicating sequence ARS1 of Saccharomyces cerevisiae. **Mol Cell Biol** 4, 2455-2466

Diffley, JFX and Stillman, B (1988) Purification of a yeast protein that binds to origins of DNA replication and a transcriptional silencer. **Proc Natl Acad Sci USA** 85, 2120-2124

Eisenberg, S, Civalier, D and Tye, B-K (1988) Specific interaction between a S. *cerevisiae* protein and a DNA element associated with certain autonomously replicating sequences. **Proc Natl Acad Sci USA** 85, 743-746

Hofmann JFX and Gasser, SM (1991) Identification and purification of a protein that binds the yeast ARS consensus sequence. **Cell** 65, 1-10

Holmes, SG and Smith, M (1989) Interaction of the H4 ARS consensus sequence and its 3' flanking domain. **Mol Cell Biol** 9, 5464-5472

Huberman, JA, Spotila, LD, Nawotka, KA, El-Assouli, SM and Davis, LR (1987) The *in vivo* replication origin of the yeast 2 mm plasmid. **Cell** 51, 473-481

Kearsey, S (1984) Structure requirements for the function of a yeast chromosomal replicator. **Cell** 37, 299-307

Kipling, D and Kearsey, SE (1990) Reversion of autonomously replicating sequence mutations in Saccharomyces cerevisiae: creation of a eucaryotic replication origin with procaryotic vector DNA. **Mol Cell Biol** 10, 265-272

Kuno, K, Murakami, S and Kuno, S (1990) Single-strand-binding factors which interact with ARS1 of *Saccharomyces cerevisiae* **Gene** 95, 73-77

Lohr, D and Torchia, T (1988) Structure of the chromosomal copy of ARS1. **Biochemistry** 27, 3961-3965

Newlon, CS (1988) Yeast chromosome replication and segregation. **Microbiol Rev** 52, 568-601

Palzkill, TG and Newlon, CS (1988) A yeast replication origin consists of multiple copies of a small conserved sequence. **Cell** 53, 441-450

Palzkill, TG, Oliver, SG and Newlon, CS (1986) DNA sequence analysis of ARS elements from chromosome III of Saccharomyces cerevisiae: identification of a new conserved sequence. **Nucleic Acid Res** 14, 6247-6263

Smith, MM. and Andrésson, OS (1983) DNA-sequences of yeast H3 and H4 histone genes from two non allelic sets encode identical H3 and H4 proteins. **J Mol Biol** 169, 641-661

Stinchcomb, DT, Struhl, K and Davis, RW (1976) Isolation and characterization of a chromosomal replicator. **Nature** 282, 39-43

Sweder, K, Rhode, PR and Campbell, JL (1988) Purification and Characterization of Proteins that Bind to Yeast ARSs. **J Biol Chem** 263, 17270-17277

Umek, RM and Kowalski, D (1988) The ease of unwinding as a determinant of initiation at yeast replication origins. **Cell** 52, 559-567

Umek, RM and Kowalski, D (1990) Thermal energy suppressses mutational defects in DNA unwinding at a yeast replication origin. **Proc Natl Acad Sci USA** 87, 2486-2490

Umek, RM, Linskens, MHK, Kowalski, D and Huberman, JA (1989) New beginnings of eucaryotic DNA replication origins. **Biochim et Biophys Acta** 1007, 1-14

Van Houten, JV and Newlon, CS (1990) Mutational Analysis of the Consensus Sequence of a Replication Origin from Yeast Chromosome III. **Mol Cell Biol** 10, 3917-3925.

Genetic Analysis of the Roles of Yeast ARS Binding Factor I

P. R. Rhode and J. L. Campbell
California Institute of Technology
Divisions of Chemistry and Biology 147-75CH
Pasadena, CA 91125 USA

Introduction

Yeast autonomously replicating sequences (ARSs) were originally identified as segments of DNA that permitted autonomous replication of colinear DNA as plasmids in yeast (Struhl *et al.*, 1979). Direct evidence that ARSs function as origins of replication on plasmids, and more important, that a subset of ARSs function as origins of replication in the chromosome, comes from 2-D gel analysis of replication intermediates (Brewer and Fangman, 1987; Huberman *et al.*, 1987, 1988). ARS1, the ARS with which we have been working for many years, is an active origin within the chromosome. Sequence comparisons between the 20 or so different ARSs that have been identified and the effect of mutations on plasmid stability have revealed two underlying similarities: A high A/T content compared to chromosomal DNA and the presence of a conserved core consensus sequence: 5' (A/T)TTTA(T/C)(A/G)TTT(A/T(T/C/G) 3' (see Campbell and Newlon, 1991 for review). Point mutations, small deletions and small substitutions abolish ARS function. Many ARSs contain multiple close matches to this sequence, though spacing and orientation varies. The core consensus sequence is a potential initiator protein binding site. DNase I footprinting patterns of the chromosomal copy of ARS1 showed that the core might be protected from digestion (Lohr and Torchia, 1988).

Although the core consensus is necessary, by itself it is not sufficient. Sequences 3' to the T-rich strand of the consensus enhance autonomous plasmid stability greatly, and in some vector contexts are absolutely required for function. Deletions into domain B (3' flanking region) severely reduce function, particularly as the deleted region gets closer to the conserved consensus. This effect is orientation dependent. Unlike other ARSs, sequences 5' to the T-rich strand of domain A in ARS1 (domain C) have also been shown to be functionally important, although less so than domain B. Various models have been proposed to account for function of DNA elements in these regions (Campbell and Newlon, 1991). All of the models have a single theme in common: That the core consensus is the primary initiation protein binding site while the 3' flanking sequences play an auxiliary role, enhancing function. We are testing this hypothesis by defining in detail the protein/DNA interactions at ARSs.

In Sweder *et al.* (1988), we described two ARS binding proteins, ABF1 and ABFIII. We have focused on ABF1 since its binding site is important for ARS function, while the ABFIII binding site is dispensable (Sweder *et al.*, 1988). Although ABF1 was first recognized as an ARS binding protein, recently it has become clear that ABF1 is an abundant, high affinity DNA binding protein that specifically recognizes the motif, RTCRYNNNNNACG, at many sites in the yeast genome. Depending upon their context, these binding sites have been defined genetically as important in transcriptional activation, mating type silencer function, and ARS activity, suggesting that ABF1 plays a global role in regulating transcription and DNA replication (see Rhode *et al.*, 1989 for discussion). In this property, ABF1 is similar to the mammalian replication/transcription factor NF1/CTF (Jones *et al.*, 1987) and perhaps to yeast *MCM1* (Jarvis *et al.*, 1989; Passmore *et al.*, 1989).

Results

Phosphorylation of ABF1. It is possible that the diverse roles of ABF1 in regulating transcription and replication are mediated by functionally distinct forms of ABF1. Previously, we have shown that in crude extracts and in the purified preparations, ABF1 is present in at least two electrophoretically distinct forms (Rhode *et al.*,1989; Sweder *et al.*, 1988). To test whether these forms represent different levels of protein modification, purified ABF1 was treated with acid phosphatase and analyzed electrophoretically. Figure 1 shows that the doublet observed in purified ABF1 corresponds to different levels of protein phosphorylation. In collaboration with S. Silve and R. Poyton, University of Colorado, we have shown that the phosphorylation state changes depending on carbon source. This is of particular interest since we have found that the phosphorylation of ABF1 correlates with the expression of several genes regulated by ABF1 binding sites in their promoters. In cells grown on glycerol or ethanol, or in stationary phase cells or cells starved for nitrogen, the phosphoprotein predominates. On glucose both forms are present in approximately equal amounts. Dephosphorylation in glucose medium is dependent on the function of the *SSN6* gene. Interestingly, *ssn6* mutants show a minichromosome maintenance defect (Schultz *et al.*, 1990). Since stationary phase and nitrogen starved cells block in G1 and since EtOH or glycerol grown cells spend a larger portion of the cell cycle growing in G1 than in the other phases, ABF1 phosphorylation may be cell cycle regulated. Studies are currently underway to test this possibility. Further resolution of the ABF1 doublet indicated that at least four phosphorylated forms are possible (Silve *et al.*, 1991). Computer analysis of the ABF1 sequence reveals at least 16 potential phosphorylation sites (Figure 1B). Further studies are required to determine if any of these sites are phosphorylated *in vivo*.

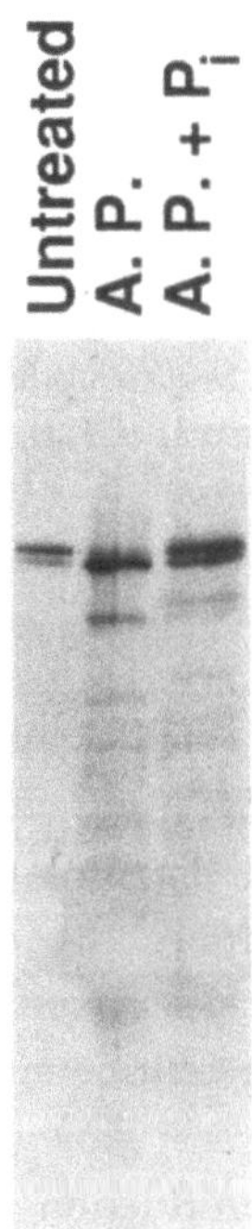

Figure 1A. Treatment of ABF1 with acid phosphatase. ABF1 (lane 1) was purified as described in Sweder *et al.* (1988) and incubated with acid phosphatase in the absence (lane 2) or in the presence (lane 3) of 0.1M Na_2HPO_4, an inhibitor of phosphatase activity. Samples were then analyzed by SDS-PAGE and Western immunoblotting with a monoclonal antibody to ABF1.

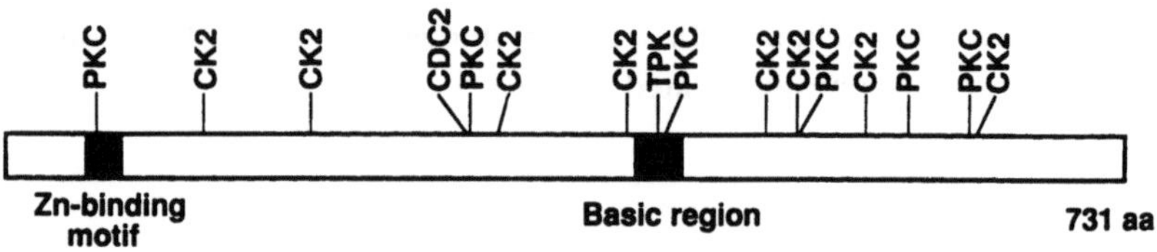

Figure 1B. Potential sites in ABF1 for protein kinases. The sequence of ABF1 was searched for potential phosphoacceptor recognition sites of protein kinases described previously (Edelman *et al.*, 1987). The bar represents the 731 aa ABF1 protein sequence; indicated above the bar are the position of serine/threonine residues that may serve as substrates for cAMP-dependent protein kinase (TPK), protein kinase C (PKS), casein kinase II (CK2), and p34^{cdc2} protein kinase (CDC2); and the shaded boxes are regions of ABF1 that may be important for DNA binding (see Discussion).

Overexpression of ABF1 is lethal. Antibodies specific to ABF1 were used to screen a λgt11 yeast genomic library and the ABF1 gene was isolated. Gene disruption experiments indicated that this gene is essential for cell viability (Rhode *et al.*, 1989). Furthermore, as shown in Figure 2, conditional overexpression of the ABF1 gene causes growth arrest. The ABF1 gene was fused to the yeast *GAL1* promoter in a centromere-containing plasmid. The *GAL1* promoter is strongly activated in yeast grown in media containing galactose. To test the effects of ABF1 overproduction on growth, yeast transformants containing the GAL1/ABF1 fusion or the vector alone were grown on media containing raffinose (uninduced condition) followed by the addition of galactose. Previous results have shown that >20-fold overproduction of ABF1 was observed in the strain carrying the GAL1/ABF1 fusion, following 4 hours of galactose induction (Rhode *et al.*, 1989). Growth of these cells was arrested after one doubling while growth in raffinose media was unaffected (Figure 2). Three to 4-fold overproduction of ABF1 on a high copy number plasmid had no effect on cell growth (data not shown). These results indicate that the cellular level of ABF1 is critical to growth. It is noteworthy that the promoter for the ABF1 gene contains five binding sites for ABF1 (Figure 3) and that ABF1 binds these sites *in vitro*, strongly suggesting some form of autoregulation. (No other promoter studied to date has more than two ABF1 sites.)

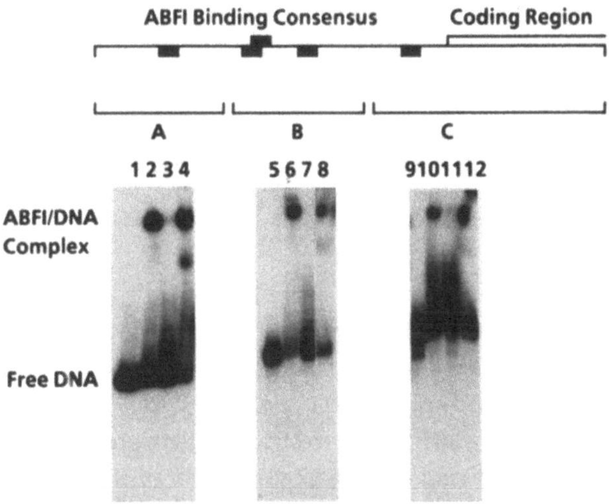

Figure 2 Overproduction of ABF1 is lethal. Strains containing the GAL1/ABF1 construct pGAL1/ABF1 (dashed lines) or the vector alone - pGAL (solid line) were grown in media containing raffinose (2%) or follow dilution into media containing raffinose (2%) or galactose (2%). Cell growth (OD_{590}) was monitored.

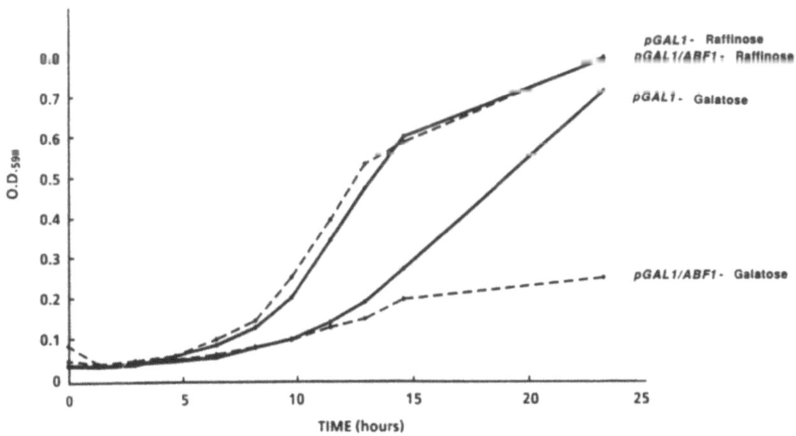

Figure 3. ABF1 binding sites in the *ABF1* promoter. Restriction fragments A

(lanes 1-4), B (lanes 5-8) and C (lanes 9-12) of the *ABF1* gene promoter region were labeled with ^{32}P and incubated with either purified ABF1 or crude extracts in the presence or absence of a specific oligonucleotide to the ABF1 binding site at ARS1 (B element). All reactions contained 1 µg of nonspecific DNA. Protein DNA complexes were analyzed by gel retardation (Sweder *et al.*, 1988). Lanes 1, 5 and 9 contain no protein; lanes 2, 6 and 10 contain crude extract; lanes 3, 7 and 11 contain crude extracts plus the ABF1 specific oligonucleotide competitor; and lanes 4, 8 and 12 contain purified ABF1.

Point mutations in ABF1. In order to examine the role of ABF1 in cell growth in more detail, we have used plasmid shuffling to isolate six temperature-sensitive *abf1* mutants (Rhode *et al.*, 1991). *abf1-1*, *-2* and *-5* mutants arrest in the first cell cycle following shift to the restrictive temperature. The terminal phenotype is unusual, showing enlarged cells with a small, aberrant bud. The sites of the mutations were determined by marker rescue and DNA sequencing.

Three of the mutants (*abf1-1*, *-2* and *-5*) map to regions of the protein thought to be important in DNA binding. Two mutants (*abf1-3*, and *-4*) map to residues between the DNA binding domains that may be important for ABF1 structure. We have analyzed ABF1 DNA binding in extracts of all of the mutants: *abf1-1* and *abf1-2* show no binding at any temperature *in vitro*; *abf1-3*, *-4*, and *-5* show temperature-sensitive DNA binding; *abf1-6* showed the same level of binding as present in wild-type extracts (Rhode *et al.*, 1991). Western analysis showed that ABF1 was present in each of the mutant extracts (not shown).

In addition, the *abf1-ts* mutants were used in analyzing functional domains of ABF1. A set of 3' deletions of the *ABF1* coding sequence were cloned into a *URA3* plasmid and introduced into an *abf1-1* and a wild-type yeast strain. These strains were used to determine whether C-terminal regions of ABF1 were required for DNA binding or complementation of the *abf1-1* temperature-sensitive phenotype. Extracts from *abf1-1* strains did not contain measurable ABF1 binding activity presumably due to the thermolability of the *abf1-1* protein. When a full length *ABF1* gene (construct 316-26) or a 500 bp deletion in the 3' end of *ABF1* (construct 316-21) were introduced into this strain, extracts showed ABF1 specific binding activity whereas when 1 kb DNA (or larger) was deleted from the 3' end of ABF1 no DNA binding was observed (Figure 4). These results indicate that the N-terminal 2/3 of the protein are required for DNA binding activity in yeast. Identical results were previously found for bacterially produced ABF1 (Halfter *et al.*, 1989). When the truncated 316-21 construct was expressed in an ABF1 strain, only ABF1/ARS1 DNA complexes corresponding to the 316-21 product and wild-type ABF1 were observed by gel retardation assays (data not shown). No intermediate complexes were observed suggesting that ABF1 does not form homodimers (or oligomers) when binding DNA. None of the C-terminal deletion constructs were able to complement the *abf1-1* temperature sensitive growth defect (Figure 4) indicating that ABF1 DNA binding activity alone (316-21 product) is not sufficient to provide for all the essential functions mediated by ABF1. Rather, the C-terminus of the protein is also required.

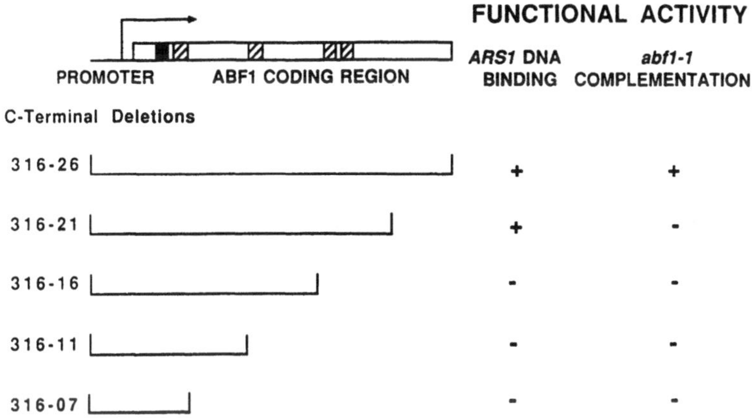

Figure 4. Complementation of *abf1-1* by truncated *ABF1* genes. A series of 3' deletions were constructed by restriction digest or exonuclease III digest. These deletions were subcloned into a *URA3-H4* ARS vector such that the normal *ABF1* promoter and transcriptional termination sequences are at the 5' and 3' ends, respectively, of the deleted *ABF1* gene. Following introduction into an *abf1-1 ura3* strain, the deletion constructs were tested for their ability to generate products that bound an oligonucleotide (element B) for the ABF1 binding site at ARS1 as assayed by gel retardation (Sweder *et al.*, 1988) and that complemented the *abf1-1* growth defect at 36°C.

We have used several methods to analyze DNA replication function in the *abf1-ts* mutants. Recently, we have carried out plasmid loss assays in *abf1-ts* mutants at a semipermissive temperature using two ARSs that have about a 10-fold difference in affinity for ABF1. ARS121 has a strong ABF1 site and ARS1 has a weak ABF1 site (Francesconi and Eisenberg, 1989). The loss rate of ARS121 is elevated 7-fold at the semipermissive temperature of 30°C, compared to 23°C. This is consistent with the loss rates *in vivo* due to mutation of the ABF1 site at ARS121 (Walker *et al.*, 1989). There is a smaller effect on ARS1, consistent with the fact that ABF1 binds less strongly and cis-acting mutations have a smaller effect at ARS1 than at ARS121. The thermolability of plasmid maintenance indicates that it is the ABF1 protein that is recognizing the ARSs, and not one of a family of proteins that recognizes the same site. In larger eukaryotes there is ample evidence that the same site is recognized by different proteins and thus previous studies on the cis-acting ABF1 site could have been explained by another protein recognizing this site, as would be consistent with results (Biswas *et al.*, 1990).

Analysis of cell cycle progression in an *abf1-ts* strain. Preliminary analysis of cell cycle progression and ^{3}H-uracil incorporation into DNA show a defect in the initiation of chromosomal DNA synthesis in a synchronized *abf1-ts* strain at the nonpermissive temperature. An asynchronous *bar1 abf1-ts* culture was grown at 25°C to early log phase and then treated with α-factor for 6 hours to arrest growth at START (G1) phase. After extensive washing to remove the α-factor, the cells were grown at either 25°C (permissive temperature) or 37°C (nonpermissive temperature for *abf1* growth). Aliquots were removed and processed for cell cycle analysis by flow cytometry. Figure 5 shows the change in the percentage of cells in G1 phase during the 5 hour time course following release from α-factor arrest. In the top panel, *abf1* cells failed to progress efficiently out of G1 phase when release from α-factor arrest at the nonpermissive temperature while cells released at the permissive temperature left G1 phase by 2 hours and continued into the rest of the cell cycle. The bottom panel shows the time course for the control *ABF1 bar1* strain. This strain progressed efficiently through the cell cycle following release from α-factor arrest at both 25°C and 37°C. Similar time courses measuring ^{3}H-uracil incorporation into nucleic acids in synchronized *abf1* strains showed that DNA synthesis was inhibited by 75% and RNA synthesis inhibited by 70% (Rhode *et al.*, 1991). In summary, the results of the plasmid loss, cell cycle progression and nucleic acid synthesis studies on the *abf1-ts* mutants suggest a direct role for ABF1 in the initiation of DNA replication.

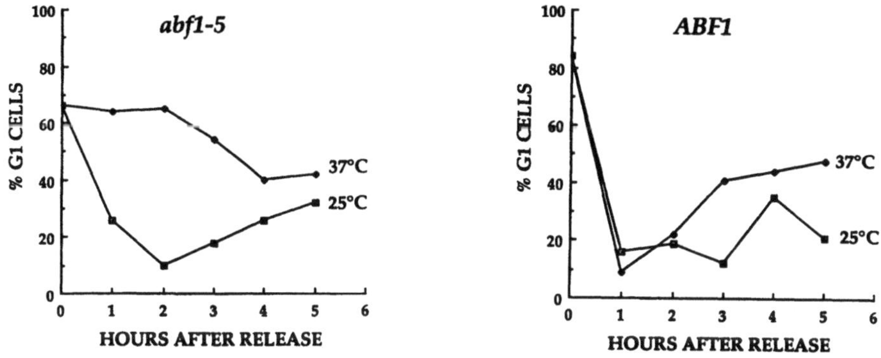

Figure 5. Flow cytometric analysis of DNA synthesis in an *abf1-ts* mutant. The *abf1-ts bar1* and *ABF1 bar1* strains were grown at 25°C to early log phase and treated with 300 ng/ml α-factor for 6 hours. The α-factor was removed by extensive washing, the cells were resuspended in fresh media and incubated at 25°C or 37°C. Aliquots were removed at the indicated times and the cells were fixed with 70% ethanol. Following RNase A treatment to remove cellular RNA, the cells were stained with propidium iodide and cellular DNA fluorescence was analyzed on a flow cytometer. The % G1 cells was determined from the DNA histograms for each sample.

Discussion

Studies of cis-acting ABF1 binding sites in the yeast genome suggest that this protein is important in regulating transcription and ARS activity. Our genetic analysis of the *ABF1* gene reported here and elsewhere (Rhode *et al.*, 1991) verify the role of ABF1 in these processes. Examination of ARS1 DNA binding activity in extracts from *abf1-ts* mutants also indicate that no other abundant ABF1-like DNA binding activities are present in these strains. Finally, analysis of cell cycle progression and DNA synthesis in conditional *abf1-ts* mutants provide further evidence for the role of ABF1 in the initiation of chromosomal replication.

Our current model for ABF1 action in replication is that this origin binding factor enhances efficient replication from chromosomal origins containing the essential ARS consensus element. Depending on the presence of other cis-acting elements and the affinity of ABF1 for its binding site, the effect of ABF1 on ARS activity may be large, as is the case for ARS121, or small, as it is for ARS1. Similar interactions between transcription factors and auxiliary cis-acting sites in viral origins of replication have been observed (see DePamphilis, 1988 for review). Several molecular mechanisms have been elucidated to explain the role of these transcription factors in enhancing the initiation of viral replication dependent on core origin sequences. These include nucleosome exclusion from the replication origin (Cheng and Kelley, 1989), facilitated interactions of the core binding factor with its binding site (Mohr *et al.*, 1991) and facilitated DNA unwinding activity of the core binding factor prior to initiation of DNA synthesis (Gutierrez *et al.*, 1990). Whether ABF1 plays a similar role at ARS elements is under current investigation.

Preliminary structure/function analysis of the *ABF1* gene suggest that ABF1 contains distinct functional domains and indicate that the DNA binding domain *per se* cannot perform all of the protein's essential functions. Examination of the protein sequence reveals two regions within the minimally defined DNA binding peptide that potentially are important for DNA binding. The previously described metal-binding motif at the N-terminus of the protein has been shown to be required for DNA binding (Halfter *et al.*, 1989). In addition, a cluster of basic amino acids is found at the C-terminus of the minimal DNA binding proteins. This region (net charge +10) is uncharacteristic of the very acidic nature of the protein as a whole (net charge -60 at pH 7). Furthermore, helical wheel analysis suggests that this region is capable of forming an amphipathic α-helix with hydrophobic amino acids (at a 7 residue repeat) on one face of the helix and basic amino acids on the other. In this respect, this region resembles the leucine-zipper motif of other DNA binding proteins (Landschulz *et al.*, 1988). How this region and the metal binding motif interact with DNA and what the role, if any, residues between these regions play is currently being examined.

Based on our initial observation, we propose that the diverse roles of ABF1 in transcription are mediated by at least two levels of regulation. First, the context of the ABF1 binding site is critical to its function. This is likely due to specific interactions with other DNA binding proteins (or other regulatory factors) and may involve different functional domains on ABF1 itself. Second, phosphorylation of ABF1 modulates its activity at a given promoter. We have found that ABF1 phosphorylation varies with different growth conditions and is correlated to the regulation of several genes whose promoter contain ABF1 binding sites. Furthermore, different ABF1-DNA complexes are observed under these conditions suggesting that interactions with other factors are dependent on the phosphorylated state of ABF1 (Silve *et al.*, 1991). Analysis of these interactions will shed further light on the role of ABF1 in transcriptional activation and repression and in the initiation of DNA replication.

References

Biswas, EE, Stefanec, MJ, Biswas, SB (1990) Mulecular cloning of a gene encoding an ARS binding factor from the yeast *Saccharomyces cerevisiae*. Proc Natl Acad Sci USA 87:6689-6692.

Brewer, BJ, Fangman, WL (1987) The localization of replication origins on ARS plasmids in *S. cerevisiae*. Cell 51:463-471.

Campbell, JL, Newlon, CS (1991) Chromosomal DNA replication. In The Molecular and Cellular Biology of the Yeast *Saccharomyces*: Genome Dynamics, Protein Synthesis and Energetics. Cold Spring Harbor Press. In press.

Cheng, L, Kelly, TJ (1989) Transcriptional activator nuclear factor 1 stimulates the replication of SV40 minichromosomes *in vivo* and *in vitro*. Cell 59:541-551.

DePamphilis, ML (1988) Transcriptional elements as components of eukaryotic origins of DNA replication. Cell 52:635-638.

Edelman, AM, Blumenthal, DK, Krebs, EG (1987) Protein serine/threonine kinases. Ann Rev Biochem 56:567-613.

Francesconi, SC, Eisenberg, S (1989) Purification and characterization of OBF1: a *Saccharomyces cerevisiae* protein that binds to autonomously replicating sequences. Mol Cell Biol 9:2906-2913.

Gutierrez, C, Guo, Z-S, Roberts, J, DePamphilis, ML (1990) Simian virus 40 origin auxiliary sequences weakly facilitate T-antigen binding but strongly facilitate DNA unwinding. Mol Cell Biol 10:1719-1728.

Halfter, H, Kavety, B, Vandekerckhove, J, Kiefer, F, Gallwitz, D (1989) Sequence, expression and mutational analysis of BAF1, a transcriptional activator and ARS1-binding protein of the yeast Saccharomyces cerevisiae. EMBO J 8:4265-4272.

Huberman, JA, Spotila, LD, Nawotka, KA, El-Assouli, SM, Davis, LR (1987) The *in vivo* replication origin of the yeast 2 micron plasmid. Cell 51:473-481.

Huberman, JA, Zhu, J, Davis, LR, Newlon, CS (1988) Close association of a DNA replication origin and an ARS element on chromosome III of the yeast *Saccharomyces cerevisiae*. Nucl Acids Res 16:6373-6484.

Jarvis, EE, Clark, KL, Sprague, GF (1989) The yeast transcription activator PRTF, a homolog of the mammalian serum response factor, is encoded by the *MCM1* gene. Genes & Develop 3:936-945.

Jones, KA, Kadonaga, JT, Rosenfield, PJ, Kelly, TJ, Tjian, R (1987) A cellular DNA-binding protein that activates eukaryotic transcription and DNA replication. Cell 48:79-89.

Landschulz, WH, Johnson, PF, McKnight, SL (1988) The leucine zipper: A hypothetical structure common to a new class of DNA binding proteins. Science 240:1759-1764.

Lohr, D, Torchia, T (1988) Structure of the chromosomal copy of yeast ARS1. Biochemistry 27:3961-3965.

Mohr, IJ, Clark, R, Sun, S, Androphy, EJ, MacPherson, P, Botchan, MR (1991) Targeting the E1 replication protein to the papillomavirus origin of replication by complex formation with the E2 transactivator. Science 250:1694-1699.

Passmore, S, Elble, R, Tye, BK (1989) A protein involved in minichromosome maintenance in yeast binds a transcriptional enhancer conserved in eukaryotes. Genes & Develop 3:921-935.

Rhode, PR, Elsasser, E, Campbell, JL (1991) The roles of yeast DNA binding protein ABF1 in DNA replication and transcription. Submitted.

Rhode, PR, Sweder, KS, Oegema, KF, Campbell, JL (1989) The gene encoding *ARS* binding factor I is essential for the viability of yeast. Genes & Develop 3:1926-1939.

Schultz, J, Marshall-Carlson, L, Carlson, M (1990) The N-terminal TPR region is the functional domain of SSN6, a nuclear phosphoprotein of Saccharomyces cerevisiae. Mol Cell Biol 10:4744-4756.

Silve, S, Rhode, PR, Campbell, J, Poyton, RO (1991) ABF1 is a phosphoprotein and plays a role in carbon source control of *COX6* transcription in *Saccharomyces cerevisiae*. Mol Cell Biol, submitted.

Struhl, K, Stinchcomb, DT, Scherer, S, Davis, RW (1979) High frequency transformation of yeast: Autonomous replication of hybrid DNA molecules. Proc Natl Acad Sci USA 76:1035-1039.

Sweder, KS, Rhode, PR, Campbell, JL (1988) Purification and characterization of proteins that bind to yeast ARSs. J Biol Chem 263:17270-17277.

Walker, SS, Francesconi, SC, Tye, BK, Eisenberg, S (1989) The OBF1 protein and its DNA-binding site are important for the function of an autonomously replicating sequence in *Saccharomyces cerevisiae*. Mol Cell Biol 9:2914-2921.

Proteins Involved in ARS Function in Yeast

B. K. Tye, V. Chang, C. Christ, R. Elble, S. Gibson, S. Passmore and H. Yan.
Section of Biochemistry, Molecular and Cell Biology, Cornell University, Ithaca, NY 14853.

Introduction

Replication of eukaryote chromosomes requires the coodinated initiation of DNA synthesis at multiple replication origins on the chromosomes. These initiation events are under cell cycle and developmental regulation that is evident in a number of well-defined biological contexts. DNA synthesis is confined within S phase of the cell cycle. During S phase, each replication origin is activated no more than once in a specified temporal order (Reynolds et al., 1989; Ferguson et al., 1991). In embryonic and somatic cells, the contrasted number of replication origins utilized reflects the demand of a replication deadline imposed by the length of the S phase in a cell division cycle (Blumenthal, et al., 1973). Thus, one may expect individual replication origins to be differentially regulated according to the cell cycle or developmental program. We are interested in finding out how differential regulation is achieved in the initiation of DNA synthesis at multiple replication origins in yeast.

The yeast genome is estimated to contain 200 to 400 replication origins that can be cloned on plasmids as autonomously replicating sequences (ARSs) (Huberman et al., 1988; Ferguson et al., 1991). The cloning of ARSs on plasmids provides a convenient means to identify gene products involved in ARS function by screening for mutants that destabilize minichromosomes containing one of these ARSs (Maine et al., 1984). The most attractive feature of this approach is that it allows one to isolate viable mutants with a defect in a specific function without relying on general phenotypes, such as cell cycle arrest, which often give no information on the execution point of the lesion. We isolated a number of yeast mutants that are defective in the maintenance of minichromosomes. Three of the mutants, *mcm1-1*, *mcm2-1* and *mcm3-1*, are of particular interest because they affect the stability of minichromosomes in an ARS-specific manner (Table 1). These mutants destabilize all minichromosomes tested (see *mcm2-1* at 37°C) but those bearing certain ARSs (such as ARS1 and ARS131) are affected much more dramatically than others (see *mcm2-1* at 23°C). Although minichromosome maintenance defects are common among mutants defective in DNA synthesis (Palmer et al., 1990), such as *cdc17*, the DNA polymerase α mutant, ARS-specificity may be unique to these *mcm* mutants (figure 1). We believe that this ARS-specificity is an indication that the defect is in the initiation step which is differentially regulated in different ARSs.

Table 1. Mitotic stabilities of minichromosomes in three ARS-specific *mcm* mutants.

	minichromosome YCp				
Strain	1	121	2B	131	131C
mcm1-1	<0.5	54	25	<0.5	39
mcm2-1 (23°C)	<0.3	67	68	<0.3	60
(35°C)	<0.3	<0.2	2	<0.3	<0.5
mcm3-1 (23°C)	<0.5	11	<0.5	<0.5	7
wildtype	91	96	97	76	96

Stabilities were measured as the percentage of cells still carrying the minichromosomes after 20 generations of nonselective growth. Stabilities were determined at 30°C unless otherwise noted.

Figure 1. Loss rates of minichromosomes YCp1 and YCp120 in different yeast strains.

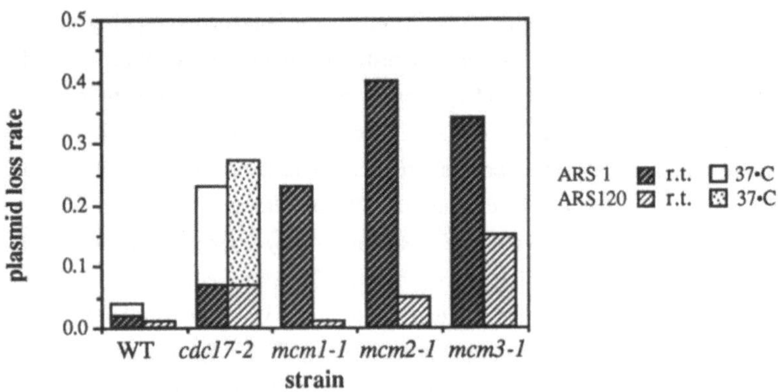

Although the *mcm1*, *mcm2* and *mcm3* mutants were isolated based solely on their *mcm* phenotype, they share a number of other phenotypes characteristic of mutants defective in DNA replication (Sinha et al., 1986). Null alleles of all three *MCM* genes are inviable (Passmore et. al, 1988; Gibson et al., 1990; Yan et al., 1991). Cells containing certain *mcm* mutant alleles have increased chromosome loss and recombination. Furthermore, these mutant strains show a cell division cycle arrest phenotype of large budded cells containing a single nucleus with a DNA content of nearly 2n. We explain this late S phase arrest as the result of leakiness in these alleles such that initiation of DNA replication takes place at enough ARSs to allow DNA synthesis to occur but not to allow complete replication of the genome. To investigate the role of these *MCM* gene products in the initiation of DNA synthesis at ARSs, we carried out functional and structural analyses of these *MCM* gene products (Christ and Tye, 1991; Yan et al., 1991). We are able to show that Mcm1, like the dnaA protein of *E. coli* (Bramhill and Kornberg, 1988; Braun et al., 1985) regulates gene expression and DNA replication by binding to gene promoters and to replication origins. Mcm2 and Mcm3 are two functionally related proteins that belong to a family of structurally similar proteins that are involved in early steps of DNA replication (Yan et al., 1991).

Mcm1, a transcription factor/replication initiator

In addition to its *mcm* defect, the original *mcm1-1* mutant is unable to mate as an α cell. As it turns out, Mcm1 is a multifunctional protein that acts as a transcription factor for mating type specific genes (Passmore et al., 1988) and also as an initiator for DNA synthesis at ARSs (Vicky Chang, unpublished results). It is identical to the previously identified transcription factor GRM/PRTF which binds cooperatively with the α1 protein to act as a transcription activator of α specific genes (Passmore et al., 1989; Jarvis et al., 1989; Ammerer, 1990) or with the α2 protein to act as a repressor of **a** specific genes (Keleher et al., 1989). Mcm1 also binds to the promoters of a number of genes involved in cell cycle progression and affects their expression (R. Elble, unpublished results). These genes include the cyclin gene *CLN3*, the p34/protein kinase gene *CDC28*, and *MCM3*. However, the effect of Mcm1 on the expression of these genes is modest, observed only in a tight *mcm1* truncation mutant but not in the original *mcm1-1* mutant, suggesting that the *mcm* defect of *mcm1-1* is not the result of an altered expression of *MCM3*.

Figure 2. Functional domains of Mcm1

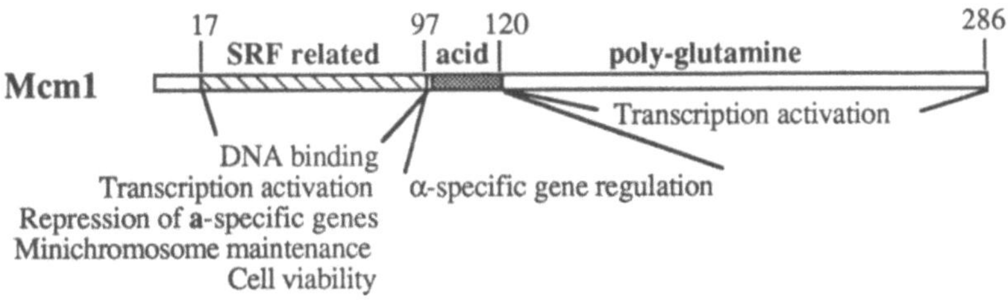

Since Mcm1 is a transcription factor, we are interested in finding out whether its action on ARSs is indirect, due to its regulation of a gene involved in ARS function, or direct as a result of its binding to ARSs. To do so, we dissected the functional domains of Mcm1 to see if we could separate the transcriptional activity of Mcm1 from its minichromosome maintenance activity. The result of this analysis is summarized in figure 2 (Christ and Tye, 1991). Mcm1 contains an N-terminal DNA binding/dimerization domain that is 70% identical to the DNA binding/dimerization domain of the human serum response factor (SRF) (Norman et al., 1988). C-terminal to this DNA binding domain is a twenty residue acidic stretch that is required for the regulation of α-specific genes. The remaining polyglutamine-rich C-terminal segment contributes 50% of the general transcriptional activity of Mcm1. Interestingly, the 80 amino acid DNA binding/dimerization domain alone is sufficient for most of the Mcm1 functions: cell viability, general transcription activation, repression of **a**-specific genes and minichromosome maintenance (table 2). Because this DNA binding domain is sufficient for both transcription activation and minichromosome maintenance, these two activities could not be uncoupled from each other by domain dissection.

Table 2. Effect of *mcm1-1* mutant protein overproduction on minichromosome stability.

Allele	Loss rate of YCp101
MCM1	0.02
mcm1-18-97	0.06
mcm1-1	0.30
+YEp24 *mcm1-1*	0.03

mcm1-18-97 allele contains only the DNA binding/dimerization domain of Mcm1. YCp101 contains ARS1.

If Mcm1 is directly involved in ARS function by binding to ARSs, we should be able to locate putative Mcm1 binding sites at or near ARSs based on the consensus derived from Mcm1 binding sites found upstream of mating type specific genes. We performed site-specific mutagenesis on at least one of these sites and showed that alterations in the Mcm1 binding site that increase or decrease its affinity for Mcm1 yields a corresponding increase or decrease in plasmid stability (V. Chang and S. Passmore, unpublished results). In addition, we noticed a possible correlation between the number and quality of potential Mcm1 binding sites in an ARS and the minichromosome stability conferred by that ARS. This notion is strengthened by two observations that suggest that the *mcm1-1* lesion results in reduced DNA binding activity. First, *mcm1-1* mutant extracts contain lowered Mcm1 DNA binding activity (Keleher et al., 1989), and second, overproduction of the mutant protein restores plasmid stability to a mutant strain (table 2). Thus, it is plausible that the ARS-specificity exhibited by the *mcm1-1* mutant is a reflection of the affinity of the mutant protein for Mcm1 binding sites at different ARSs.

Mcm2 and Mcm3 are structurally related

Although the *mcm1*, *mcm2* and *mcm3* mutants share a number of similar phenotypes, the sequences of the Mcm2 and Mcm3 proteins indicate that these two proteins have a special relationship to each other. Both proteins are large (890 a.a. for Mcm2 and 971 a.a. for Mcm3) compared to Mcm1, and share striking sequence similarities (figure 3; Yan et al., 1991). The greatest similarities between these proteins concentrate in three regions, I, II and III, with region II containing the best and most extensive homology (50% identity in 200 amino acids). A search with the amino acid sequences of these proteins against a protein data base maintained by Dr. Mark Goebl revealed that this sequence similarity is shared by three other known proteins: Cdc46 of *S. cerevisiae* (Hennessy and Botstein, 1991) and Cdc21 and Nda4 of *S. pombe* (S. Kearsey, see accompanying chapter). All of these proteins have been implicated in the early steps of DNA replication. The *S. pombe cdc21* mutant arrests in late S phase but has an execution point before the hydroxyurea block (see Kearsey et al., same volume). Cdc46 has been shown to be required for DNA synthesis. It is mobilized from the cytoplasm to the nucleus as mitosis is completed and persists there until the next round of DNA synthesis is initiated (Hennessy et al., 1990). Interestingly, such a cell cycle dependent nuclear localization property has been postulated for a "licensing factor" (Blow and Laskey, 1988) that may control DNA replication in the cell cycle.

Although all four proteins are strikingly similar in structure, they are distinct proteins required to perform unique functions essential for growth, since each gene is essential for viability. In fact, aside from their common domains, there are features unique to individual proteins that may provide clues to their functions. For example, N-terminal to region I in Mcm2, there is a putative C4 zinc finger motif that is characteristic of DNA binding domains (Evans and Hollengerg, 1988). Mutagenesis of two of the cysteine residues resulted in lethality suggesting that they are essential for Mcm2 function (Yan et al., 1991). This zinc finger motif is not conserved in the other members of this protein family suggesting that if only one of these proteins binds to DNA, Mcm2 is the likely candidate.

Figure 3. Alignment of the Mcm2 and Mcm3 protein sequences.

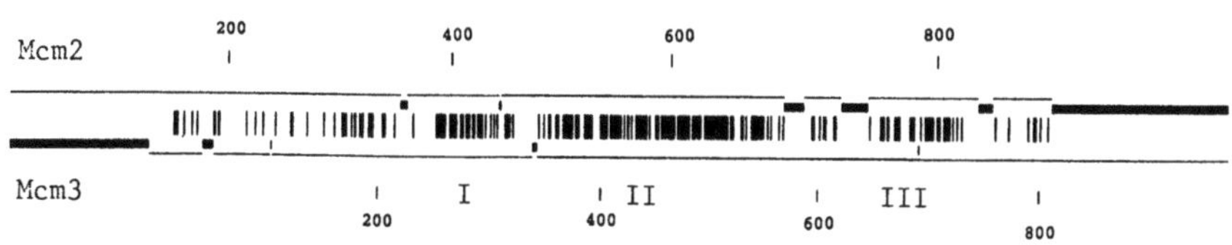

Horizontal lines represent amino acid sequences of Mcm2 and Mcm3. Bars indicate gaps in the alignment. Vertical lines indicate positions of identity.

Mcm2 and Mcm3 are functionally related.

The structural similarities between Mcm2 and Mcm3 led us to examine the functional relationship between these two proteins. Results of this genetic analysis are summarized in figure 4. We found that the *mcm2-1 mcm3-1* double mutant is lethal at a temperature that is permissive for each of the single mutants. Overproduction of Mcm2 suppresses the *mcm3-1* mutation while overproduction of the Mcm3 protein exacerbates the *mcm2-1* mutation. This latter observation is perhaps not too surprising since overproduction of Mcm3 in wildtype strains results in an ARS-specific minichromosome maintenance defect (Gibson et al., 1990). Such defects due to the overproduction of Mcm3 could be explained if Mcm3 is a component of a multisubunit complex, with inactive complexes resulting from stoichiometric imbalance. It would be interesting to find out if Mcm2 and/or other members of this protein family are subunits of a multicomponent complex, interacting through their common functional domains.

Figure 4. Functional Relationship between Mcm2 and Mcm3

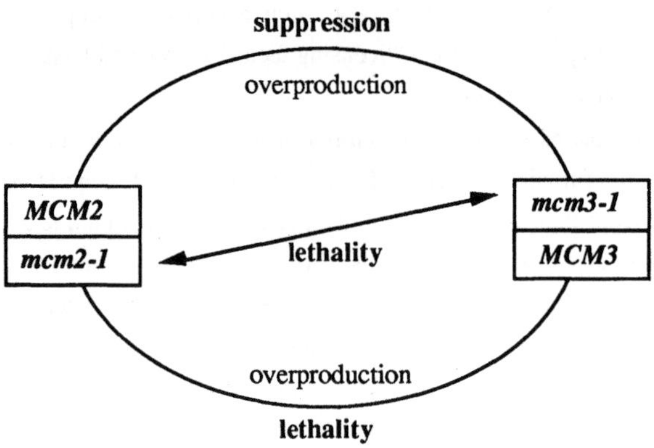

Another look into ARS-specificity

The phenotypic similarities between the three *mcm* mutants and the structure/function relationship between Mcm2 and Mcm3 suggest that the Mcm proteins perform similar functions at ARSs. Our initial motivation to study the *MCM* genes was based on the premise that the ARS-specific minichromosome maintenance phenotype is indicative of an involvement in the regulation of ARS function. We felt that a direct comparison of the ARS specificities of the *mcm* mutants may help us to gain insight into the role of these *MCM* gene products in ARS function.

We further defined ARS-specificity in minichromosome maintenance in the *mcm* mutants by using a larger sample of well characterized ARSs (Yan et al., 1991). The loss rates of seven minichromosomes, each carrying *CEN5* and an ARS element, were measured in wildtype and mutant strains at room temperature. To compare the effects of each mutation on a particular minichromosome, the relative loss rates of each minichromosome in mutant and wildtype are presented in a bar graph in figure 5. Two conclusions can be drawn from this study. First, there is no correlation between the relative strengths of the ARSs in the wildtypeand *mcm* mutant strains. This is illustrated by the fact that all minichromosomes are stable in the wildtype strain but to varying degrees. We observed that there is no correlation

Figure 5. Comparison of plasmid loss rates in mutant and wildtype.

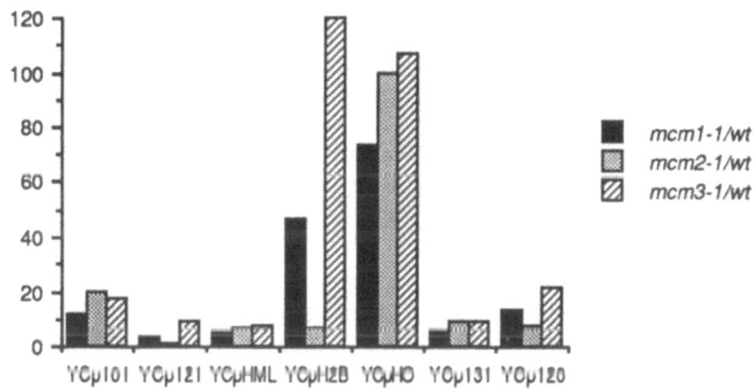

Loss rate in WT	0.020	0.020	0.050	0.003	0.003	0.040	0.007

between the hierarchy of ARSs, based on plasmid stabilities, in wildtype cells and in the mutant strains (e.g. compare YCpHO and YCpH2B in wildtype and *mcm2-1*). Thus, the *mcm* phenotype of these mutants is probably not due to an amplification of preexisting differences in ARSs as a result of indirect effects, such as effects on growth rates or the length of the phases of the cell cycle etc.. Second, although the *mcm* mutants affect the stability of all ARSs tested, they exert their most dramatic effects on different sets of ARSs. The best example is illustrated by the dramatic difference in the effects of *mcm2-1* and *mcm3-1* on YCpH2B. Thus, it appears that the *mcm* mutants have similar but nonidentical ARS-specificities.

Models for the action of Mcm1, Mcm2 and Mcm3

We are interested in understanding the mechanism for the initiation of DNA replication that leads to the regulation of origin usage in different stages of the cell cycle and of development. Emerging evidence suggests that the strategy used in the regulation of DNA replication at the large number of replication origins in the genome is perhaps not very different from that used in the regulation of gene expression at promoters of diverse genes. Multiple factors, acting in specific combinations, determine the probability, tissue specificity and cell cycle timing of DNA synthesis initiation events in a manner similar to the regulation of gene expression. In fact, throughout this meeting, a recurring theme reported in prokaryotes and eukaryotes is that replication initiators are often transcription factors. The dnaA protein, which is the replication initiator for OriC in *E. coli*, also acts as a transcription factor for its own expression (Braun et al., 1985) and the expression of a number of genes. In yeast, two transcription factors, ABF1 (Walker et al., 1990; also see J. Campbell and J. Diffley in same volume) and Mcm1, appear to play a role in replication initiation at ARSs. In mammalian systems, the transcription factors NFI and oct-1 have been shown to be involved in viral DNA replication (Cheng and Kelly, 1989; Verrijzer et al., 1990). The report by N. Heinz that the first identified mammalian chromosomal origin, at the *dhfr* gene, contains binding sites for the proliferation-specific transcription factors Ap1 and NFIII/OTF1, adds further support to the belief that transcription factors may be involved in the initation of DNA replication (Dailey et.al., 1990).

Functional domain analysis of three transcription factors, NFI, NFIII and Mcm1, suggests that the DNA binding domain of these proteins is sufficient for their role in replication initiation at replication origins. This finding suggests that the mode of action of these proteins at replication origins may be different from that at promoters, requiring the binding of target sites but requiring no physical interactions with the initiation complex. This view is consistent with a simple prerequisite for replication initiation, which is to melt or unwind the origin region. The melting of replication origins could be achieved by the binding of any of a number of proteins. In turn, differential regulation may be accomplished by combinatorial actions of these proteins.

The minichromosome maintenance phenotype of the *mcm1-1*, *mcm2-1* and *mcm3-1* mutants suggests that it is unlikely that the *MCM* gene products act only at a subset of ARSs. Rather, if they do act directly at ARSs, they contribute to the activation of all ARSs, but each to different extents. The homologous domain shared by Mcm2, Mcm3, Cdc46 and Cdc21 suggests that these proteins, that are involved in the early steps of DNA replication, either they interact with a common target, or they interact with one another through this homologous domain. A plausible model for the action of the *MCM* gene products is that Mcm1, and the

Mcm2/Mcm3 complex bind to ARSs independently of each other to signal a replication initiation event. The contribution of Mcm1, Mcm2 and Mcm3 to ARS activity depends on the number and/or the quality of their binding sites at ARSs. The reports by Dr. S. Eisenberg that ABF1/OBF1 binds to different ARSs with different affinities (Eisenberg et al., 1988) and by Dr. J. Campbell that this differential affinity to ARSs translates into different plasmid stabilities in the *abf1*^{ts} mutants (this volume) show that ABF1 is another example of such a replication initiator. These many different ARS binding factors acting in a combinatorial manner provide diversity in the regulation of initiation events at ARSs.

References

Ammerer, G. (1990) Genes & Dev. 4, 299-312.
Blumenthal, A.B., Kriegstein, H.G., and Hogness, D.S. (1973) Cold Spr. Harbor Symp. Quant. Biol. 38, 205.
Bramhill, D. and Kornberg, A. (1988) Cell. 54, 915.
Braun, R.E., O'Day, K., Wright, A. (1985) Cell. 40, 159.
Christ, C. and Tye, B. K. (1991) Genes & Dev. 5. In press.
Dailey, L., Caddle, M., Heintz, N. and Heintz, N. (1990) Molec. & Cell. Biol. 10, 6225.
Eisenberg, S., Civalier, C. and Tye, B.K. (1988) Proc. Natl. Acad. Sci. U.S.A. 85, 743.
Ferguson, B.M., Brewer, B.J., Reynolds, A.E., Fangman, W.L. (1991) Cell, in press.
Evans, R.M., and Hollenberg. (1988) Cell 52, 1.
Gibson, S., Surosky, R. and Tye, B.-K. (1990) Mol. Cell.Biol. 10, 5707.
Hennessy, K.M., Clark, C.D., and Botstein, D. 1990. Genes & Devel. 4, 2252.
Huberman, J.A., Zhu , J., Davis, R. and Newlon, C. (1988) Nucleic Acids Res. 16, 6373.
Jarvis, E.E., Clark, K.L. and Sprague, G.F. (1989) Genes & Dev. 3, 936.
Keleher, C.A., Passmore, S. and Johnson, A.D. (1989) Mol. Cell. Biol. 9, 5228.
Maine, G.T., Sinha, P., and Tye, B.K. (1984) Genetics. 106, 365.
Norman, C., Runswick, M., Pollock, R., and Treisman, R. (1988) Cell. 55, 989-1003.
Palmer, R.E., Hogan, E., and Koshland, D. (1990) Genetics. 125, 763.
Passmore, S., Maine, G.T., Elble, R., Christ, C., and Tye, B.K. (1988) J. Mol. Biol. 204, 593
Passmore, S., Elble, R. and Tye, B.K. (1989) Genes & Dev. 3, 921.
Reynolds, A.E., McCarroll, R., Newlon, C. and Fangman, W. (1989) Mol. & Cell. Biol. 9, 4488.
Sinha, P., Chang, V. and Tye, B.K. (1986) J. Mol. Biol. 192, 805.
Yan, H., Gibson, S., and Tye, B.K. (1991) Genes & Devel. In press.
Walker, S.S., Francesconi, S.C., and Eisenberg, S. (1990) Proc. Natl. Acad. Sci. 87, 4665.

ARS Binding Factors from *Saccharomyces cerevisiae*

John F.X. Diffley[1] and Bruce Stillman[2]

[1]Imperial Cancer Research Fund
Clare Hall Laboratories
Blanche Lane
South Mimms, Potters Bar
Herts. EN6 3LD
UK
[2]Cold Spring Harbor Laboratory
Cold Spring Harbor, NY 11724
USA

Two proteins, designated ARS Binding Factors (ABF) 1 and 2, were identified several years ago as proteins that interact specifically with the yeast replication origin, ARS1 (Diffley and Stillman, 1988a). Subsequent analysis has revealed that ABF1 plays a key role in cell growth while ABF2 is involved in the maintenance of the mitochondrial genome. In this article, we summarize some of these results.

ABF1

ABF1 was purified as a polypeptide with an apparent molecular weight of 135kDa (Diffley and Stillman 1988a, Sweder et al 1988). ABF1 binds to a single site within ARS1 (Diffley and Stillman, 1988a, Sweder et al., 1988) and to sites within several other ARSs including three of the four ARSs associated with the silent mating cassettes *HML* and *HMR* (silencers) (Shore et al., 1987, Diffley and Stillman, 1988a, Buchman et al., 1988). The ABF1 binding sites at several ARSs have been shown to be important for ARS function on plasmids. Furthermore, the ABF1 binding site within at least one of the silencers (*HMRE*) is important for both ARS and silencer function (Brand et al. 1987). Additionally, ABF1 binding sites have been found upstream of a large number of yeast genes and these binding sites have been shown in several cases to correspond to UAS elements (see Dorsman et al., 1990, Della Setta et al., 1990). Therefore, ABF1 is an important, multifunctional DNA binding protein.

To unravel the various functions of ABF1 *in vivo*, we have raised a polyclonal antiserum against purified ABF1 protein and used this antiserum to identify and isolate the *ABF1* gene from a λgt11 yeast genomic DNA expression library (Diffley and Stillman 1989). The predicted ABF1 coding sequence is closely related to another multifunctional DNA binding protein from yeast, RAP1 (Diffley and Stillman 1989). The *ABF1* gene is essential for viability (Diffley and Stillman,1989, Rhode et al., 1989, Halfter et al., 1989) and, furthermore, overproduction of ABF1 in yeast using the GAL1 promoter is lethal (J.D. and B.S. unpublished data) indicating that intracellular levels of ABF1 are critical.

The Domain Structure of ABF1

We have begun to approach the *in vivo* function of ABF1 by identifying domains within the protein involved in different processes. First, we have attempted to identify the region of the protein that contacts the DNA recognition sequence. Previous results using the reversible sulfhydryl modifying reagent p-(hydroxymercuri) benzenesulfonate (PMPS) (Diffley and Stillman, 1989) and *in vitro* mutagenesis (Halfter et al., 1989) indicated that DNA binding by ABF1 required at least one of the three Cys residues located near the N-terminus of the protein. Furthermore, reversal of PMPS inhibition was shown to require zinc, indicating that this region of the protein formed an essential metal binding domain (Diffley and Stillman, 1989).

To analyze a large number of ABF1 mutations, we have used *in vitro* transcription and translation to generate mutant proteins which could subsequently be assayed for sequence-specific DNA binding activity. The results from these experiments (summarized in Fig.1) indicate that two separate regions of the ABF1 protein are required for high affinity, sequence-specific DNA binding; one located near the N-terminus of the protein (N-DOMAIN) and a second located nearly 100 amino acids away in the middle of the coding sequence (C-DOMAIN). Furthermore, the spacing between these two domains is not important since removal of the protein sequence between them does not affect its affinity for DNA (Diffley and Stillman, manuscript in preparation).

In other proteins with bipartite DNA binding domains like the b-zip proteins and the helix-loop-helix proteins, one of the subdomains does not actually contact DNA but, instead, is required for dimerization, with the dimer being the active DNA binding entity. This, however, is not the case with ABF1, since it binds to its recognition sequence as a monomer (Diffley and Stillman, manuscript in preparation). To address their role in DNA binding, we have overexpressed each subdomain in *E.coli* and purified them to apparent homogeneity (Fig.2A). Each of these purified polypeptides has intrinsic, low affinity DNA binding. Furthermore, one of these domains (C-DOMAIN) retains at least some of the sequence specific contacts of the intact protein (Diffley and Stillman, manuscript in preparation). Consistent with the fact that the central domain does not contain any cysteine residues, DNA binding by this domain is insensitive to PMPS inhibition while DNA binding by a protein containing the entire DNA binding domain (Fig.2B) or just the amino terminal domain (N-DOMAIN - Diffley and Stillman, manuscript in preparation) is sensitive to PMPS inhibition.

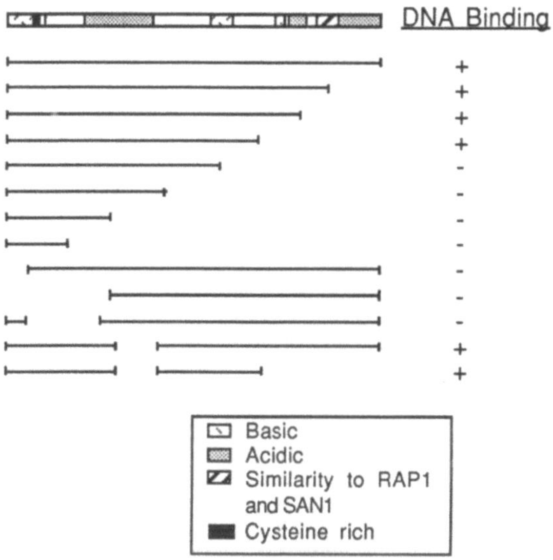

Fig.1. Summary of DNA Binding Activity of ABF1 Deletion Mutants. The indicated ABF1 deletion mutants were transcribed in vitro using T3 RNA polymerase, translated in a wheat germ lysate and assayed for ARS1 specific DNA binding activity essentially as described (Diffley and Stillman, 1988a) except that electrophoresis was in 10% acrylamide gels and binding reactions contained 200μg of boiled, sonicated salmon sperm DNA. Regions of the ABF1 protein are indicated in the legend.

We have employed a modified "plasmid shuffle" strategy (Sikorski and Boeke, 1991), summarized in Fig. 3A to begin to analyze some of the proposed *in vivo* functions of ABF1. In this strategy, we have constructed haploid yeast strains with three key features. First, the chromosomal copy of the *ABF1* gene has been replaced with the selectable marker, *TRP1*. Because the *ABF1* gene is essential for viability, it must be supplied to these cells. Thus, the second key feature is that these strains harbor a plasmid containing *URA3* as a selectable marker and the full length *ABF1* gene driven by its natural

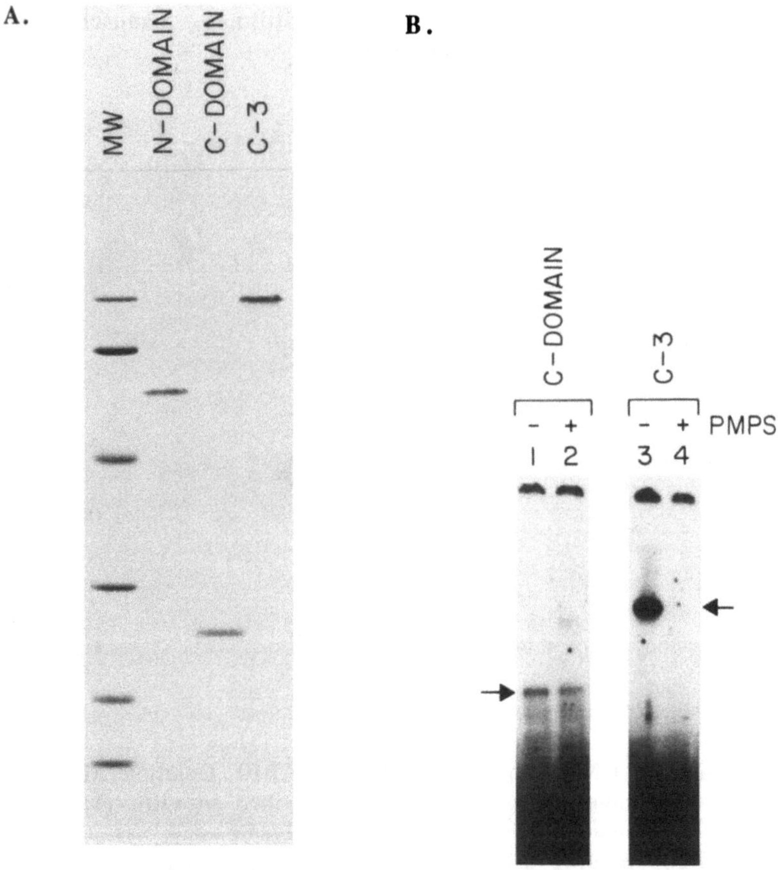

Fig.2 A) SDS polyacrylamide gel electrophoresis of DNA binding subdomains purified from *E.coli* extracts. 5 µg of each purified polypeptide was subjected to electrophoresis in 15% SDS polyacrylamide gels and stained with Coomassie blue. These polypeptides contain amino acids 1-250 (N-DOMAIN), 313-485(C-DOMAIN) and 1-485 (C-3). The molecular weight markers are, from top to bottom, 97, 66, 43, 31, 21 and 14kDa. **B)** Sensitivity of C-DOMAIN to the sulfhydryl reagent PMPS. Aliquots of purified C-DOMAIN and C-3 were dialyzed into buffer containing 10mM Tris HCl pH 7.5 and 1mM EDTA, treated with 1mg/ml PMPS on ice for 30 minutes and assayed for ARS1 specific DNA binding activity essentially as described (Diffley and Stillman, 1988a, 1989).

promoter. The choice of *URA3* as selectable marker is important because this marker can be selected for (by conversion to uracil prototrophy) and against (by growth on the poisonous uracil precursor analog, 5-fluoroorotic acid - 5-FOA). The third key feature of these strains is that they are *MATα* and contain either a wild type *HMRE* silencer or a mutated silencer in which the binding site for RAP1 has been eliminated with a linker insertion (e^- silencer). Since the ABF1 and RAP1 binding sites at *HMRE* are normally redundant, mutations in

A.

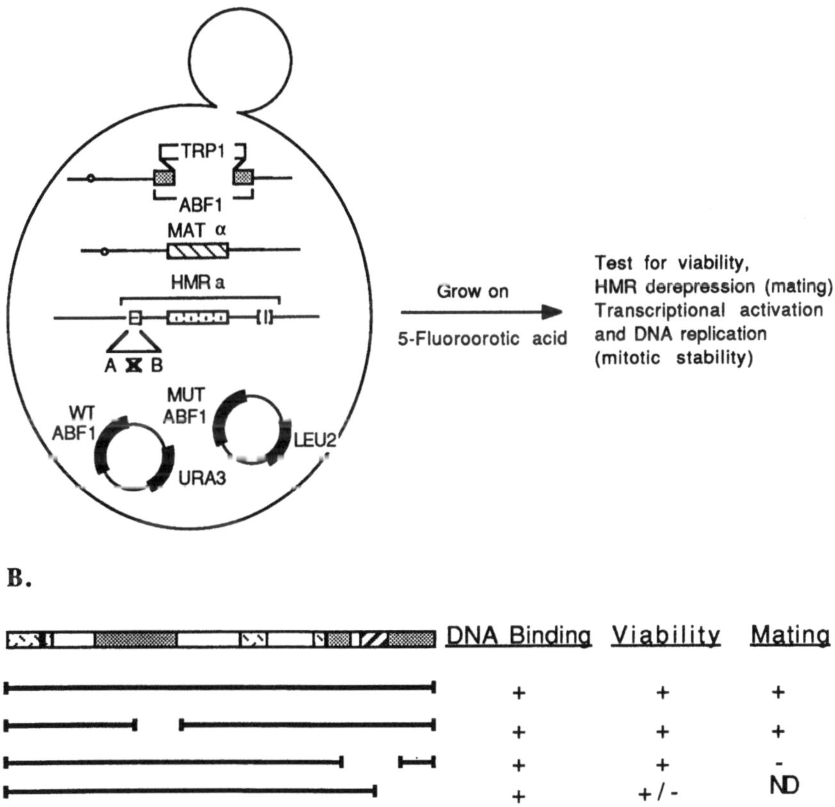

Fig. 3. A) Strategy and **B)** Results of the Plasmid Shuffle. Details are described in the text. The legend to B is the same as Fig.1. ND in not determined.

ABF1 which specifically affect transcriptional repression of *HMR* might be predicted to have a much greater effect in the absence of the RAP1 site than in its presence. Such defects can be measured indirectly by measuring the mating eficiency of such strains (derepression of *HMR* - which contains *MATa* information - in *MATα* cells will result in cells behaving as *MATa/α* diploids that cannot mate).

Into these strains, separate plasmids containing specifically engineered mutations in the *ABF1* gene and a separate selectable marker (*LEU2*) are introduced. If the mutation introduced does not eliminate an essential function of ABF1, cells will be able to grow on 5-FOA, generating a new strain driven solely by the mutant *ABF1* gene. These mutant strains can then be analyzed for ABF1-specific functions like mitotic stability of ARS1-containing plasmids, mating ability and transcription of ABF1-dependent genes.

Results from our initial analysis of ABF1 deletion mutants are shown in Fig.3B. Since we anticipate that sequence-specific DNA binding will be required for all of the *in vivo* functions of ABF1, we have initially targeted regions that we know are not involved in DNA binding for analysis. Thus, the ABF1 deletion mutants shown in Fig.3B all bind to DNA with wild type affinity and encode structurally stable proteins. Any *in vivo* defects exhibited by these mutants, therefore, must be due to the loss of some other important function of ABF1.

Deletion of the region of ABF1 between the DNA binding domains has very little effect. Strains in which all ABF1 functions must be performed by this mutant are healthy and nearly normal, growing only marginally slower than the wild type. Interestingly, this region contains several of the asparagine rich tracts found in ABF1 suggesting that these tracts may not serve any critical function.

Deletions in the C-terminus of the ABF1 protein, however, have several potentially important effects. From analysis of the ABF1 coding sequence, we noted that a short region in the C-terminus of the ABF1 protein is related, not only to RAP1, but also to *SAN1*, a suppressor of certain weak mutant alleles of *SIR4* (a gene required for *HMR* and *HML* repression)(Diffley and Stillman, 1989) suggesting that this region may be involved in transcriptional repression of the mating type loci. Strains in which this region of the *ABF1* gene have been removed are defective for mating in an e^- silencer (approximately 1% wild type level), which is dependent upon the ABF1 binding site but not in an e^+ silencer (approximately 30% wild type level) which is not dependent upon the ABF1 binding site. This strain produces stable ABF1 protein *in vivo* and ARS1 plasmids are maintained efficiently (data not shown) suggesting a specific defect in mating type repression. Finally, deletion of the extreme C-terminus of the ABF1 gene causes a severe reduction in growth rate, suggesting that this region is important for cell growth. Future work will focus on identifying proteins that interact with ABF1 in transcriptional activation, repression and DNA replication.

ABF2

ABF2 is an abundant, small, basic protein that exhibits both specific and non-specific DNA binding (Diffley and Stillman, 1988a and b). The protein has been purified as a polypeptide of approximately 20kDa (Diffley and Stillman, in press) and a polyclonal antiserum to the purified protein has been raised. By indirect immunofluorescence, ABF2 roughly colocalizes with cytoplasmic mitochondrial DNA, while subcellular fractionation experiments indicate that, in addition to being mitochondrial, a significant amount of ABF2 is also found in the nucleus (Diffley and Stillman, in press).

A.

MNSYSLLTRSFHESSKPLFNLASTLLKASKRTQLRNELIKQGPKRPTSAYFLY
LQDHRSQFVKENPTLRPAEISKIAGEKWQNLEADIKEKYISERKKLYSEYQKAKKE
FDEKLPPKKPAGPFIKYANEVRSQVFAQHPDKSQLDLMKIIGDKWQSLDQSIKDK
YIQEYKKAIQEYNARYPLN.

B.

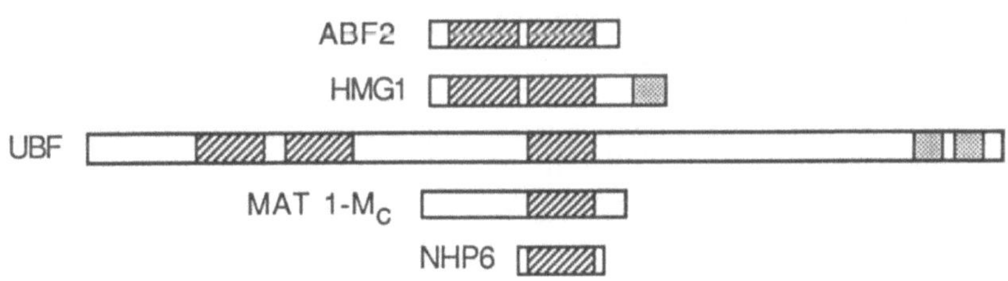

Fig. 4 ABF2 and Other Proteins are Related to the High Mobility Group Protein, HMG1. **A)** The predicted amino acid sequence of the ABF2 protein. Note that the mitochondrial signal sequence which is cleaved off in the mature protein is shown in boldface and the two HMG boxes in ABF2 are underlined. **B)** The positions of the HMG boxes found within several proteins (hatched boxes) is compared. Stippled boxes are acidic regions found in some of these proteins. The proteins included are described in the text.

We have used the amino terminal sequence of the ABF2 protein to design oligonucleotides to clone the *ABF2* gene using the polymerase chain reaction. The sequence of this gene was determined and analyzed (Diffley and Stillman, in press) and the predicted amino acid sequence of the ABF2 protein is shown in Fig. 4A. Consistent with its apparent mitochondrial location, ABF2 contains a leader of 26 amino acids that conforms well to a mitochondrial signal sequence and is cleaved of in the mature protein. Analysis of the rest of the ABF2 coding sequence has revealed that it is closely related to the vertebrate non-histone chromosomal high mobility group protein, HMG1. The HMG1 protein consisits of three domains; two closely related domains heretofore referred to as HMG boxes and a C-terminal domain composed entirely of acidic amino acid residues. ABF2 contains two HMG boxes closely related to HMG1 but lacks the acidic tail. In addition to HMG1 and ABF2, several other proteins also contain HMG boxes (Fig. 4B) such as the RNA polymerase I transcription factor UBF (Jantzen et al, 1990), the *Schizosaccharomyces pombe* mating type gene MAT1-M_c (Kelly et al., 1988) and the *S.cerevisiae* chromosomal protein NHP6 (Kolodrubetz and Burgum, 1990).

The *ABF2* gene is not essential for viability, but is required for the stable maintenance of the mitochondrial genome (Diffley and Stillman, in press). Estimates of the amount of intracellular ABF2 indicate that there are at least 250,000 copies of the protein *in vivo*, or, one for every 15 bp of mitochondrial DNA. Taken together with the fact that ABF2 binds to DNA by wrapping and, presumably, therefore, compacting DNA (Diffley and Stillman, unpublished data), these results suggest that ABF2 may be the "mitochondrial nucleosome", serving to package and organize the mitochondrial genome. Footprinting analysis on mitochondrial promoters (Diffley and Stillman, in press) suggests that ABF2 may also have a more direct role in mediating mitochondrial transcription. Finally, the presence of a significant amount of processed ABF2 in the nucleus suggests that it may also be involved in some type of feedback mechanism, informing the nucleus as to the amount of mitochondrial DNA.

References

Brand,A.H., Micklem, G. and Nasmyth, K. (1987) *Cell* **51**:709-719.

Buchman, A.R., Kimmerly, W.J., Rine, J. and Kornberg, R.D. (1988) *Mol. Cell. Biol.* **8**:210-225.

Della Setta, F., Treich, I., Buhler, J.-M. and Sentenac, A. (1990) *J. Biol. Chem.* **265**:15168-15175.

Diffley, J.F.X. and Stillman, B. (1988a) *Proc. Natl. Acad. Sci. (USA)* **85**:2120-2124.

Diffley, J.F.X. and Stillman, B. (1988b) in *Cancer Cells vol. 6, Eucaryotic DNA Replication*, B. Stillman and T.J. Kelly, Eds. (Cold Spring Harbor Laboratory Press, Cold Spring Harbor, NY) pp. 235-243.

Diffley, J.F.X. and Stillman, B. (1989) *Science* **246**:1034-1038.

Dorsman, J.C., van Heeswijk, W.C. and Grivell, L.A. (1990) *Nucleic Acids Research* **18**:2769-2776

Halfter, H., Kavety, B., Vandekerckhove, J., Kiefer, F. and Gallwitz, D. (1989) *EMBO J.* **8**:4265-4272.

Jantzen, H.-M., Admon, A., Bell, S.P. and Tjian, R. (1990) *Nature* **344**:830-836.

Kelly, M., Burke, J., Smith, M., Klar, A. and Beach, D. (1988) *EMBO J.* **7**:1537-1547.

Kolodrubetz, D. and Burgum, A. (1990) *J. Biol. Chem.* **265**:3234-3239.

Rhode, P.R., Sweder, K.S., Oegama, K.F. and Campbell, J.L. (1989) *Genes and Development* **3**:1926-1939.

Rose, M.D., Winston, F. and Hieter, P. *Methods in Yeast Genetics.* (1989) Cold Spring Harbor Laboratory Press, Cold Spring Harbor, NY

Shore, D., Stillman, D.J., Brand, A.H. and Nasmyth, K. (1987) *EMBO J.* **6**:461-467.

Shore, D. and Nasmyth, K. (1987) *Cell* **51**:721-732.

Sikorski, R.S. and Boeke, J.D. (1991) in *Methods in Enzymology vol. 194 Guide to Yeast Genetics and Molecular Biology*, C. Guthrie and G.R. Fink Eds. (Academic Press, New York) pp.302-318

Sweder, K.S., Rhode, P.R. and Campbell, J.L. (1988) *J. Biol. Chem.* **263**:17270-17277.

Mutations Within the Minimal Replicon of Plasmid pPS10 Increase Its Host Range

Rafael Giraldo, Margarita Martín, María Elena Fernández-Tresguerres, Concepción Nieto, and Ramón Díaz

Centro de Investigaciones Biológicas

Velázquez 144

28006 Madrid

Spain

ABSTRACT

Mutations in the basic replicon of the *Pseudomonas* plasmid pPS10 permit its establishment in *E.coli*. Some of the mutations affect the synthesis and/or activity of RepA, the replication protein of the plasmid, and one of the mutants, pMM141, is further analysed. Replication of pMM141 in *E.coli* requires functional *oriV* and *repA* sequences, and depends on the DnaA protein of the host. pMM141 can replicate in *Pseudomonas*. Preliminary analysis of the replication of this mutant in extracts of *E.coli*, shows that the plasmid replication protein RepA as well as the host proteins DnaA, DnaB, DnaG, HU, SSB, DNA gyrase and DnaK are required . *In vitro* replication of pMM141 can occur in the absence of transcription by host RNA polymerase. Reconstitution experiments locate the mutation in a 0.3 kb fragment of the basic replicon of pMM141 that includes the 5′ end of the *repA* gene and that contains the coding region for the leucine zipper motif of RepA (Giraldo et al, 1989). It is proposed that the broad host-range character of the mutant is due to the adjustment of RepA interactions to the ones required for initiation of replication in *E.coli.*

INTRODUCTION

The study of the factors and molecular interactions which determine the host range character of plasmids of Gram-negative bacteria has attracted considerable interest due to their clinical and environmental implications and to the use of plasmids as cloning vectors (Thomas, 1989). The usual approach to investigate plasmid host-range determinants, has been to analyze replication of broad-host range replicons. *In vitro* replication systems have been important tools in these studies (Diaz and Staudenbauer, 1982 ; Sherzinger et al, 1984; Haring and Scherzinger 1989; Pinkney et al, 1988) .

During the course of our studies with a plasmid isolated from *Pseudomonas savastanoi*, pPS10, we realized that this replicon failed to give stable transformants in *E.coli:* only tiny and non-viable colonies ("abortive establishment") could be observed in some of the transformation experiments (Nieto, 1989). This observation prompted us to study the possible reasons of this deficient establishment. It was eventually found that, after *in vitro* mutagenesis with hydroxilamine, variants of mini-pPS10 plasmids that transform efficiently *E.coli* could be easily isolated. This result indicates the possibility to isolate by direct selection host-range mutants of pPS10 and opens new ways of studying plasmid host-range determinants.

The analysis of the host-mutants of pPS10 is greatly facilitated by the available information on the pPS10 replicon: i) the sequence of a 1823 bp region that contains the basic replicon of the plasmid has been obtained; in this sequence, the origin of replication, *oriV*, and the gene, *repA*, of a protein that can activate *in trans* replication from *oriV* sequences, have been identified (Nieto,1988 ; Nieto et al, submitted), ii) a leucine zipper motif has been identified in the N- terminal region of RepA, and a consensus motif, based in the comparison with other plasmid replication proteins, has been established (Giraldo et al, 1989), iii) a mutation that introduces a site for *Cla*I in a dnaA box present in *oriV* and others that adjust the sequence of this box to the consensus described in *E.coli*, have been obtained (Giraldo, 1991; Nieto et al, submitted), iv) different miniplasmid and cloning vectors based on the pPS10 replicon are available (Nieto et al, 1990).

Here we describe the isolation and preliminary characterization of a broad host range mutant of the pPS10 mini derivative pRG14, that is affected in the replication protein. The possible involvement of the leucine zipper motif of RepA in this host range phenotype is discussed.

MATERIAL AND METHODS

Bacterial strains and plasmids.

The strains of *E.coli* K12 used here were: C600EL (from K. Nordström), that is a restriction deficient strain, BT1000 (Weshler et al,1973) that is a *polA1* mutant , WM1490 (from W. Messer) that is a *dnaA*-null strain and WM1502 (from W. Messer) that is a *dnaA*$^+$ strain isogenic to

WM1490. Two restriction deficient strains of *Pseudomonas* were used: PAO1024 (from K. Nordström) that is a strain of *P.aeruginosa,* and KT2440 (Bagdasarian and Timmis,1982) that is a strain of *P.putida* . Plasmids used were: pRG15, a wild-type mini pPS10 and its mutants pRG14 and pRG16 (Giraldo, 1991; Nieto et al, submitted): in pRG16 the 9 bp that define the dnaA box (5´ATATCCACA3´) are substituted by 9 bp (5´AATCGATGA3´) that include a *Cla*I restriction site, and in pRG14 the A in the 5´ position of this box is subtituted by T to match the consensus found in *E.coli;* pCN51 is a shuttle vector containing the pPS10 and pBR322 replicons (Nieto et al, 1990); pCN510 is a derivative of pCN51 lacking the *Eco*RI fragment that contains the *repA* gene of pPS10 (this work); pRG21 is a pBR322 recombinant carrying a *Bam*HI-*Eco*RI *oriV* fragment of pRG14 (Giraldo, 1991); pMMBrepA, is broad host range recombinant carrying *repA* under control of P*tac* (Nieto, 1989) ; pLSK5 (Messer et al, 1988) is a pMB9 based vector carrying the *dnaA* gene of *E.coli* under control of P*tac* ; pTWY310 (from C. Thomas), is a pBR322 recombinant carrying the *dnaA* gene of *P.aeruginosa*, and pHI206 (from H. Ingmar), is a broad host range recombinant, carrying the *dnaA* gene of *P.putida* under the control of the P*tac*.

Experimental procedures

General manipulations with DNA (Sambrook *et al.*, 1989) and *In vitro* mutagenesis of plasmid DNA with hydroxylamine (Humphreys et al,1976) were as described. Transformation of *E.coli* cells with plasmid DNA was done according to Lederberg and Cohen (1974) and transformation of *P.aeruginosa* cells as described by Bagdasarian and Timmis (1982). Preparation of extracts for *in vitro* replication and the *in vitro* replication assays were basically as described (Diaz and Staudenbauer, 1982).

RESULTS

Preliminary analysis: the failure of pPS10 to replicate in *E.coli* is not due to low level of repA or to a poor dnaA box.

The fact that the pPS10-pBR322 shuttle vector pCN51 can be established in *E.coli* roles out as reasons of the "abortive" establishment" of pPS10, expression of plasmid-lethal functions or uncontrolled "run-away" replication. The poor expression of *repA-lacZ* fusions in *E.coli*, and the mismatch of the dnaA box present in the origin of replication of pPS10 with the consensus dnaA box of *E.coli*, (Nieto et al, submitted), indicate two possible factors that could limit the establishment of pPS10 in *E.coli* : i) low level of RepA, and ii) poor accesibility of the DnaA protein of the host to the origin of replication. The availability of a conditional overproducer of the RepA protein of pPS10, pMMBrepA, and of a mini-pPS10 replicon, pRG14, with the dnaA box adjusted to the consensus dnaA box of *E.coli*, made it possible to evaluate the contribution of these factors. However, neither the increased level of RepA protein promoted by pMMBrepA, nor the adjustment of the dnaA box (pRG14 mutant), made possible a full establishment of the pPS10 replicon in *E.coli*; this failure was not relieved either by over-expression of the DnaA protein of *E.coli* promoted by pLSK5 or by the presence in the *E.coli* cells of pTWY310 or pHI206, recombinants that supply respectively DnaA proteins of

P.aeruginosa or *P.putida* (data not shown). It was tentatively concluded, that the faillure of pPS10 to transform *E.coli* was not due to low levels of RepA or to a poor dnaA box. Additional factors, as specific protein-protein and /or protein-DNA interactions, could be limiting.

Host range mutants of pPS10: isolation of pMM141

An attempt was made to test if variants of pPS10 able to transform *E.coli* could be isolated. For this purpose, the wild-type pPS10 miniplasmid pRG15 and the mutants pRG14 (dnaA box adjusted to the *E.coli* consensus) and pRG16 (dnaA box substituted by a 9 base sequence that includes a *ClaI* site), were mutagenized *in vitro* with hydroxylamine and used to transform C600EL, a restriction deficient strain of *E.coli*. Data in Table 1 show that host-range mutants of pRG14

			Transformants/µg	
Plasmid	dnaA box	NH_2OH	PAO1024	C600EL
pRG15	ATATCCACA	-	10^4	0
id	id	+	id	270
pRG14	TTATCCACA	-	id	0
id	id	+	id	270
pRG16	AATCGATGA	-	id	0
id	id	+	id	75*

Table 1. Induction by hydoxylamine of host-mutants of pPS10 in different mini-pPS10 plasmids. * indicates small size colonies after 24 hours of incubation.

and pRG15 can be easily isolated but that host-range mutants of pRG16 appeared less frequently and in addition they grew poorly in the selection plates. This last result indicates that specific *oriV* sequences are required for efficient replication of the host mutants in *E.coli* and it makes highly improbable that the establishment of the mutants in this host were due to activation of a new and cryptic replicon. One of these mutants derived from pRG14, pMM141, was selected for a more extensive characterization.

In vivo replication studies with pMM141

Data in Table 2 show that pMM141 transforms efficiently *P. aeruginosa* PAO1024 and *P.putida* KT2440, which indicates, that pMM141 is a broad-host range mutant of pPS10. Furthermore, it was found that

Strain	Relevant genotype	Transformants/µg	
		pRG14	pMM141
C600EL	res-	0	106
BT1000	polA1	0	106
PAO1024	res-	104	104
KT2440	res-	10	10

Table 2. Stablishment by transformation of pRG14 and pMM141 in different hosts. C600EL and BT1000 are *E.coli* K12 strains; PAO1024 a and KT2440 are strains of *P.aeruginosa* and *P.putida* respectively.

pRG21, a pBR322-*oriV* recombinant, can be established as an autonomous replicon in cells of BT1000 (*polA*1) carrying pMM141 (data not shown). These results indicate the broad-host range character of the mutant and strongly suggest that the mutation present in pMM141 affects the synthesis and/or activity of the replication protein RepA, the trans-acting replication factor of pPS10. The fact that pMM141 can replicate in a *polA1* strain, made also possible to map the broad-host-range mutation substituting fragments of the pBR322-pPS10 vector pCN51 by the equivalent fragments of pMM141 and looking for the ability of the pCN51-like recombinants to transform BT1000. The mutation was located, in this way, inside an *Eco*RI-*Sph*I fragment of pMM141 that includes the 5′ end of *repA* (see Fig. 1).

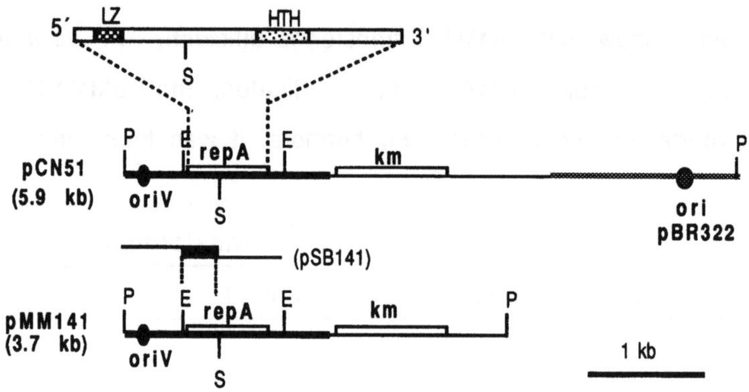

Fig.1. comparative maps of the pCN51 and pMM141 replicons used to map by reconstitution experiments the mutation of pMM141. *Pvu*II-*Sph*I and *Eco*RI-*Eco*R1 fragments used in the reconstitutions and the overlap that define the location of the mutation are indicated. Enlarged is shown the *repA* gene with indications of the relative positions of the leucine zipper(LZ) and helix-turn-helix (HTH) motifs. *Eco*RI fragment of pMM141 incorporated in pSB141 is indicated. basic replicon of pPS10, region containing the kanamycin resistance gene (km) , region derived from pBR322.

pSB141 is one of the pCN51-like recombinants carrying the broad-host range mutation of pMM141. Inversion or deletion of the *Eco*RI fragment of pSB141 that contains the mutated *repA* region, destroys its ability to transform BT1000 (data not shown).

The presence of a dnaA box in *oriV* indicates the involvement of the DnaA protein of the host in replication of pPS10. To evaluate this indication we compared the efficiency of transformation of pMM141 in a*dnaA*-null strain, WM1490 and in the *dnaA* $^+$ isogenic strain WM1502, using the efficiency of transformation of the DnaA independent replicon RFS1010 as control. It was found that pMM141 transforms the DnaA$^+$ strain at least 1000 fold more efficiently than the DnaA-null strain (data not shown), which indicates that the DnaA protein is required for replication of pMM141 in *E.coli* . The requirement of the DnaA protein for *in vivo* replication of pMM141 in *E.coli* is consistent with the inhibitory effect of antibodies against DnaA on the *in vitro* replication of pMM141 (next section).

Replication of pMM141 in cell-free extracts of *E.coli*

Data shown in Fig. 2 , lane 3, indicates that pMM141 can replicate in plasmid-free extracts of *E.coli* prepared from cells of C600EL containing pSB141. *In vitro* replication of pMM141 yields the supercoiled monomeric form of the plasmid (m ccc) and accumulates dimers (d) and partially replicated molecules (i) that migrate between monomers and dimers. The *in vitro* replication proccess does not require transcription by host RNA polymerase and can be inhibited by antibodies against DnaA, DnaB, DnaG, HU, SSB and DnaK proteins (Fig. 2 lanes 3-9) . Extracts prepared from cells containing a derivative of pPS10, pCN510, carrying a deletion of the *Eco*RI fragment in which is located the *repA* gene, do not support replication of pMM141 (Fig.2 lane 2).

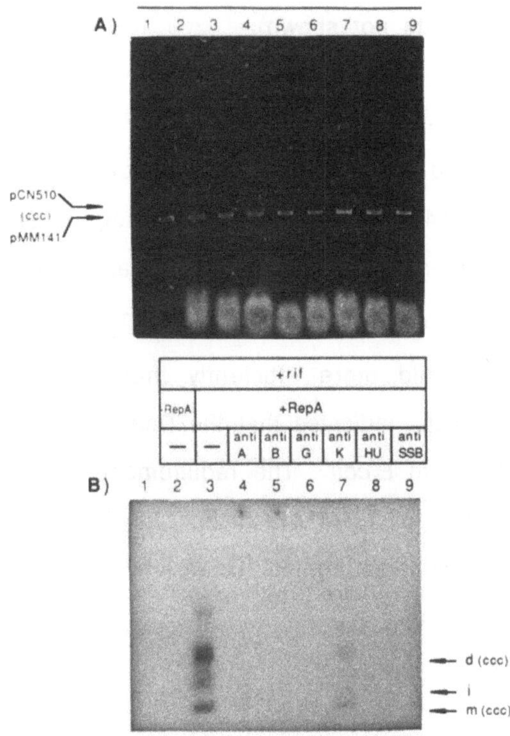

Fig.2. *In vitro* replication of plasmid pMM141. A. samples run in agarose gel and stained with ethidium bromide. Lane 1: sample of the pMM141 preparation used as DNA substrate in the *in vitro* assays. Lanes 2-9 : *in vitro* replication samples. B. Autoradiograms corresponding to the samples shown in A. rif: rifampicin; anti A,B...:indicate treatment with antibodies against DnaA, DnaB....; d: dimers; i: replicating intermediates; m: monomers.

DISCUSSION

The isolation and characterization of the broad host range mutant pMM141 presented here indicate that modifications in the activity of the RepA protein can extend the host range of the pPS10 replicon to *E.coli*.

Replication of pPS10 in *E.coli* depends on replication proteins of the host that act in the initial stages of DNA replication (Bramhill and Kornrberg, 1988). This indicates that broad-host range character of the mutant is not due to dispensability of the replication machinery of the host during the initial stages, as it is the case in RSF1010 (Scherzinger et al, 1984; Haring and Scherzinger, 1989). The situation seems to be similar to that of RK2 where it has been proposed that the differential interactions of two forms of the replication protein TrfA with sequences of the origin of replication and/or with components of the replication machinery of the host, are important determinants of the broad-host range character of the plasmid (Durland and Helinski, 1979; Pinkney et al, 1988; Thomas, personal communication). It is interesting to note that the mutation of pMM141 is located in a fragment that contains the 5′ end of *repA* that codes for the leucine zipper motif of RepA (Giraldo et al., 1989); this makes very attractive the possibility that modifications of particular residues of this motif could adjust interactions of RepA in *oriV* to those required for initiation of replication in the new host.

It is possible that mutations in other regions of the RepA protein or in the *oriV* sequences can also adjust replication of pPS10 to *E.coli*. It is also possible that modifications in particular components of the replication machinery of the host could favor the establishment of the wild-type replicon in *E.coli*.

It is interesting to underline the participation of DnaK in replication of pPS10, a protein that can modulate relevant interactions in nucleoprotein structures initiating DNA replication (Echols,1990).

The strategy reported here has been attempted without success to establish in *Pseudomonas* plasmids as ColE1 and R1 that have a host range restricted to *Enterobacteria.*

REFERENCES

Bagdasarian M, Timmis KN (1982) Host-vector systems for gene cloning in *Pseudomonas*. Curr top Microbiol Immunol 96:47-67

Bramhill D, Kornberg A (1988) A model for initiation at origins of DNA replication. Cell 54: 915-918

Diaz R, Staudenbauer WL (1982) Replication of the broad-host range plasmid RSF1010 in cell-free extracts of *Escherichia coli* and *Pseudomonas aeruginosa*. Nucl Acids Res 10:4687-4702

Durland RH, Helinski DR (1979) Replication of an origin-containing derivative of plasmid RK2 dependent on a plasmid function provided *in trans*. Proc Natn Acad Sci USA 76:1648-1652

Echols H (1990) Nucleoprotein structures in initiation of DNA replication, transcription and site-specific recombination. J Biol Chem 265:14697-14700

Giraldo R, Nieto C, Fernández-Tresguerres ME, Diaz R (1989) Bacterial zipper. Nature 342:866

Giraldo R (1991) Estudios genéticos y bioquímicos sobre la participación de proteinas iniciadoras en la replicación del DNA de los plasmidos de bacterias Gram-negativas R1 y pPS10. Doctoral Thesis, Universidad Complutense, Madrid

Haring V and Scherzinger E (1989) Replication proteins of the IncQ plasmid RSF1010. in "Promiscuous plasmids of Gram-negative bacteria". Thomas, CM (ed). Academic Press. London, pp.95-124

Humphreys GO, Willshaw GA, Smith HR, Andersson ES (1976) Mutagenesis of plasmid DNA with hydroxylamine, isolation of mutants of multi-copy plasmids. Mol Gen Genet 145:101-108

Lederberg EM, Cohen SN (1974) Transformation os *Salmonella typhimurium* by plasmid deoxyribonucleic acid. J Bacteriol 119:1072-1074

Messer W, Seufert W, Schaefer C, Gielow A, Harmann H, Wende M (1988) Functions of the DnaA protein of *E.coli* in replication and transcription. Biochem Biophys Acta 951:351-358

Nieto C (1988) Desarrollo de vectores de clonaje a partir del plásmido pPS10 de *Pseudomonas savastanoi* y caracterización de su

replicón básico. Doctoral Thesis. Universidad Autónoma, Madrid.
Nieto C, Fernandez-Tresguerres ME, Sanchez N, Vicente M, Diaz R (1990) Cloning vectors derived from a naturally occurring plasmid of *Pseudomonas savastanoi*, specifically tailored for the genetic manicuplations in *Pseudomonas*. Gene 87:145-149
Pinkney M, Diaz R, Lanka E, Thomas CM (1988) Replication of mini RK2 plasmid in extracts of *Escherichia coli* requires plasmid encoded protein TrfA and host-encoded proteins DnaA, B,G, DNA gyrase and DNA polymerase III. J Mol Biol 203:927-938
Scherzinger E, Bagdasarian MM, Scholz P, Lurz R, Rückert B, Bagdasarian M (1984) Replication of the broad host range plasmid RSF1010: Requiriments of three plasmid-encoded proteins. Proc Natn Acad Sci USA 81:654-658
Sambrook J, Fritsch EF, Maniatis T (1989) Molecular cloning, a laboratory manual. Cold Spring Harbor, New York
Thomas CM (ed) (1989) Promiscuous plasmids of Gram-negative bacteria. Academic Press, London
Wechsler JA, Nusslein V, Otto B, Klein A, Bonhoeffer F, Hezzman R, Gloger L, Schaller H (1973) Isolation and characterization of thermosensitive *Escherichia coli* mutants defective in deoxyribonucleic acid replication. J Bacteriol 113:1381-1388

ACKNOWLEDGMENTS

This work was supported by Grant BIO88-0294 of the Spanish CAICYT. Part of this work was done during a short visit of R.G. to the lab. of C Thomas (University of Birminghmam, UK), and of R.D. to the Dept. of H. Schuster (MPI, Berlin, FRG). These visits were founded by scientific interchange schemes between Spain-UK and Spain-FRG respectively. We greatefully acknowledge W. Messer, C. Thomas and E. Lanka for strains and plasmids. Antibodies used in the *in vitro* replication studies were kind gifts of E. Lanka, M. Kohiyama and V. Dyke. C.N. and R.G. were supported by fellowships from the Spanish MEC. We thank G. de Torrontegui for a chritical reading of the manuscript. The technical assistance of C. Pardo and A. Serrano is greatefully aknowledged.

Human Type I DNA Topoisomerase: Regulation of Expression and Function

R. Knippers, Ch. Lambrecht, H. Romig, A. Richter and N. Kunze
Division of Biology
University of Konstanz
D-7750 Konstanz, FRG

The replication of DNA requires two fundamental steps, namely an unwinding of the parental DNA double helix and the utilization of the emerging single strand sequences as templates for the synthesis of progeny DNA strands. The unwinding of the double helix is catalyzed by DNA helicases breaking the hydrogen bonds between complementary DNA strands at the expense of ATP hydrolysis (reviewed by Matson and Kaiser-Rogers, 1990). Inevitably, this reaction causes topological strain in circular or long linear DNA molecules. This strain is due to alterations in the helical twist and writhe of the double helix and must be relaxed by DNA topoisomerases as an essential condition for the progression of the replication fork.

Topoisomerases release torsional strain by a three-step-reaction cycle:(i) strand breakage and covalent linkage of the enzyme to ends of the broken DNA; (ii) relaxation of the torsional strain by strand rotation or strand passage at the site of the discontinuity in the DNA and (iii) the release of the enzyme coupled to the resealing of the DNA break.

Type I and type II DNA topoisomerases are distinguished by a number of important properties, including the fact that the type I enzymes cut one strand of the double helix whereas the type II topoisomerases cut both strands (reviewed by Vosberg, 1985; Wang, 1985).

Genetic analysis in yeast established that type I and type II topoisomerases can mutually substitute several of their replicative functions but type II topoisomerase has an essential role in the termination phase of the replication cycle including the segregation of the intertwined daughter molecules (Thrash et al, 1985; Brill et al., 1987; Uemura et al., 1987). This conclusion was supported by studies investigating the replication of circular SV40 DNA *in vitro* (Yang et al., 1987). However, it seems also to be generally accepted that topoisomerase I appears to provide the major relaxation function (swivel) during replicative chain elongation in yeast as well as in higher eukaryotes. This conclusion is mainly based on the effects of the specific inhibitory drug camptothecin which blocks the toposiomerase I reaction cycle at the step of covalent protein-DNA linkage (Liu,1989). This leads *in vivo* to a rapid inhibition of replicative chain elongation due to strand breakages in replicating chromosomes (Snapka, 1986; Avemann et al., 1988; Hsiang et al., 1989). Conversely, in the presence of topoisomerase II specific drugs like the podophyllotoxin derivative VM26, SV40 DNA replication proceeds to a terminal stage and leads to an accumulation of late replicative intermediates and catenated DNA molecules (Richter et al., 1987; Snapka et al., 1988; Fields-Berry and DePamphilis, 1989).

Thus, topoisomerase I appears to be a genuine replication elongation factor even though it performs numerous additional functions in other genetic processes such as transcription and, possibly, recombination (Vosberg,1985; Wang,1985).

In this communication, we present a brief overview on our current research concerning the regulation of the replication function of topoisomerase I. This topic has two aspects. First, we investigate how the necessary amount of topoisomerase I is guaranteed in rapidly proliferating cells. We have elucidated the organization of the human topoisomerase I gene and begun studying the regulation of its promoter. Second, we investigate how topoisomerase I gains access to replicating chromatin, and how it is directed to those sites on chromatin where its function is needed. For these latter experiments, we used SV40 minichromosomes as a model system. The advantage of this system is that SV40 minichromosomes possess many characteristics also found in cellular chromatin, yet are small enough to be amenable to biochemical manipulations.

Genetics of topoisomerase I

Topoisomerase I is needed in all genetically active cells. In cultured proliferating cells, the amounts of specific mRNA and of active enzyme essentially double over the cell cycle and provide the new cells with sufficient quantities of topoisomerase I for transcription and replication (Fig.1; Heck et al., 1988; Romig and Richter, 1990 a).

Similar results were obtained studying liver cells after partial hepatectomy (Duguet et al., 1983) and other cell types (summarized in Romig and Richter, 1990a). A much stronger activation of topoisomerase I gene transcription was detected after phorbol ester treatment (Hwong et al., 1989) and after infection with adenovirus (Romig and Richter, 1990b).

To better understand these regulatory phenomena and to elucidate the genetics of this physiologically important enzyme we have investigated the gene structure of the human topoisomerase I including its promoter elements.

We determined three loci in the human genome containing topoisomerase I sequences. Two loci, one on chromosome 1 and a second one on chromosome 22, contain processed truncated pseudogenes (Yang et al., 1990) whereas the third locus, hTOP1, on chromosome 20 carries the active gene which is composed of 21 exons spread out over more than 85 kb of genomic DNA (Kunze et al., 1991) (Fig.2).

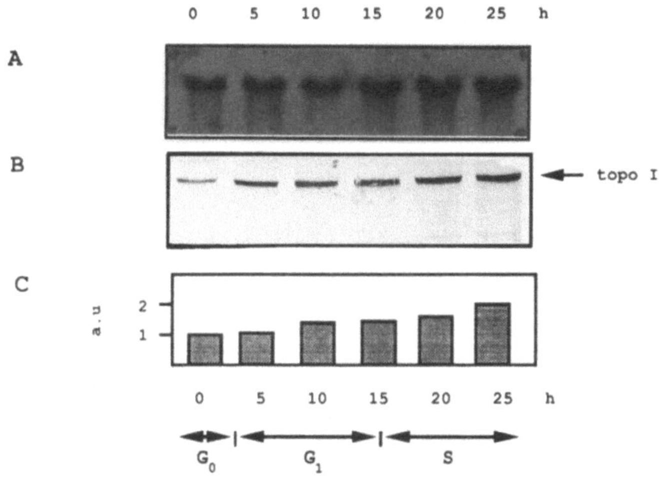

Figure 1: Expression of DNA topoisomerase I in human embryonic lung fibroblast (Hel) cells. (A) Expression of topoisomerase I mRNA. Total cellular RNA was prepared from cells growth arrested by serum starvation (0 h) and from cells 5, 10, 15, 20 and 25 hours after stimulation with serum. The RNA was separated on 0.8% agarose gels (Maniatis et al., 1982), transfered to nitrocellulose filters and hybridized with a ^{32}P-labeled topoisomerase I cDNA probe (Oddou et al., 1988). (B) Expression of topoisomerase I protein. Nuclear extracts were prepared from growth arrested cells and from cells 5, 10, 15, 20 and 25 hours after serum stimulation. 150 µg of protein were separated on 12.5% polyacrylamide gels, transfered to nitrocellulose filters and immuno-stained with topoisomerase I specific antibodies. (C) Expression of topoisomerase I activity. Nuclear extracts were prepared from growth arrested cells and from cells 5, 10, 15, 20 and 25 hours after serum stimulation. The relaxation activity of topoisomerase I was determined after serial dilutions of the extracts with supercoiled SV40 DNA as substrate and quantitated as described (Romig and Richter, 1990a). The total activity present in extracts of resting cells was set as 1, and the topoisomerase I activity present in extracts prepared 5, 10, 15, 20 and 25 hours after serum stimulation is given in arbitrary units. G_0-phase as well as the durations of G_1- and S-phase of the cell line used are given in below.

The hTOP1 upstream region is characterized by structural features typical for household gene promoters with multiple transcriptional start sites and GC-rich sequence elements (Kunze et al., 1990) some of which may be binding sites for the transcription factor Sp1. Interspersed within the promoter region are motifs known as transcription factor binding sites such as an cAMP response element and binding sites for the Oct1 transcription factor. These sequence elements may be required for gene activation in response to external stimuli (Fig.3). This possibility is currently under investigation.

In addition to transcriptional regulation, enzymatic activity may also be determined by other mechanisms. It is known, for example, that topoisomerase I is activated by phosphorylation of a single serine residue (Durban et al.,1985; Kaiserman et al., 1988) but the role of this reaction for the replicative function of topoisomerase I is unknown.

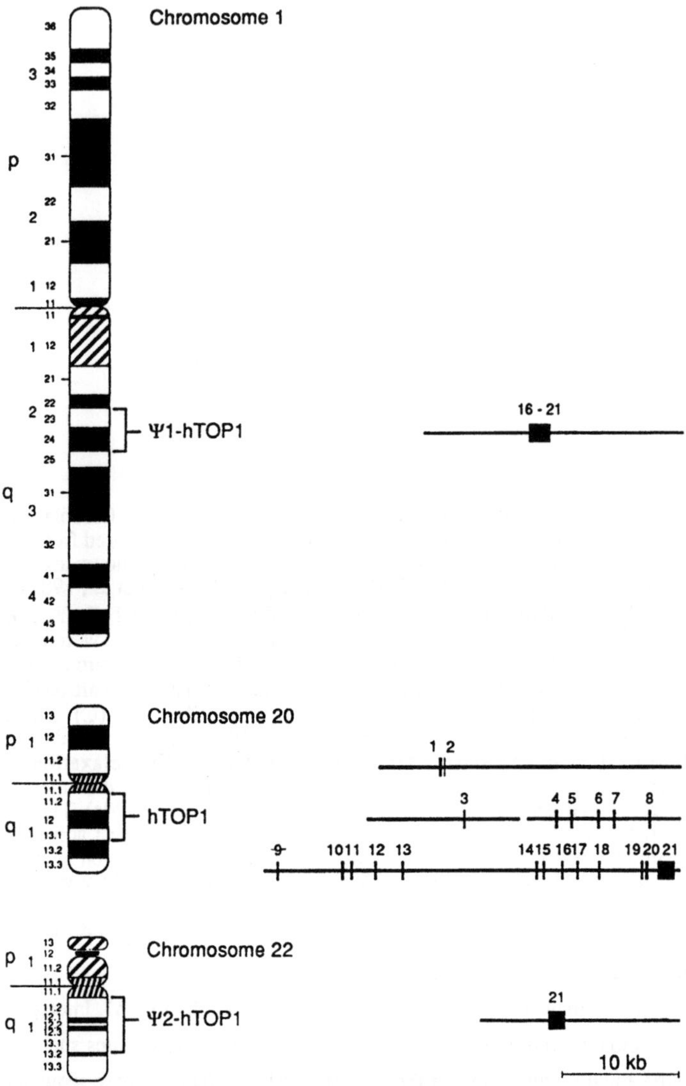

Figure 2: Chromosomal assignment and genomic structure of the human topoisomerase I gene and two pseudogenes. *Left*: Chromosomal localisation of the regions harbouring topoisomerase I sequences as revealed by *in situ* hybridisation to metaphase spreads of human chromosomes and Southern blot hybridization to DNA isolated from rodent-human hybrid cell lines (Kunze et al., 1989). *Right*: Genomic organisation of the loci shown on the left. The funtional topoisomerase I gene (hTOP1) consists of 21 exons (black bars) spread over at least 85 kb of genomic DNA (Kunze et al., 1991). The exact sizes of intron 2, 3 and 8 are unknown (indicated by interruption of the horizontal line). The human pseudogene 1 (Ψ1-hTOP1) consists of sequences homologous to exon 16 through exon 21 of the functional gene without intervening sequences. Pseudogene 2 (Ψ2-hTOP1) contains only sequences homologous to exon 21 of the functional gene (Yang et al., 1990).

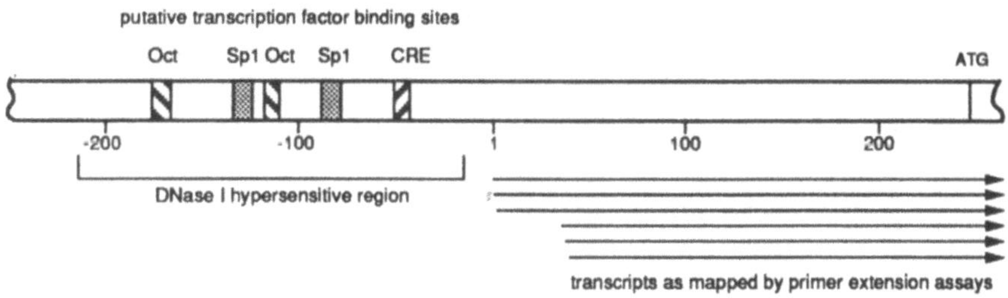

Figure 3: Schematic presentation of the promoter region of the human topoisomerase I gene. An about 500 bp long region upstream of the translation start site is shown. Transcription starts at multiple sites (indicated by horizontal arrows) and the most 5´transcription start site is designated as +1. The nucleotide sequence of the region preceding the transcriptional start site is GC-rich (67%), contains a high frequency of potential methylation sites and is hypersensitive to DNase I digestion. (Kunze et al., 1990). It contains several sequence motifs known as transcription factor binding sites. Oct: octamer protein binding site; Sp1: Sp1 binding site; CRE: cAMP responsive element. ATG: translation start point of topoisomerase I.

Topoisomerase I on replicating chromatin.

As described above, topoisomerase I expression is activated in rapidly proliferating cells. Thus, the enzyme is present in high amounts and effectively functions as a replication elongation factor. The question arises of how topoisomerase I gains access to replicating chromatin. We investigate this point using SV40 chromatin.

Replicating and non-replicating SV40 minichromosomes were differentially marked by (^{3}H)-thymidine pulse and (^{14}C)-thymidine long term labelling, respectively. To determine the distribution of topoisomerase I between these structures we used the drug camptothecin and accumulated reaction intermediates as covalent enzyme-DNA complexes. These complexes were first treated with sodium dodecylsulfate disrupting all noncovalent protein-DNA interactions and then precipitated via potassium dodecylsulfate micelles which trap covalent protein-DNA complexes but not free DNA (Fig.4).

Immunoprecipitations using specific antibodies showed that these protein-DNA complexes carried indeed topoisomerase I (Fig.5). Thus, camptothecin-induced covalent enzyme-DNA complexes are formed in much higher amounts with replicating compared to non-replicating chromatin suggesting that replicating chromatin carries much more topoisomerase I (Champoux, 1988).

Since nonreplicating minichromosomes are present in the cell in a vast excess over replicating intermediates the latter must be rather specific targets for topoisomerase I. With other words, the enzyme must recognize structural features which are specific for replicating chromatin or are induced by the replication process itself.

It has been shown by many earlier experiments that replicating chromatin is less densely packed than non-replicating chromatin (reviewed by van Holde,1988). Therefore, replicating chromatin may possess regions of protein-free DNA where topoisomerase I could bind and act. But this explanation is probably not sufficient. We could show that topoisomerase I recognizes and cuts with comparable efficiencies at the same sites in protein-free DNA as well as in

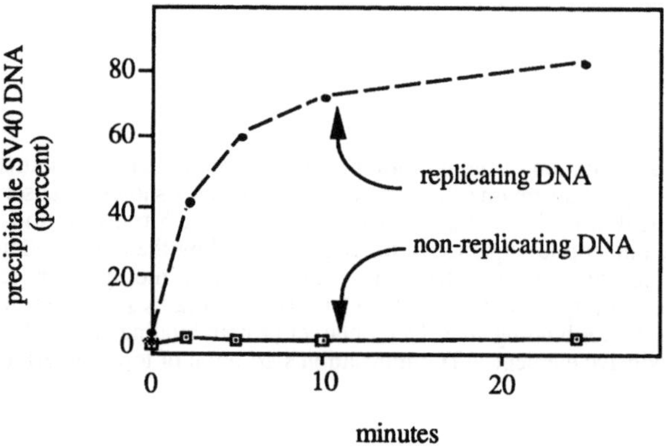

Figure 4: Formation of covalent protein-DNA complexes in SV40 infected cells. Non-replicating SV40 DNA was labeled with 0.1 µCi/ml ^{14}C-TdR (30Ci/mmole) in DME medium from 24 to 36 hours after infection. Replicating SV40 DNA was pulse labeled for 2 minutes with 100 µCi/ml ^{3}H-TdR (3000 Ci/mmole) 36 hours post infection at the time of maximal DNA synthesis. The pulse labeling reaction was either stopped immediatelly (0 min), or a chase reaction was performed in the presence of 100 mM camptothecin in DME for 2.5, 5, 10 and 25 minutes. The cells were denatured by the addition of 1% sodium dodecylsulfate in TE buffer and the total SV40 DNA was recovered in Hirt supernatants (Hirt, 1967). Remaining protein-DNA complexes present in these supernatants were precipitated on ice by the addition of potassium chloride to a final concentration of 250 mM. The precipitate formed was pelleted (2000 g, 2 min, room temperature) and washed three times in 250 µl of TE containing 250 mM potassium chloride. The final pellet was dissolved in 100 µl of 0.1 M HCl and the radiocativity was determined by liquid scintillation counting in 1 ml aquasol. In parallel samples the trichloroacetic acid precipitable total radioactivity was determined. The amount of potassium dodecylsulfate precipitable protein-DNA complexes is given as percentage of the total amount of SV40 DNA present in the sample. Open symbols: non-replicating SV40 DNA. Closed symbols: replicating SV40 DNA.

nucleosome-associated DNA (Fig.6). Thus, the action of topoisomerase I is not restricted to nucleosome-free DNA stretches in chromatin. At least in principle, an altered nucleosomal organisation or the presence of nucleosome-free regions immediatelly in front of the replication forks (Avemann et al., 1988) is no sufficient explanation for the presence of topoisomerase I on replicating DNA molecules.

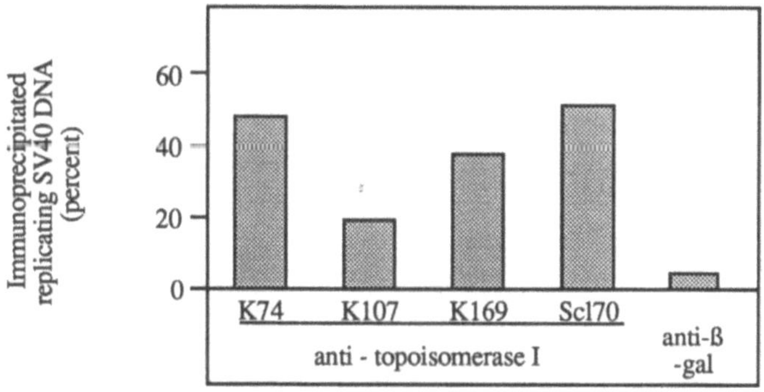

Figure 5: Immunoprecipitation of complexes of topoisomerase I covalently bound to replicating SV40 DNA. Replicating SV40 DNA was pulse labeled as described in the legend to Figure 4 and a chase reaction was performed for 25 minutes in the presence of camptothecin. The cells were then denatured by the addition of 1% sodium dodecylsulfate in TE buffer and the total SV40 DNA was recovered from Hirt supernatants. SV40 DNA as well as protein-SV40-DNA complexes were separated from free protein by sucrose gradient centrifugation (Richter et al.,1987). Immunoprecipitation was performed with antibodies against calf thymus topoisomerase I as described. The amount of immunoprecipitable topoisomerase-DNA complexes is given as percentage of the total amount of replicating SV40 DNA present in the sample. K74: affinity purified rabbit antibody against topoisomerase I; K107: rabbit serum against topoisomerase I; K169: rabbit serum against a topoisomerase I fragment expressed in E.coli from a cDNA sub-fragment (Oddou et al., 1988); Scl70: serum from a patient suffering from sytemic sclerosis recognizing topoisomerase I (Verheijen et al., 1990); anti-ß-gal: rabbit serum against E.coli ß-galactosidase.

Another possibility to explain the preferential interaction of topoisomerase I with replicating molecules is that the enzyme could be a component of the multienzyme replication complex which assembles soon after initiation of a replication cycle. However, earlier experiments do not support this possibility. It has been shown that camptothecin treatment of replicating viral and cellular DNA causes many double strand DNA breaks unless the progression of the replication fork is inhibited by aphidicolin which blocks replicative DNA polymerases (Avemann et al., 1988; Hsiang et al., 1989). A simple explanation is that double strand breaks occur when the replication forks run into single strand discontinuities generated by topoisomerase I acting at sites ahead of the fork (Fig.7).

These considerations suggest that topoisomerase I is probably not part of the multienzyme replication complex but is bound to DNA sites on the unreplicated part of the molecule. In fact, biochemical assays have shown that the enzyme interacts well with different topological forms of DNA but is strongly activated when superhelical tension is induced (e.g. Caserta et al., 1990). In replicating chromatin, an overwinding of the helix ahead of the advancing replication fork may play an important role in the activation process. This possibility was investigated in model studies. As shown in Fig.8, purified topoisomerase I, when added *in vitro* to isolated non-replicating SV40 minichromosomes, forms covalent enzyme-DNA complexes in the presence of camptothecin. However, 2 - 3 times more topoisomerase I-DNA complexes are

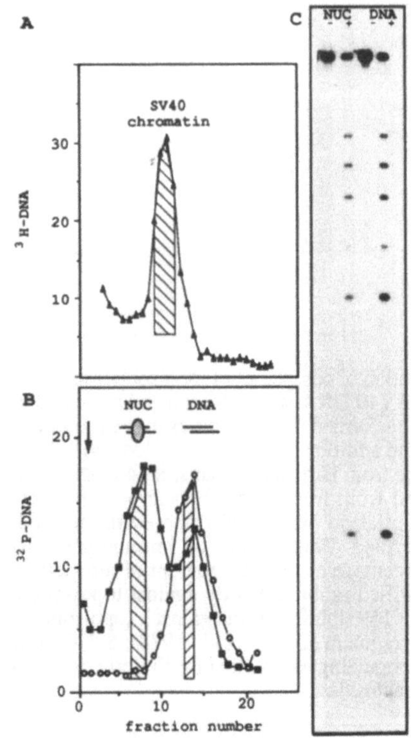

Figure 6: Topoisomerase I activity on DNA and DNA assembled into histone octamer complexes. (A) Purification of SV40 chromatin. SV40 minichromosomes labeled with 0.1 µCi/ml ^{3}H-TdR for 20 hours were extracted from nuclei of infected TC7 cells 48 hours after infection as described (Sundin and Varshavsky, 1981). Extracted minichromosomes were purified on 7.5% to 30% sucrose gradients (Richter et al., 1987) containing 450 mM sodium chloride. SV40 chromatin prepared in this way was essentially free of topoisomerase I. The SV40 chromatin was used as nucleosome donor in histone transfer experiments (see Fig.6B and C) and as substrate for topoisomerase I in the reaction described in the legend of Figure 8. (B) Purification of mononucleosomes. Mononucleosomes were assembled on a 154 bp long DNA fragment by the histone transfer method described by Drew and Travers (1985) with isolated SV40 minichromosomes as nucleosome donors. After the histone transfer reaction the samples were adjusted to 400 mM sodium chloride and 0.1% Nonidet P40 and the mononucleosomes (NUC) formed were separated from excess donor chromatin (arrow) and free DNA (DNA) by centrifugation. (C) Topoisomerase I cleavage reaction. The cleavage reactions were performed in a total of 25 µl of cleavage buffer (20 mM Tris HCl pH 7.5, 75 mM KCl, 5 mM $MgCl_2$, 1 mM DTT, 0.05 mg/ml BSA) for 15 min at 20°C in the presence of 20 µM camptothecin and 10 ng of the labeled 154 bp DNA fragment (DNA) or the same amount of DNA assembled into mononucleosomes (NUC) in the presence (+) or absence (-) of 20 ng of purified topoisomerase I. After incubation for 15 min at 20°C 200 µl of 0.5% sodium dodecylsulfate in TE was added to stop the reaction. The samples were treated with proteinase K and the DNA was purified and separated on 4% polyacrylamide sequencing gels.

formed after pretreating the minichromosomes with intercalating ethidium bromide. This drug is known to induce alterations of the twist angle and superhelicity. Comparabel structural changes are expected to occur in front of the progressing replication fork and may create the sites recognized by topoisomerase I.

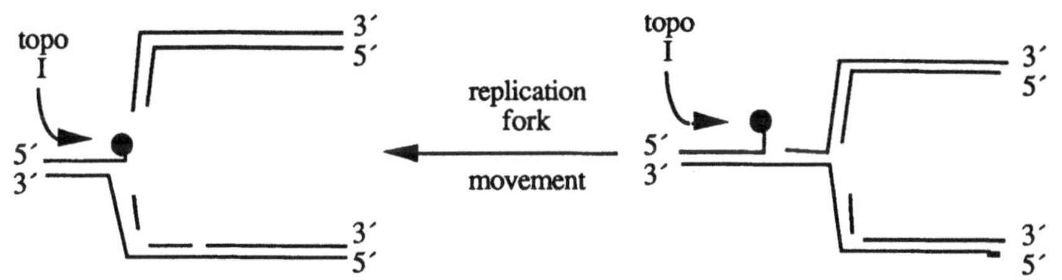

Figure 7: Topoisomerase I creates "double strand" breaks in replicating DNA molecules. Camptothecin stabilizes the cleavable complex formed between topoisomerase I and DNA. When such protein-DNA complexes were formed in front of an advancing replication fork (left side), the collision of fork and protein-DNA complex will result in breakage and arrest of the replication fork (right side).Arrow: direction of replication.

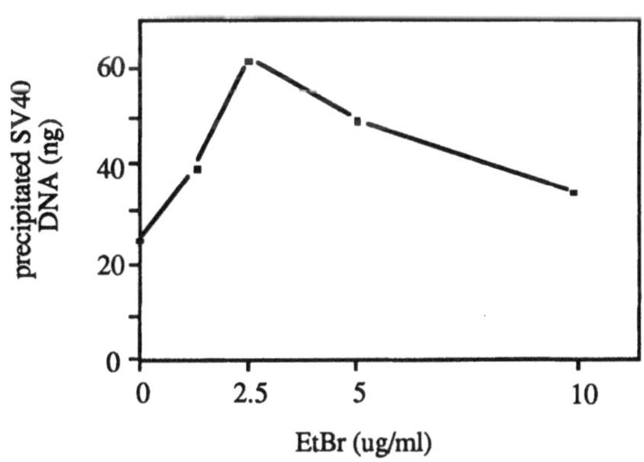

Figure 8: Ethidium bromide stimulates the formation of covalent complexes between topoisomerase I and SV40 chromatin *in vitro*. SV40 chromatin was labeled and purified as described in the legend to Figure 6A. The isolated SV40 chromatin was incubated with 20 ng of purified topoisomerase I for 15 min at 20°C in the presence of 20 μM camptothecin and increasing amounts of ethidium bromide (EtBr). The reaction was terminated by the addition of 200 μl of 0.5% sodium dodecylsulfate in TE and the amount of potassium dodecylsulfate precipitable SV40 DNA was determined as described in the legend to Figure 4.

It could be argued that ethidium bromide causes a dissociation of nucleosomes providing for stretches of free DNA to which topoisomerase I may preferentially bind. However, recent experiments have shown that histone octamers readily assemble on positively supercoiled DNA forming stable chromatin-like structures (Clark and Felsenfeld,1991).
In summary, we propose that topoisomerase I resides in replicating chromatin at sites ahead of the fork where it is activated by local topological changes induced by the advancing replication fork. But this is certainly not the only location of topoisomerase I in replicating chromatin because we found a covalent binding of topoisomerase I with DNA strands, pulse-labeled with (^{3}H)-thymidine for very short times (data not shown), indicating that the enzyme also occurs and acts behind the fork on newly synthesized DNA strands. This could simply mean that this part of the molecule is highly accesssible for the enzyme but a more interesting possiblity is that the "post-fork" topoisomerase I functions in the assembly of new nucleosomes on freshly replicated DNA. Indeed, many experiments have established that topoisomerase I is an indispensable cofactor for chromatin assembly reactions performed under *in vitro* conditions.

Conclusions

The available evidence suggests that topoisomerase I provides the swivel mechanism needed for the continued unwinding of the parental DNA strands during replicative chain elongation. The enzyme is probably located at several sites on the unreplicated part of replicative intermediates and acts when it encounters the positive superhelical tension generated by the advancing replication fork. It is, however, important to realize that chromatin rather than free DNA is the *in vivo* substrate for the cellular replication enzymes. Thus, the replicative unwinding of DNA occurs in the context of the constrained negative superhelicity of the DNA in chromatin complexes. And the precise topology of overwound DNA in front of a replication fork advancing through an array of nucleosomes has yet to be determined. Furthermore, chromatin assembly behind the fork requires that DNA becomes wrapped around histone cores in a negative direction leading to local and transient positive supercoiling. This superhelical tension may be directly relaxed by rotation of the replicating DNA strand. However, this could be impeded by the multienzyme replicative complex. In this case, the activity of topoisomerase I will be necessary to obtain the constrained negative superhelicity of DNA on nucleosomes. Clearly, the concert of topological changes around a replication fork is complex and has yet to be carefully analyzed.

Acknowledgement

This work was supported by Deutsche Forschungsgemeinschaft (SFB 156)

References

Avemann K, Knippers R, Koller T, Sogo JM (1988) Camptothecin, a specific inhibitor of type I DNA topoisomerase, induces DNA breakages at replication forks. Mol Cell Biol 8: 3026-3034

Brill SJ, DiNardo S, Voelkel-Meiman K, Sternglanz R (1987) Need for DNA topoisomerase activity as swivel for DNA replication and for transcription of ribosomal RNA. Nature 326: 414-416

Caserta M, Amadei A, Camilloni G, DI Mauro E (1990) Regulation of the function of eukaryotic topoisomerase I: analysis of the binding step and the catalytic constants of topoisomerization as a function of DNA topology. Biochemistry 29: 8152-8157

Champoux JJ (1988) Topoisomerase I is preferentially associated with isolated replicating simian virus 40 molecules after treatment of infected cells with camptothecin. J Virol 62: 3675-3683

Clark DJ, Felsenfeld,G (1991) Formation of nucleosomes on positively supercoiled DNA. EMBO J 10: 387-395

Drew HR, Travers AA (1985) DNA bending and its relation to nucleosome positioning. J Mol Biol 186: 773-790

Duguet M, Lavenot C, Harper F, Mirambeau G, De Recondo AM (1983) DNA topoisomerases from rat liver: physiological variations. Nucleic Acids Res 11: 1059-1075

Durban E, Goodenough M, Mills J, Busch H (1985) Topoisomerase I phosphorylation in vitro and in rapidly growing Novikoff hepatoma cells EMBO J 4: 2921-2926

Fields-Berry, DePamphilis ML (1989) Sequences that promote formation of catenated intertwines during termination of DNA replication. Nucl Acids Res 17: 3261-3273

Heck MMS, Hittelman WN, Earnshaw WC (1988) Differential expression of DNA topoisomerases I and II during the eukaryotic cell cycle. Proc Natl Acad Sci USA 85: 1086-1090

Hirt B (1967) Isolation of viral genomes from eukaryotic cells. J Mol Biol 26: 365-369

Hsiang YH, Lihou MG, Liu LF (1989) Arrest of replication forks by drug-stabilized topoisomerase I-DNA cleavable complexes as a mechanism of cell killing by camptothecin. Cancer Res 49: 5077-5082

Hwong C-L, Chen M-S, Hwang J (1989) Phorbol ester transiently increases topoisomerase I mRNA levels in human skin fibroblasts. J Biol Chem 264: 14923-14926

Kaiserman HD, Ingebritsen TS, Benbow RM (1988) Regulation of Xenopus laevis topoisomerase I activity by phosphorylation *in vitro*. Biochemistry 27: 3216-3222

Kunze N, Yang GC, Jiang ZY, Hameister H, Adolph S, Wiedorn KH, Richter A, Knippers R (1989) Localization of the active type I DNA topoisomerase gene to human chromosome 20q11.2-13.1, and two pseudogenes on chromosomes 1q23-24 and 22q11.2-13.1. Hum Genet 84: 6-10

Kunze N, Klein M, Richter A , Knippers R (1990) Structural characterization of the human DNA topoisomerase I gene promoter. Eur J Biochem 194: 323-330

Kunze N, Yang G, Dölberg M, Sundarp R, Knippers R, Richter A (1991) Structure of the human type I DNA topoisomerase gene. J Biol Chem in press

Liu LF (1989) DNA topoisomerase poisons as antitumor drugs. Ann Rev Biochem 58: 351-375

Maniatis T, Fritsch EF, Sambrook J (eds) (1982) Molecular Cloning. Cold Spring Harbor Press New York

Matson SW, Kaiser-Rogers KA (1990) DNA helicases. Ann Rev Biochem 59: 289-329

Oddou P, Schmidt U, Knippers R, Richter A. (1988) Monoclonal antibodies neutralizing mammalian DNA topoisomerase I activity. Eur.J.Biochem. 177: 523-529

Richter A, Strausfeld U, Knippers R (1987) Effects of VM26 (teniposide), a specific inhibitor of type II DNA topoisomerase, on SV40 DNA replication in vivo. Nucleic Acids Res 15: 3455-3468

Romig H, Richter A (1990a) Expression of the topoisomerase I gene in serum stimulated human fibroblasts . Biochim Biophys Acta 1048: 274-280

Romig H, Richter A (1990b) Expression of the type I DNA topoisomerase gene in adenovirus-5 infected human cells. Nucleic Acids Res 18: 801-808

Snapka RM (1986) Topoisomerase inhibitors can selectively interfere with different stages of simian virus 40 DNA replication. Mol Cell Biol 6: 4221-4227

Snapka RM, Powelson MA, Strayer JM (1988) Swivelling and decatenation of replicating Simian virus 40 genomes *in vivo*. Mol Cell Biol 8: 515-521

Sundin O, Varshavsky A (1981) Arrest of segregation leads to accumulation of highly intertwined catenated dimers. Cell 25: 659-669

Thrash C, Bankier AT, Barrel BG, Sternglanz R (1985) Cloning, characterization, and sequence of the yeast DNA topoisomerase I gene. Proc Natl Acad Sc USA 82: 4374-4378

Uemura T, Morino K, Uzawa S, Shiozaki K, Yanagida M (1987) Cloning and sequencing of Schizosaccaromyces pombe DNA topoisomerase I gene, and effect of gene disruption. Nucleic Acids Res 15: 9727-9739

Van Holde KE (1988) Chromatin. Springer Verlag Berlin Heidelberg New York

Verheijen R, van den Hoogen F, Beijer R, Richter A, Penner E, Habets WJ, van Venrooij WJ (1990) A recombinant topoisomerase I used for autoantibody detection in sera from patients with systemic sclerosis. Clin.exp.Immunol 80: 38-43

Vosberg HP (1985) DNA topoisomerases: enzymes that control DNA conformation. Cur Topics Microbiol Immunol 114: 19-102

Wang JC (1985) DNA topoisomerases. Ann Rev Biochem 54: 665-697

Yang G, Kunze N, Baumgärtner B, Jiang Z, Sapp M, Knippers R, Richter A (1990) Molecular structures of two human DNA topisomerase I retrosequences. Gene 91: 247-253

Yang L, Wold MS, Li JJ, Kelly TJ, Liu LF (1987) Roles of DNA topoisomerases in simian virus 40 DNA replication *in vitro*. Proc Natl Acad Sc USA 84: 950-954

Holoenzyme Forms of Yeast DNA Polymerases δ and ε

Peter M.J. Burgers
Department of Biochemistry and Molecular Biophysics
Washington University School of Medicine
660 S. Euclid Avenue, Box 8231
St. Louis, MO 63110

Introduction

A detailed understanding of the mechanism of chromosome replication in the eukaryotic cell and of the fidelity of this process requires a comprehensive biochemical and genetic characterization of the DNA polymerases and their accessory factors. Genetic manipulation in the yeast *Saccharomyces cerevisiae* has shown that at least three nuclear DNA polymerases are essential for cell viability. These are the products of the *POL1* (also designated *CDC17*), *POL2*, and *POL3* (*CDC2*) genes (Johnson. et al, 1985, Morrison et al., 1990, Sitney et al., 1989, Boulet et al., 1989). Extensive structural and functional similarity between yeast and mammalian cells for many components of the replication apparatus including the replicative DNA polymerases has prompted us to propose a unified nomenclature for the eukaryotic DNA polymerases (Burgers et al., 1990). The catalytic subunit of DNA polymerase α, previously called DNA polymerase I in yeast, is encoded by the yeast *POL1* gene (Johnson et al., 1985). Polα is a 4-subunit enzyme and contains an associated DNA primase activity. One still unsolved problem in the biochemistry of Polα is whether this enzyme possesses a proofreading exonuclease activity (reviewed in Burgers, 1989). Whereas the 4-subunit yeast enzyme does not have a detectable 3'-5'-exonuclease activity, the purified monomeric polymerase polypeptide exhibited such an exonuclease activity in one, but not in another study (Kunkel et al., 1989, Brooke et al., 1991). Polδ, previously designated PolIII in yeast, is the product of the yeast *POL3* gene (Sitney et al., 1989, Boulet et al., 1989). The 2-subunit enzyme has a 3'-5'-exonuclease activity capable of proofreading *in vitro* (Bauer, G.A. et al., 1988). The catalytic subunit of Polε, PolII, is the product of the yeast *POL2* gene. This 4-subunit enzyme also has a proofreading exonuclease activity (Hamatake et al., 1990).

The biochemical function of Polδ follows mainly from *in vitro* SV40 DNA replication studies (Challberg and Kelly, 1989, Stillman, 1989, for reviews). The proliferating cell nuclear antigen (PCNA), which is a processivity factor for Polδ, is required for *in vitro* SV40 DNA replication. In its absence, early replicative intermediates accumulate and leading strand DNA synthesis is abolished (Prelich and Stillman, 1988). From these initial studies it was concluded that Polδ was required for leading strand DNA synthesis. Later Polδ was actually shown to be involved in SV40 DNA replication (Lee et al., 1989, Weinberg and Kelly, 1989). For Polε, however, no biochemical function during SV40 DNA replication has yet been established. In this paper we show that both Polδ and Polε form holoenzyme complexes at the primer terminus of primed single-stranded DNA templates in a manner compatible with a function for these enzymes at the leading and lagging strand, respectively. Formation of these holoenzyme complexes is dependent upon the presence of the polymerase, PCNA, ATP, and an additional complex of accessory proteins: Replication Factor C (RF-C).

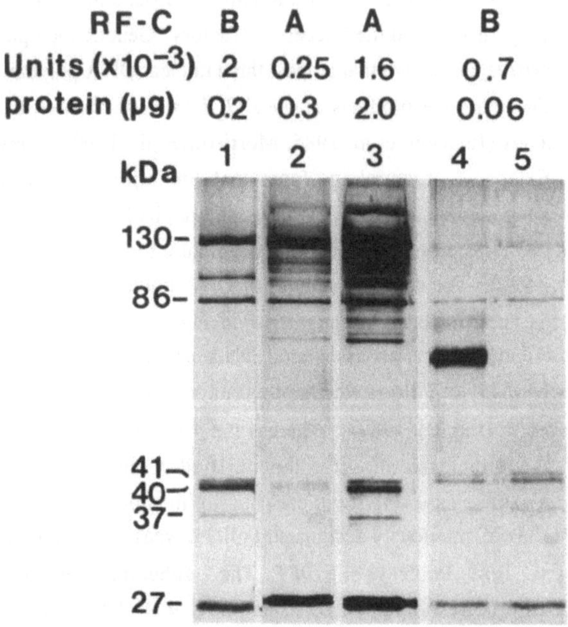

Figure 1. SDS-PAGE Analysis of RF-C

Superose 6 peak activity fractions of RF-C, form A or B as indicated, was analyzed by 10% SDS-PAGE. Form B of RF-C (60 ng) was incubated with (lane 4) or without (lane 5) 500 ng of calf intestinal alkaline phosphatase. The phosphatase migrates as a 65 kDa protein in this gel system. Protein was visualized by silver staining.

Results and Discussion

Purification and Subunit Characterization of Yeast Replication Factor C. PCNA interacts in a protein-protein fashion with Polδ, thereby increasing the processivity of the polymerase on poly(dA).oligo(dT) (Tan et al., 1986, Bauer and Burgers, 1988). PCNA, however, failed to significantly increase the processivity of Polδ on primed natural single-stranded (SS) DNA and as a result the synthetic activity of Polδ on such template-primers was minimal. Coating of the SS DNA with a single-stranded DNA binding protein, either from *E. coli* (SSB) or from yeast (ySSB, or yRF-A) had little effect, whether PCNA was present or not. This lack of activity provided an assay for an additional yeast replication factor. Because of the considerable structural and functional homology between yeast and mammalian cells for replication proteins we surmised that this factor would presumably be analogous to mammalian Replication Factor C (RF-C), also called Activator 1. Mammalian RF-C forms a complex with PCNA at the primer terminus using the energy of ATP hydrolysis and this complex promotes efficient DNA synthesis by Polδ (Tsurimoto and Stillman, 1990, Lee et al.,1991). Thus, our assay for the purification of yRF-C was based upon its requirement in the PCNA and ATP dependent replication by Polδ of SSB coated SS mp18 DNA primed with a single oligonucleotide primer.

RF-C was purified about 25,000 fold by chromatography on Affigel Blue, heparin agarose, SS DNA cellulose, MonoS, MonoQ, and, finally, gel filtration on Superose 6. During SS DNA cellulose chromatography two forms of RF-C were separated, which had identical hydrodynamic and enzymatic properties. They were designated RF-C, form A, and RF-C, form B. Both forms showed identical subunits at 130, 86, 41, 40, 37, and 27 kDa (Figure 1). None of these subunits were related through phosphorylation (Figure 1, lanes 4,5) or proteolysis. The latter was shown by N-chlorosuccinimide fingerprinting (data not shown). Form A had about 1/10 of the specific activity of form B, which was probably due to the substoichiometric presence of the 41, 40, and 37 kDa subunit in this form (Figure 1, lanes 1-3).

Interaction of RF-C with ATP and with PCNA. RF-C has an ATPase activity which is stimulated about 15-20 fold in the presence of single-stranded mp18 phage DNA (Table 1). Priming of this SS phage DNA with either 1 or about 5 primers caused little or no stimulation of the ATPase activity (Table 1). These results do not indicate that RF-C alone is a primer terminus dependent ATPase. In the presence of PCNA, however, entirely different results were obtained for the unprimed vs. the primed DNA templates. Saturating levels of PCNA caused minimal stimulation of the RF-C mediated ATPase on unprimed DNA, but caused a large increase in this ATPase when the phage DNA contained one single primer, and an even larger increase when about 5 primers were present on the DNA (Table 1). These results strongly indicate that in the presence of PCNA, RF-C becomes a primer terminus dependent ATPase.

Table 1. ATPase Activity of RF-C

DNA	Primers/circle	ATPase (pmol/min) - PCNA	ATPase (pmol/min) + PCNA	ATPase ratio
None	-	0.4	-	
SS mp18	0	6.6	7.7	1.2
SS mp18	1	6.9	11.8	1.7
SS mp18	5	7.3	22.9	3.1

ATPase assays (30 ml) were carried out in 40 mM Tris-HCl, 8 mM $MgCl_2$, 0.1 mg/ml BSA, 1 mM DTT, 50 mM NaCl, 50 mM [a-^{32}P] ATP, 0.1 mg of SS mp18 DNA, either unprimed or primed with a single oligonucleotide primer, or with about 5 primers generated by DNase I digestion of DS mp18 DNA, 20 ng of RF-C, with or without 100 ng of PCNA. Aliquots were analyzed by PEI cellulose chromatography to quantitate % of ADP formed.

Formation of a Holoenzyme Form of Polδ. ATP is required to form a Polδ holoenzyme with PCNA and RF-C. dATP but no other dNTP or rNTP substituted for ATP in the activation process. ATPγS did not activate. Rather, it potently inhibited the activation by ATP. Likewise, AMP-PNP and dAMP-PNP , which have a β–γ-imido diphosphate linkage, also did not activate, indicating that ATP hydrolysis is required for activation. dAMP-PNP, however, is a good analog of dATP for incorporation into DNA by Polδ and Polε. Replication of SSB coated singly primed SS mp18 DNA by Polδ with PCNA and RF-C was not dependent on ATP because of the presence of dATP in the dNTP mix. However, upon substitution of dATP by dAMP-PNP (the mix of dTTP, dCTP, dGTP, and dAMP-PNP is called dNTP-PNP's) replication became absolutely dependent on ATP or dATP.

The different stages in the formation of a stable holoenzyme form of Polδ were determined by the isolation of DNA-bound complexes by gel filtration on Biogel A-5m. DNA and DNA-bound proteins elute in the void volume fractions of the column whereas unbound proteins and nucleotides elute much later in the included fractions. When SSB coated singly primed SS mp18 DNA was incubated with PCNA, RF-C, and Polδ, either each separately or all three factors together, but without ATP added, only RF-C eluted in the void volume of the column. This was indicated by the observation that Polδ and PCNA, but not RF-C had to be added back to the void volume fractions of the column in order to replicate the phage DNA in these fractions. Upon preincubation of the DNA with RF-C and PCNA in the presence of ATP, however, a stable DNA.PCNA.RF-C complex could be isolated by gel filtration. This complex contained both RF-C and PCNA, as only Polδ needed to be added to the void volume fractions for efficient replication of the phage DNA (Figure 2, lanes 1-3). In this second step Polδ bound to the isolated DNA.PCNA.RF-C complex in the absence of ATP to form a stable

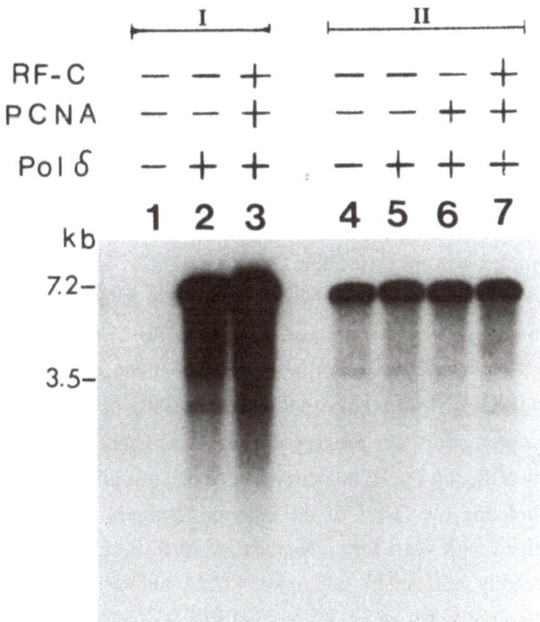

Figure 2. Binding of Polδ to a PCNA.RF-C Complex

Complexes were formed in gel filtration buffer [30 mM Tris-HCl 7.5, 5% glycerol, 8 mM $MgCl_2$, 0.1 mM EDTA, 2 mM DTT, 0.1 mg/ml BSA, 40 mM NaCl] in a 80 µl incubation containing singly primed SS mp18 DNA (2 mg) coated with SSB (17 mg), 0.1 mM ATP, 400 ng of RF-C, and 200 ng of PCNA at room temperature for 30 sec. The mixture was cooled on ice and filtered through a 1.1 ml Biogel A-5m column. Void volume fractions were collected and small aliquots incubated with [a-^{32}P] dNTP's with or without addition of factors for 4 min at 37°C (lanes 1-3). The rest of the void volume fractions was incubated with 50 ng of Polδ in the gel filtration buffer without any other additions and filtered again as described above. Aliquots of the void volume fractions from this second column were then incubated in replication assays as described above. Products were analyzed by 1% alkaline agarose gel electrophoresis.

holoenzyme complex at the primer terminus in quantitative yields. This was shown by carrying out a second gel filtration step (Figure 2, lanes 4-7). The DNA isolated in the void volume fractions of the second gel filtration column replicated efficiently. In addition, no increase in replication was observed when all the replication factors were added back, indicating that all primer termini were occupied with holoenzyme complexes.

Formation of a Holoenzyme Form of Polε. In contrast to Polδ, Polε is extremely processive on the model template-primer system poly(dA).oligo(dT). Still, a slight increase in processivity was detected upon addition of PCNA, indicating that an interaction between the two should exist (Hamatake et al., 1990). However, gel filtration experiments as described above showed that only small levels of DNA-bound Polε eluted in the void volume fractions. In addition, the presence of these small levels of Polε in the void volume fractions was independent of PCNA, RF-C, and ATP. These experiments indicate that if Polε formed a holoenzyme with ATP, PCNA, and RF-C, this complex was not very stable.

The efficiency of DNA synthesis by Polε on SSB coated singly primed SS mp18 DNA was very high and basically unaffected by the addition of PCNA, RF-C, and ATP. Reasoning, however, that the high efficiency of DNA synthesis by Polε alone could mask a possible interaction with PCNA and RF-C, we sought to inhibit the activity of Polε alone in order to measure such a possible interaction. In prokaryotic DNA replication studies it has been established that the activity of *E. coli* Pol III holoenzyme is much less affected by moderate salt levels than that of non-holoenzyme forms of the enzyme (Burgers and Kornberg, 1982). Similar results were obtained with yeast Polε. Replication of SSB coated singly primed SS mp18 DNA by Polε was readily inhibited by salt, unless PCNA and RF-C as well as ATP were included into the assay (Figure 3). Together with the gel filtration results, these experiments show that Polε forms a holoenzyme with PCNA and RF-C, albeit it less stable than Polδ holoenzyme.

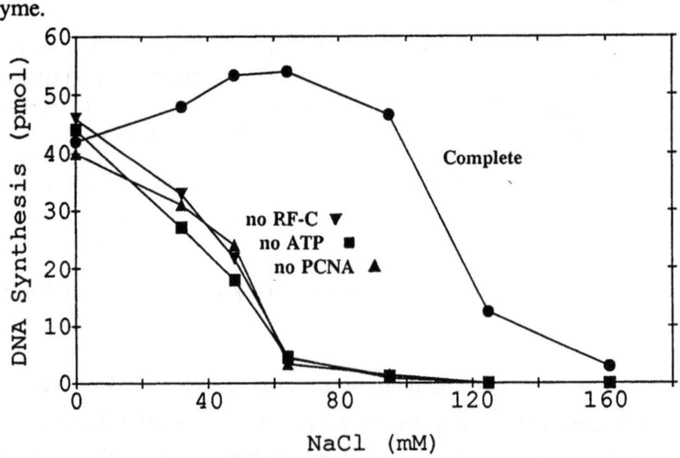

Figure 3. A Holoenzyme Form of Polε Revealed by Salt

Replication of singly primed SS mp18 DNA for 10 min at 37°C was carried out as described in Table 1, except that [^{3}H] dNTP-PNP's were added, NaCl was omitted from the initial reaction and added as indicated, 0.5 mM ATP was added as indicated, and each assay contained 90 ng of Polε.

<u>Activities of Polδ and Polε Holoenzyme</u>. A fiber autoradiographic analysis of replicating DNA isolated from yeast cells growing at 30°C has shown that the rate of progression of the replication fork is about 60 nt/sec (Rivin and Fangman, 1980). The rate of DNA replication by the *in vitro* reconstituted holoenzyme forms of Polδ and Polε was also determined at 30°C. The rate of DNA synthesis by Polδ holoenzyme was 110 nt/sec, regardless whether the DNA was coated with *E. coli* or yeast SSB (Figure 4).

The elongation by Polε holoenzyme was puntuated by dissociation and reassociation events, which occurred every few hundred nucleotides and showed up as strong replicational pause sites (Figure 5). The mean rate of DNA replication by Polε holoenzyme was about 20 nt/sec, but the actual rate during processive DNA synthesis was about 50 nt/sec. Thus, both holoenzyme forms have synthetic rates commensurate with the *in vivo* measured rate of fork movement.

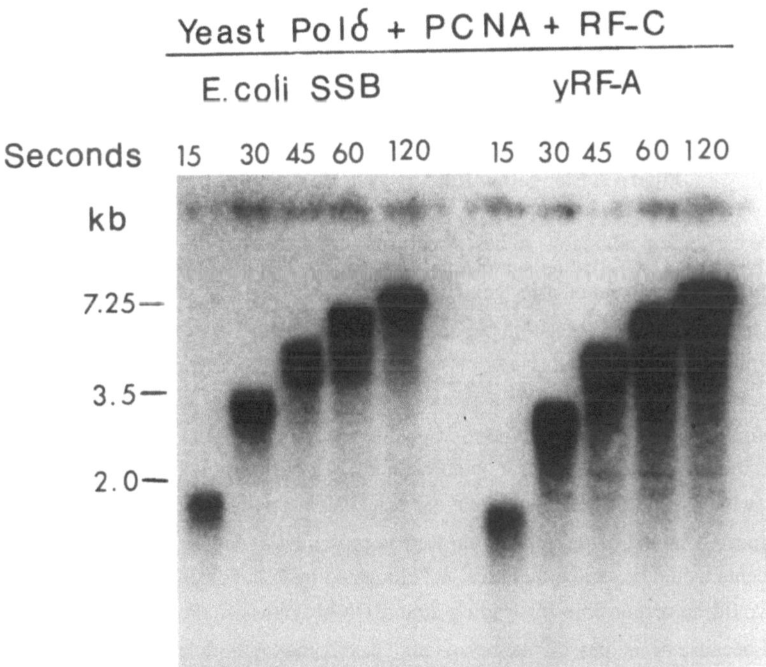

Figure 4. Elongation Rate by Polδ Holoenzyme

Mixtures were assembled as described in Figure 2, except that Polδ was included in the initial preincubation at 30°C. Reactions were started by addition of [^{32}P] dNTP's. Aliquots were analyzed by electrophoresis on a 1% alkaline agarose gel.

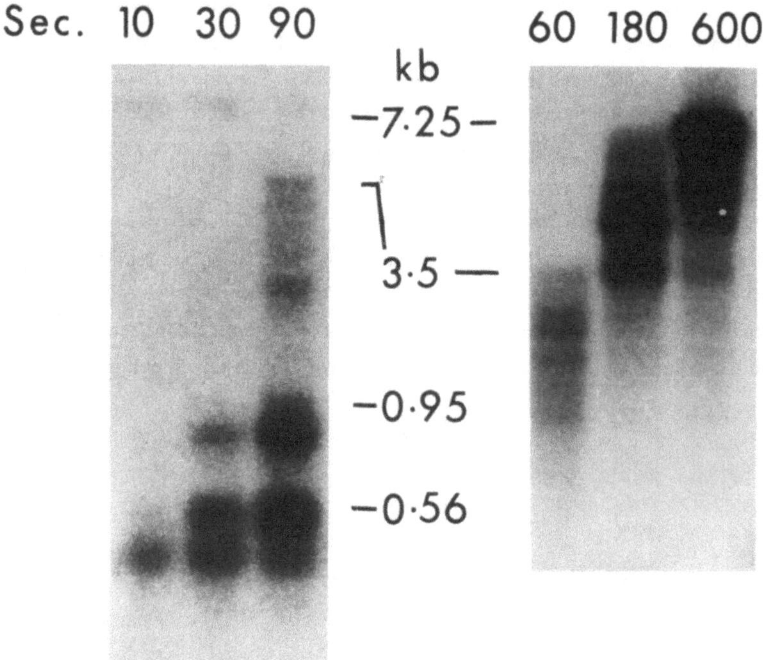

Figure 5. Elongation Rate by Polε Holoenzyme

Reactions were carried out analogous to described in Figure 4. Early time-points (10, 30, 90 sec) were analyzed on a 2% alkaline agarose gel (left panel), and later time-points (1, 3, 10 min) from a separate experiment on a 1% gel (right panel).

Model of the Eukaryotic Replication Fork. What are the in vivo functions of Polδ and Polε? Genetic analysis has shown that both are essential for yeast cell growth, suggesting that both are required for nuclear DNA replication. Considering that high fidelity chromosome replication is essential for the cell and that both Polδ and Polε, but possibly not Polα carry out proofreading, it is possible that Polδ and Polε are responsible for the bulk of chromosomal DNA replication. A model of the replication fork is presented in Figure 6. In this model Okazaki fragments would be started by Polα , but elongated by Polε holoenzyme, while Polδ holoenzyme would be responsible for leading strand DNA synthesis. Polε is placed at the lagging strand because of its dissociative behaviour. In this model, both lagging and leading strand DNA replication are dependent upon PCNA and RF-C for elongation. In addition coordination between the two strands could for instance be mediated by PCNA, which is known to exist as a dimer (Tan et al., 1986, Bauer and Burgers, 1988) . Support for this or a similar type of model comes from the recent observations by Nethanel and Kaufman (1990) that, during SV40 DNA replication *in vitro* in isolated nuclei, extremely small lagging strand fragments (<35 nt) were synthesized in a BuPhdGTP sensitive fashion, whereas these small fragments were chased into Okazaki size fragments (~250 nt) in a BuPhdGTP resistant fashion. BuPhdGTP is a potent inhibitor of Polα, but not of Polδ or Polε.

This model is testable in yeast because it makes the following predictions: 1) protein-protein interactions between Polα and Polε likely exist; 2) mutations in the exonuclease domains of Polδ and Polε should have similar phenotypic effects. Gap filling during Okazaki fragment maturation and telomer replication, both essential functions, could also be specialized operations to which yeast may have dedicated one specific DNA polymerase to the exclusion of others. A combination of genetic and biochemical approaches may be successful in assigning precise tasks to these three DNA polymerases.

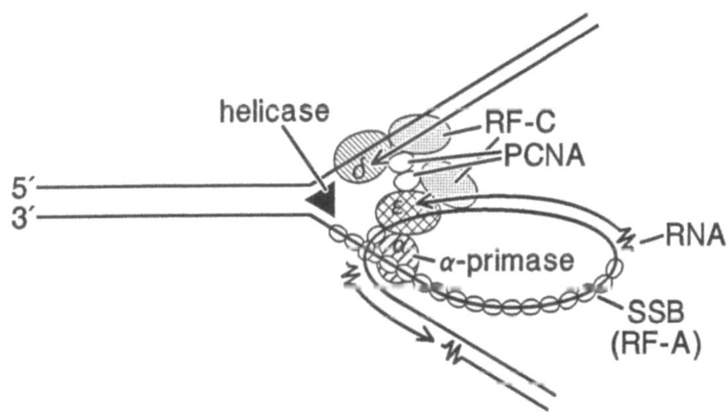

Figure 6. Model of the Eukaryotic Replication Fork

Acknowledgements

I thank Bonnie Yoder for excellent technical assistance. This work was supported in part by grant GM32431 from the National Institute of Health during a tenure as an Established Investigator of the American Heart Association (860179).

References

Bauer, G.A., and Burgers, P.M.J. 1988, The yeast analog of mammalian cyclin/proliferating-cell nuclear antigen interacts with mammalian DNA polymerase δ. Proc. Natl. Acad. Sci. USA **85**:7506-7510.

Bauer, G.A., Heller, H., and Burgers, P.M.J. 1988, DNA polymerase III from *Saccharomyces cerevisiae*. I. Purification and characterization. J. Biol. Chem. **263**:917-924.

Boulet, A., et al.,1989, Structure and function of the *Saccharomyces cerevisiae CDC2* gene encoding the large subunit of DNA polymerase III. EMBO J. **8**:1849-1854.

Brooke, R.G., et al., 1991, Purification and characterization of the 180- and 86- kilodalton subunits of the *Saccharomyces cerevisiae* DNA primase-DNA polymerase protein complex. J. Biol. Chem. **266**:3005-3015.

Burgers, P.M.J., 1989, Eukaryotic DNA polymerases α and δ: conserved properties and interactions, from yeast to mammalian cells. Progr. Nucleic Acids Res Mol. Biol. **37**:235-280.

Burgers, P.M.J., and Kornberg, A.,1982, ATP activation of DNA polymerase III holoenzyme of *Escherichia coli*. J. Biol. Chem. **257**:11474-11478.

Burgers, P.M.J., et al., 1990, Revised nomenclature for eukaryotic DNA polymerases. Eur. J. Biochem. **191**:617-618.

Challberg, M.D., and Kelly, T.J., 1989, Animal virus DNA replication. Annu. Rev. Biochem. **58**:671-717.

Hamatake, R.K., et al., 1990, Purification and characterization of DNA polymerase II from the yeast *Saccharomyces cerevisiae*: identification of the catalytic core and a possible holoenzyme form of the enzyme. J. Biol. Chem. **265**:4072-4083.

Johnson, L.M., et al., 1985, Isolation of the gene encoding yeast DNA polymerase I. Cell **43**:369-377.

Kunkel, T.A., et al., 1989, Fidelity of DNA polymerase I and the DNA polymerase I-DNA primase complex from *Saccharomyces cerevisiae*. Mol. Cell. Biol. **9**:4447-4458.

Lee, S-H., Eki, T., and Hurwitz, J., 1989, Synthesis of DNA containing the simian virus 40 origin of replication by the combined action of DNA polymerases α and δ . Proc. Natl. Acad. Sci. USA **86**:7361-7365.

Lee, S-H., Kwong, A.D., Pan, Z-Q., and Hurwitz, J., 1991, Studies on the activator 1 protein complex, an accessory factor for PCNA-dependent DNA polymerase δ. J. Biol. Chem. **266**:594-602.

Morrison, A., et al.,1990, A third essential DNA polymerase in *S. cerevisiae*. Cell **62**:1143-1151.

Nethanel, T., and Kaufman, G., 1990, Two DNA polymerases may be required for synthesis of the lagging DNA strand of simian virus 40. Virology **64**:5912-5918.

Prelich, G., and Stillman, B., 1988, Coordinated leading and lagging strand DNA synthesis during SV40 DNA replication *in vitro* requires PCNA. Cell **53**:117-126.

Rivin, C.J., and Fangman, W.L., 1980, Replication fork rate and origin activity during the S phase of *Saccharomyces cerevisiae*. J. Cell Biol. **85**:108-115.

Sitney, K.C., Budd, M.E., and Campbell, J.L., 1989, DNA polymerase III, a second essential DNA polymerase, is encoded by the *S. cerevisiae CDC2* gene. Cell **56**:599-605.

Stillman, B., 1989, Initiation of eukaryotic DNA replication in vitro. Annu. Rev. Cell Biol. **5**:197-245.

Tan, C-K., Castillo, C., So., A.G., and Downey, K.M., 1986, An auxiliary protein for DNA polymerase δ from fetal calf thymus. J. Biol. Chem. **261**:12310-12316.

Tsurimoto, T., and Stillman, B., 1990, Functions of replication factor C and proliferating-cell nuclear antigen: functional similarity of DNA accessory proteins from human cells and bacteriophage T4. Proc. Natl. Acad. Sci. USA **87**:1023-1027

Weinberg, D.H., and Kelly, T.J., 1989, Requirement for two DNA polymerases in the replication of simian virus 40 DNA *in vitro*. Proc. Natl. Acad. Sci. USA **86**:9742-9746.

The POL1 and POL3 DNA Synthesis Genes in Fission Yeast *Schizosaccharomyces pombe*

G. Baldacci, D. Bouvier, V. Damagnez, G. Pignède and A-M. de Recondo
C.N.R.S. UPR 272,
Laboratoire de Biologie Moléculaire de la Réplication
94800 Villejuif
France

ABSTRACT

We isolated and sequenced the *S. pombe* genes encoding the catalytic subunits of DNA polymerases α (*POL1*) and δ (*POL3*) by sequence homology with *S. cerevisiae* gene probes. Both genes contain a short intron in the 5' region of the open reading frame. About 300 bp 5' to *POL1* is located the *RPL7b* gene which encodes a predicted ribosomal protein and is divergently transcribed. The codon usage of *POL1* and *POL3* is in agreement with both genes being weakly expressed; on the contrary, *RPL7b* shows a codon usage characteristic of highly expressed genes. The pol α and pol δ predicted peptides were compared with known sequences of DNA polymerases. Both polypeptides contain the seven conserved amino acid blocks diagnostic of a super-family of DNA polymerases. Moreover, pol α shows additional blocks conserved only in α-type eukaryotic DNA polymerases known to date, and pol δ shows a high overall identity percentage with the homologous *S. cerevisiae* protein indicating an important structural inter-species conservation. We also raised antibodies against two fusion proteins containing *E. coli* β-galactosidase and part of either pol α or pol δ. These antibodies inhibit *in vitro* two distinct DNA polymerase activities present in *S. pombe* cell extracts.

INTRODUCTION

Yeasts are good model systems for the study of DNA replication in eukaryotic cells because of their accessibility to molecular, genetic and biochemical analyses (Guthrie and Fink, 1991). Presently, the budding yeast *Saccharomyces cerevisiae* is the most studied species among yeasts. As far as DNA polymerases are concerned, different groups have characterized in *S. cerevisiae* the genes encoding: i) the catalytic subunits of DNA polymerases α, δ, ε and γ

(Pizzagalli et al., 1988; Boulet et al., 1989; Morrison et al. 1990; Foury, 1989), ii) the primase subunits (Plevani et al., 1987; Foiani et al. 1989) and the p86 subunit of DNA polymerase α (Brooke et al. 1991), iii) the p79 subunit of DNA polymerase ε (Araki et al., 1991).

In mammalian cells only the gene encoding the catalytic subunit of DNA polymerase α has been isolated and sequenced (Wong et al., 1988). However, enzymatic activities showing features specific of the other types of DNA polymerases have been detected (for reviews see Burgers, 1989; Thömmes and Hübscher, 1990; Bambara and Jessee, 1991). The alignment of primary structures of known DNA polymerases allowed the identification of conserved amino acid regions which seem related to the catalytic site (Wang et al., 1989; Blanco et al., 1991).

Replicative genes in *S. cerevisiae* are periodically expressed near the G1/S phase of the cell cycle: a hexamer element located in the upstream promoter sequences of these genes is involved in the regulation of transcription *via* interaction with a regulatory protein (Lowndes et al., 1991). In cycling mammalian cells, the mRNAs of pol α are constitutively present in the different phases of the cell cycle, with only a moderate elevation prior to the S phase (Wahl et al., 1988).

We have chosen to study DNA replication in *S. pombe* as an eukaryotic model system because i) several genetic, molecular and morphological features are similar to those of animal cells (Nasim et al., 1989) and ii) chromosome III from *S. pombe* is able to replicate in mouse cells (Allshire et al., 1987). By heterologous molecular hybridization with natural or synthetic gene probes we have isolated and sequenced the *S. pombe* genes encoding the catalytic subunits of DNA polymerases α and δ. We have studied their transcription and raised antibodies against fusion proteins produced in *E. coli*. We also studied the transcription of these genes during the cell cycle to determine whether their regulation resembles that observed in *S. cerevisiae* or that observed in animal cells (Damagnez et al., 1991a; Damagnez et al., 1991b; Bouvier et al., submitted; Pignède et al., 1991).

MATERIALS AND METHODS

Current techniques of molecular biology used for the work reported in this paper are described in Sambrook et al., 1991. The immunological procedures are described in Harlow and Lane, 1988.

Molecular and cellular biology techniques adapted to *S. pombe* are described in Moreno et al., 1991.

Strains, culture media, growth conditions, chemical and biochemical products, specific experimental conditions used in performing experiments, and computer alignments and analyses reported in this paper are described in Damagnez et al., 1991a; Damagnez et al., 1991b; Pignède et al., 1991.

RESULTS AND DISCUSSION

To search for genes encoding DNA synthesis proteins in *S. pombe* we performed heterologous molecular hybridizations of natural or synthetic radioactive gene probes corresponding to known replicative genes from eukaryotes with *S. pombe* DNA (Figure 1).

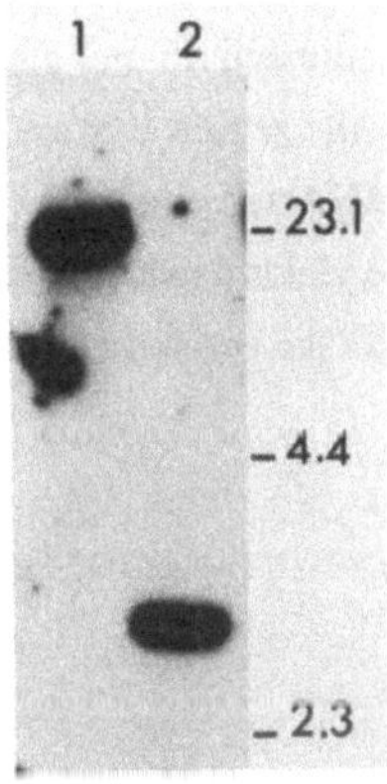

Figure 1. Autoradiogram of the hybridization of a ^{32}P-labeled *S. cerevisiae POL3* gene probe to 4 µg *S. pombe* DNA digested with HindIII (lane 1) and HaeIII (lane 2). Sizes in kb were determined by comparison with DNA fragments of known molecular weight. Hybridization conditions were in 5xSSC at 55°C. Washes were performed at 55-60°C in 0.1xSSC.

As a second step, the same probes were used for the screening of a *S. pombe* genomic library constructed in the lambda phage replacement vector EMBL4. Recombinant phages were isolated and the *S. pombe* DNA fragments were sequenced on both strands. Using this strategy we isolated the *S. pombe* DNA regions hybridizing with the *S. cerevisiae POL1* and *POL3* genes encoding the catalytic subunits of DNA polymerases α and δ, respectively (Damagnez et al., 1991a; Pignède et al., 1991). We also isolated the unique region of *S. pombe* DNA hybridizing with a probe corresponding to the p86 subunit of DNA polymerase α: this work is in progress. However, we have been unable to detect clear hybridization signals with probes corresponding to *S. cerevisiae* primases and to the mitochondrial DNA polymerase.

Identification of *POL1* and *POL3* gene products was based on the comparison of the predicted peptides with known DNA polymerases α and δ, respectively. However, since we constructed a genomic DNA library, we looked for the presence of introns in the rescued genes. We established their number and their approximate location by S1 Nuclease treatment of hybrids between the genes and the corresponding mRNAs. Because the steady-state amounts of *POL1* and *POL3* mRNAs are low in wild type cells, we cloned the respective genes in an expression vector under the control of the constitutive *adh* promoter (McLeod et al., 1987). Large amounts of transcripts were obtained from transformed cells. RNA was hybridized with the same DNA fragment used for the construction of the recombinant expression vector and submitted to S1 Nuclease digestion experiments. The two genes show the presence of an intron located in the 5' part of the coding sequence (Damagnez et al., 1991a; Figure 2). To precisely identify the intron positions, we first localized the consensus sequences involved in splicing in *S. pombe* (Mertins and Gallwitz, 1987) at the sites indicated in our clones by S1 Nuclease experiments. Oligonucleotides complementary to the region located 3' to the putative introns were then synthesized, hybridized with total mRNA and extended by reverse trancription in the presence of dideoxynucleotides. This technique yields a ladder of radioactive bands corresponding to the mRNA sequence, which directly shows the two base pairs joined as the result of intron

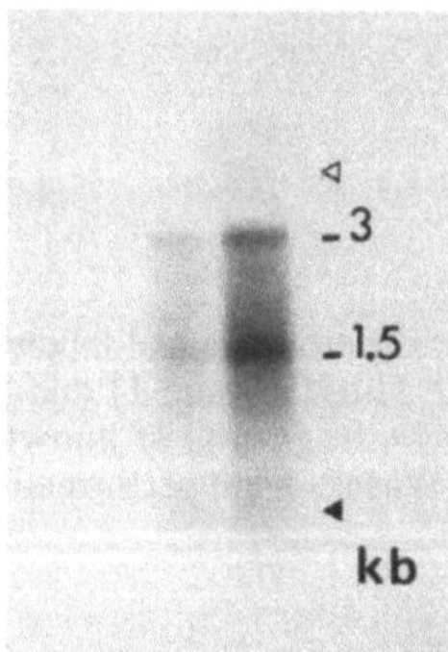

Figure 2. Autoradiogram of the hybridization of a ^{32}P-labeled 4.5 kb ScaI *S. pombe* DNA fragment containing the *POL3* gene and surrounding sequences, with S1 Nuclease-resistant hybrids between the DNA fragment and RNA transcribed from this region. 140 μg RNA extracted form *S. pombe* cells transformed with an expression vector containing the 4.5 kb ScaI fragment under the control of *adh* promoter was co-precipitated with 5 μg of the 4.5 kb ScaI fragment, hybridized 16 hours in 80% formamide and digested 1 hour (left) or 2 hours (right) at 25°C with 200 units of S1 Nuclease. Reaction products were electrophoresed in an alkaline 1.2% agarose gel and blotted onto nitrocellulose. A double amount of reaction products was loaded in the right lane. The open triangle corresponds to the undigested 4.5 kb fragment. The closed triangle indicates a fragment protected by a transcript external to *POL3* coding sequence. Sizes were determined by comparison with DNA fragments of known molecular weight.

splicing. The spliced introns of both *POL1* and *POL3* genes show 5'-, branch site- and 3'-internal intron sequences corresponding to known *S. pombe* splicing consensus sequences (Damagnez et al., 1991a; Pignède et al., 1991). Interestingly, additional potential introns deduced from the presence and the relative distances of consensus sequences could be present inside these genes. However, in our hands the S1 Nuclease protection experiments clearly indicate the presence of a unique intron. In both *POL1* and *POL3* the intron actually excised is the most 5'-located among potential introns. In *S. pombe* single introns are preferentially located in the 5' part of genes, possibly due to a role in splicing of the proximity of the 5' mRNA end. However, some *S. pombe* genes contain more than one intron (Russel, 1989), and it is possible that elements other than the consensus sequences ensure the correct maturation of pre-mRNAs. Alternatively, one could think that the excision of potential additional introns existing in *POL1* and *POL3* is prevented in a specific way.

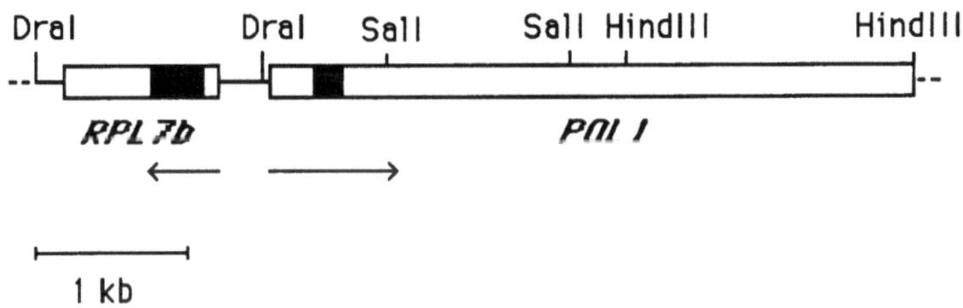

Figure 3. Partial restriction map of the *S. pombe* chromosome II region containing the *RPL7b* and *POL1* genes. Genes are indicated with boxes; black areas correspond to introns. Arrows indicate the direction of transcription.

The physical organization of the *S. pombe* DNA region containing *POL1* on Chromosome II is shown in Figure 3. The *RPL7b* gene, encoding the predicted ribosomal protein L7b, was detected because of its proximity to *POL1* (Damagnez et al., 1991b). An interesting aspect of the short region separating *POL1* and *RPL7b* is the initiation of divergent transcripts. Since the steady-state amounts of respective mRNAs are different, this region should contain the sequence elements involved in the divergent transcription and in the different regulation of these genes. Preliminary results show that a short sequence present 5' to the *RPL7b* gene is able to strongly increase the transcription of a reporter gene inserted in an *S. pombe* expression vector. In fact, the comparison of the codons used in these genes (Table 1 and Table 2) fits with the observation that *RPL7b* is an actively transcribed gene and that *POL1* is weakly expressed

TABLE 1

F	TTT	71.7	S	TCT	36.9	Y	TAT	71.4	C	TGT	76.9
F	TTC	28.3	S	TCC	11.3	Y	TAC	28.6	C	TGC	23.1
L	TTA	29.6	S	TCA	19.9	*	TAA	0.0	*	TGA	100.0
L	TTG	20.0	S	TCG	12.1	*	TAG	0.0	W	TGG	100.0
L	CTT	23.5	P	CCT	41.9	H	CAT	66.7	R	CGT	27.5
L	CTC	7.0	P	CCC	16.1	H	CAC	33.3	R	CGC	10.0
L	CTA	8.7	P	CCA	32.3	Q	CAA	73.0	R	CGA	22.5
L	CTG	11.3	P	CCG	9.7	Q	CAG	27.0	R	CGG	5.0
I	ATT	69.0	T	ACT	36.4	N	AAT	70.4	S	AGT	14.9
I	ATC	14.9	T	ACC	16.7	N	AAC	29.6	S	AGC	5.0
I	ATA	16.1	T	ACA	31.8	K	AAA	60.6	R	AGA	26.3
M	ATG	100.0	T	ACG	15.2	K	AAG	39.4	R	AGG	8.8
V	GTT	45.9	A	GCT	55.8	D	GAT	67.8	G	GGT	34.4
V	GTC	16.2	A	GCC	16.3	D	GAC	32.2	G	GGC	15.6
V	GTA	28.4	A	GCA	17.4	E	GAA	76.3	G	GGA	39.1
V	GTG	9.5	A	GCG	10.5	E	GAG	23.7	G	GGG	10.9

POL1 codon usage. Total number of codons = 1406. Codon-usage frequencies are expressed as a percentage value among each degenerate codon set.

TABLE 2

F	TTT	54.5	S	TCT	66.7	Y	TAT	37.5	C	TGT	0.0
F	TTC	45.5	S	TCC	16.7	Y	TAC	62.5	C	TGC	0.0
L	TTA	10.0	S	TCA	0.0	*	TAA	100.0	*	TGA	0.0
L	TTG	40.0	S	TCG	8.3	*	TAG	0.0	W	TGG	100.0
L	CTT	25.0	P	CCT	66.7	H	CAT	80.0	R	CGT	83.3
L	CTC	25.0	P	CCC	33.3	H	CAC	20.0	R	CGC	11.1
L	CTA	0.0	P	CCA	0.0	Q	CAA	72.7	R	CGA	0.0
L	CTG	0.0	P	CCG	0.0	Q	CAG	27.3	R	CGG	0.0
I	ATT	59.1	T	ACT	80.0	N	AAT	25.0	S	AGT	0.0
I	ATC	36.4	T	ACC	20.0	N	AAC	75.0	S	AGC	8.3
I	ATA	4.5	T	ACA	0.0	K	AAA	22.2	R	AGA	5.6
M	ATG	100.0	T	ACG	0.0	K	AAG	77.8	R	AGG	0.0
V	GTT	62.5	A	GCT	73.1	D	GAT	100.0	G	GGT	57.1
V	GTC	31.3	A	GCC	23.1	D	GAC	0.0	G	GGC	35.7
V	GTA	6.3	A	GCA	3.8	E	GAA	47.6	G	GGA	7.1
V	GTG	0.0	A	GCG	0.0	E	GAG	52.4	G	GGG	0.0

RPL7b codon usage. Total number of codons = 251. Codon-usage frequencies are expressed as a percentage value among each degenerate codon set.

(Russel, 1989). Similarly, *POL3*, which is a weakly transcribed gene as determined by northern blot analysis, shows a codon usage typical of poorly expressed genes (data not shown). An interesting question is whether the presence of a divergent transcript in the 5' region is common to other replicative genes. Our study of the *POL3* promoter region indicates the presence of a transcript encoded by the region 5' to the gene. Indeed, the 0.65 kb S1 Nuclease-resistant band in Figure 2 is detected by the part of the probe located 5' from the *POL3* gene (data not shown). Additional work is required to identify in untransformed cells the full size of the transcript, the direction of transcription and the peptide it encodes. It is remarkable that a RNA molecule is divergently transcribed 5' from the human PCNA gene (proliferating cell nuclear antigen) encoding an auxiliary protein which interacts *in vitro* with DNA polymerase δ (Rizzo et al., 1990). The presence of regions responsible for divergent transcription has been detected at several sites of eukaryotic and prokaryotic chromosomes (Beck and Warren, 1988). The physiological meaning of such an organization is still unclear; however, it has been proposed that a strong promoter could facilitate the transcription of a gene under the control of a weak promoter, by increasing the local concentration of RNA polymerase II.

Once established the gene and mRNA sequences, we could compare the predicted peptides encoded by *S. pombe POL1* and *POL3* with other DNA polymerases. These comparisons showed the presence, in the predicted peptides encoded by *S. pombe POL1* and *POL3*, of seven diagnostic amino acid blocks, conserved in a group of DNA polymerases including human DNA polymerase α and *S. cerevisiae* DNA polymerases α and δ (Figure 4). The overall identity percentages were: i) *S. pombe*/human pol α, 34%; ii) *S. pombe*/*S. cerevisiae* pol α,

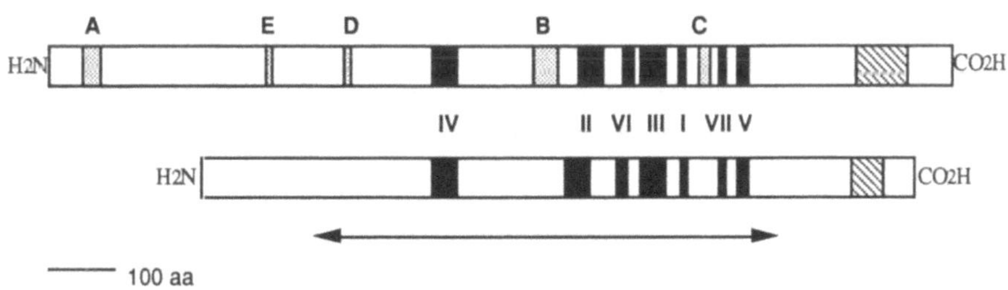

Figure 4. Schematic representation of *S. pombe* DNA polymerases α (top) and δ (bottom). Roman numbers from I to VII indicate amino acid (black) blocks conserved in a super-family of DNA polymerases. Letters indicate amino acid (stippled) blocks conserved in human, budding and fission yeast DNA polymerase α. In the most C-terminal (hatched) blocks the positions of cysteine residues are conserved. The double arrow corresponds to the region showing 60% identity between *S. pombe* and *S. cerevisiae* DNA polymerase δ.

TABLE 3

POL α		
human	[1281]	C 2 C 23 C 4 C 32 C 4 C 17 C 2 C
S. pombe	[1256]	C 2 C 23 C 1 C 30 C 4 C 13 C 4 C
S. cerevisiae	[1287]	C 2 C 23 C 2 C 30 C 4 C 13 C 4 C
POL δ		
S. pombe	[991]	C 2 C 11 C 2 C 28 C 2 C 9 C 4 C
S. cerevisiae	[1005]	C 2 C 11 C 2 C 28 C 2 C 9 C 4 C
POL ε		
S. cerevisiae	[2108]	C 2 C 18 C 2 C 30 C 2 C 11 C 1 C
REV3		
S. cerevisiae	[1308]	C 2 C 12 C 2 C 28 C 2 C 18 C 4 C

Conservation of cysteine residues in C-terminal regions of eukaryotic DNA polymerases. Numbers correspond to amino acid residues separating cysteine residues (C). The position of the first cysteine residue in each protein sequence is indicated in square brackets.

37%; iii) *S. pombe/S. cerevisiae* pol δ, 52%. Cysteine residues at distances compatible with the formation of zinc fingers (structures which could interact with DNA) have been detected in the C-terminal region of eukaryotic DNA polymerases and of the peptide predicted from the *S. cerevisiae REV3* gene (Morrison et al., 1989). Table 3 shows the alignment of cysteine residues in the C-terminal region of the predicted peptides encoded by *S. pombe POL1* and *POL3* genes and of other eukaryotic DNA polymerases known to date. Taken together, the following arguments indicate that *S. pombe POL1* and *POL3* respectively encode the catalytic subunits of DNA polymerase α and δ: i) the hybridization with the *S. cerevisiae POL1* or *POL3* gene probes, ii) the presence of the seven amino acid blocks diagnostic of DNA polymerases, iii) the overall identity percentages of alignments between *S. pombe POL1* and *POL3* predicted peptides with eukaryotic DNA polymerases α and δ, and iv) the conservation of cysteine residues in the C-terminal regions.

In order to raise antibodies against *S. pombe* pol α, we constructed a recombinant expression vector containing part of the *POL1* gene fused with the *E. coli* β-galactosidase gene. The fusion protein was purified from *E. coli* cell extracts and injected into rabbits. The antibodies were purified from rabbit sera and used for immunoblots with *S. pombe* cell extracts. Anti-pol α antibodies detect a protein showing an apparent molecular of about 170 kDa weight in SDS-PAGE, and some bands of lower molecular weight which, by two-dimensional peptide mapping, appear as proteolysis products of the 170 kDa protein (Figure 5). This protein is undetectable in *S. pombe* stationary phase cells and becomes apparent when cells are stimulated to proliferate (Figure 5). With a similar strategy, we obtained antibodies against a β-galactosidase/pol δ fusion protein. These antibodies detect a protein showing an apparent molecular weight in SDS-PAGE of about 123 kDa in *S. pombe* cell extracts and inhibit *in vitro* a DNA polymerase activity distinct from pol α. They also detect a 140 kDa protein in extracts from mammalian cells (Pignède et al., 1991). The availability of homologous gene probes and of specific antibodies allowed us to trace pol α and pol δ mRNAs and proteins along the cell cycle. Preliminary results indicate that, in both cases, the mRNAs and the proteins are

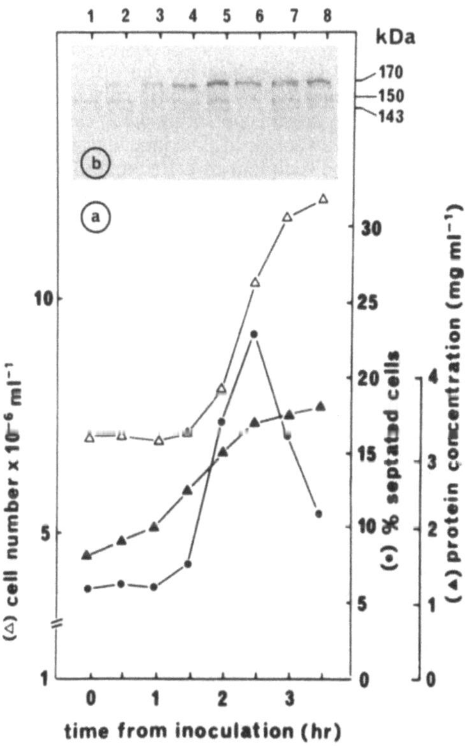

Figure 5. Stimulation of *S. pombe* stationary phase cells to proliferate. Stationary phase cells were collected, refed with fresh medium and allowed to proliferate. Extracts were prepared from 10^8 cells every 30 min and immunoblotted with anti-pol α antibodies. a: parallel evolution of cell plate index, cell number and protein content of each extract; b: immunoblots of the corresponding extracts.

constitutively present during the different steps of the cell cycle. However, moderate increase (about three times) of the transcripts and proteins appear in late G1 phase.

In conclusion, i) the primary structure of the catalytic subunit of DNA polymerase α from *S. pombe* shows the presence of amino acid blocks present only in the two other eukaryotic pol α known to date; ii) the comparison between the budding and fission yeast catalytic subunits of DNA polymerase δ shows an important structural conservation, also confirmed by the ability of antibodies raised against *S. pombe* pol δ to detect a protein in mammalian cell extracts; iii) the regulation of transcripts during the cell cycle is closer to that observed in animal cells than in *S. cerevisiae* cells.

AKNOWLEDGMENTS

We are grateful to P. Plevani and G. Lucchini for the gift of plasmids containing the *S. cerevisiae POL1*, *PRI1* and *PRI2* genes, to G. Faye for the plasmid containing the *S. cerevisiae POL3* gene and to D. Hinkle for the plasmid containing the *S. cerevisiae* gene encoding the p86 subunit of pol alpha. We thank J. Tillit for skillful technical assistance. V.D. is the recipient of a M.R.T. fellowship and G.P. is the recipient of a Beckman/IFSBM fellowship. This work was supported in part by grant 6704 to G.B. from the A.R.C. and by a grant to A.M. de R. from the L.N.F.C.

REFERENCES

Allshire RC, Cranston G, Gosden JR, Maule JC, Hastie NE, Fantes PA (1987) A fission yeast chromosome can replicate autonomously in mouse cells. Cell 50:391-403

Araki H, Hamatake RK, Johnston LH, Sugino A (1991) *DPB2*, the gene encoding DNA polymerase II subunit B, is required for chromosome replication in *Saccharomyces cerevisiae*. Proc Natl Acad Sci USA 88:4601-4605

Bambara RA, Jessee CB (1991) Properties of DNA polymerases delta and epsilon, and their roles in eukaryotic DNA replication. Biochim Biophys Acta 1088:11-24

Beck CF, Warren RAJ (1988) Divergent promoters, a common form of gene organization. Microbiol Rev 52:318-326

Blanco L, Bernad A, Blasco MA, Salas M (1991) A general structure for DNA-dependent DNA polymerases. Gene 100:27-38

Boulet A, Simon M, Faye G, Bauer GA, Burgers PMJ (1989) Structure and function of the *Saccharomyces cerevisiae CDC2* gene encoding the large subunit of DNA polymerase III. EMBO J 8:1849-1854

Brooke RG, Singhal R, Hinkle DC, Dumas LB (1991) Purification and characterization of the 180- and 86-kilodalton subunits of the *Saccharomyces cerevisiae* DNA primase-DNA polymerase protein complex. J Biol Chem 266:3005-3015

Burgers PMJ (1989) Eukaryotic DNA polymerases alpha and delta: conserved properties and interactions, from yeast to mammalian cells. Progr Nucleic Acids Res Mol Biol 37:235-281

Damagnez V, Tillit J, De Recondo A-M, Baldacci G (1991a) The *POL1* gene from the fission yeast, *Schizosaccharomyces pombe*, shows conserved amino acid blocks specific for eukaryotic DNA polymerases alpha. Mol Gen Genet 226:182-189

Damagnez V, De Recondo A-M, Baldacci G (1991b) Identification of a gene encoding the predicted ribosomal protein L7b divergently transcribed from POL1 in fission yeast *Schizosaccharomyces pombe*. Nucleic Acids Res 19:1089-1104

Foiani M, Santocanale C, Plevani P, Lucchini G (1989) A single essential gene, *PRI2*, encodes the large subunit of DNA primase in *Saccharomyces cerevisiae*. Mol Cell Biol 9:3081-3087

Foury F (1989) Cloning and sequencing of the nuclear gene *MIP1* encoding the catalytic subunit of the yeast mitochondrial DNA polymerase. J Biol Chem 264:20552-20560

Guthrie C, Fink GR (eds) (1991) Guide to yeast genetics and molecular biology. Methods in Enzymology 194. Academic Press Inc. San Diego California

Harlow E, Lane D (1988) Antibodies. Cold Spring Harbor Laboratory Press Cold Spring Harbor New York

Lowndes NF, Johnson AL, Johnston LH (1991) Coordination of expression of DNA synthesis genes in budding yeast by a cell-cycle regulated *trans* factor. Nature 350:247-250

McLeod M, Stein M, Beach D (1987) The product of of the *mei3*+ gene, expressed under control of the mating-type locus, induces meiosis and sporulation in fission yeast. EMBO J 6:729-736

Mertins P, Gallwitz D (1987) Nuclear pre-mRNA splicing in the fission yeast *Schizosaccharomyces pombe* strictly requires an intron-contained, conserved sequence element. EMBO J 6:1757-1763

Moreno S, Klar A, Nurse P (1991) Molecular genetic analysis of fission yeast *Schizosaccharomyces pombe*. In: Guide to yeast genetics and molecular biology. Methods in Enzymology 194:795-823. Academic Press Inc. San Diego California

Morrison A, Araki H, Clark AB, Hamatake RK, Sugino A (1990) A third essential DNA polymerase in *S. cerevisiae*. Cell 62:1143-1151

Morrison A, Christensen RB, Alley J, Beck AK, Bernstine EG, Lemontt JF, Lawrence CW (1989) *REV3*, a *Saccharomyces cerevisiae* gene whose function is required for induced mutagenesis, is predicted to encode a nonessential DNA polymerase. J Bacteriol 171: 5659-5667

Nasim A, Young P, Johnson BF (eds) (1989) Molecular Biology of the Fission Yeast. Academic Press Inc. San Diego California

Pignède G, Bouvier D, De Recondo A-M, Baldacci G (1991) Characterization of the *POL3* gene product from *S. pombe* indicates inter-species conservation of the catalytic subunit of DNA polymerase delta. J Mol Biol in press

Pizzagalli A, Valsasnini P, Plevani P, Lucchini G (1988) DNA polymerase I gene of *Saccharomyces cerevisiae* : Nucleotide sequence, mapping of a temperature-sensitive mutation, and protein homology with other DNA polymerases. Proc Natl Acad Sci USA 85:3772-3776

Plevani P, Francesconi S, Lucchini G (1987) The nucleotide sequence of the *PRI1* gene related to DNA primase in *Saccharomyces cerevisiae*. Nucleic Acids Res 15:7975-7989

Rizzo MG, Ottavio L, Travali S, Chang C-D, Kaminska B, Baserga R (1990) The promoter of the human Proliferating Cell Nuclear Antigen (PCNA) gene is bidirectional. Exp Cell Res, 188:286-293

Russel P (1989) Gene cloning and expression in fission yeast. In Molecular Biology of the Fission Yeast (Nasim A, Young P, Johnson BF eds) 243-271. Academic Press Inc. San Diego California

Sambrook J, Fritsch EF, Maniatis T (1989) Molecular Cloning. Cold Spring Harbor Laboratory Press Cold Spring Harbor New York

Thömmes P, Hübscher U (1990) Eukaryotic DNA replication. Eur J Biochem 194:699-712

Wahl AF, Geis AM, Spain BH, Wong SW, Korn D, Wang TS-F (1988) Gene expression of human DNA polymerase alpha during cell proliferation and the cell cycle. Mol Cell Biol 8:5016-5025

Wang TS-F, Wong SW, Korn D (1989) Human DNA polymerase alpha: predicted functional domains and relationships with viral DNA polymerases. FASEB J 3:14-21.

Wong SW, Wahl AF, Yuan P-M, Arai N, Pearson BE, Arai K-I, Korn D, Hunkapiller MW, Wang TSF (1988) Human DNA polymerase alpha gene expression is cell proliferation dependant and its primary structure is similar to both prokaryotic and eukaryotic replicative DNA polymerases. EMBO J 7:37-47

DNA Polymerase δ (POL3) of *Saccharomyces cerevisiae*

G. Faye, F. Fabre, M. Simon, L. Giot, A. Boulet, P. Bénit and A. Vassal

Institut Curie-Biologie, Bât. 110, Centre Universitaire, 91405 Orsay, France

Introduction

There are three DNA polymerases in yeast which are essential for cell proliferation. Whereas DNA polymerase α (POL1) might be involved in the initiation of nuclear DNA replication (Pizzagalli et al., 1988), the specific functions of DNA polymerase ε (POL2) (Morrison et al., 1990) and DNA polymerase δ (POL3) (Boulet et al., 1989) during the S phase, are not clearly established. Moreover, it is not known whether these DNA polymerases are implicated into other processes such as DNA reparation or recombination.

We are presenting some informations concerning the properties and functions of DNA polymerase δ.

1. Proofreading 3'-5' exonuclease activity in DNA polymerase δ

The biochemically defined DNA polymerase III (Burgers et al., 1990) is composed of two subunits, one of 124 (DNA polymerase δ = POL3) and the other of 55 kd. This complex possesses a proofreading 3'-5' exonuclease activity (Bauer et al., 1988). Two lines of evidences suggest that this activity is a quality of the DNA polymerase δ protein. First, the DNA polymerase III complex isolated from cdc2 temperature sensitive mutants (mutants in the POL3 gene), grown at non permissive temperature, lacks both the polymerase and exonuclease

ExoI

			Mutations: 1DV, 1DA, 2EA
Sc POL3	(δ)	315	LRIMSF***D***I***E***CAGRIGVFP
Sc POL2	(ε)	284	PVVMAF***D***I***E***TTKPPLKFP
EBV		290	YQALAF***D***I***E***CLGEEG-FP

ExoI'

		Mutation: 3EA
Sc POL3	371	SMIFSHAT***E***EEM***L***S
Sc POL2	347	FTIFNEND***E***VAL***L***Q
EBV	348	VEVYEFPS***E***LDM***L***Y

ExoII

		Mutation: 4DA
Sc POL3	396	DV***I***IGY***N***TTN***FD***IPY***L***LNRAK
Sc POL2	372	TV***I***STF***N***GDF***FD***WPF***I***HNRSK
EBV	373	EI***V***TGY***N***VAN***FD***WPY***I***LDRAR

Figure 1. 3'-5' exonuclease domains. Amino-acid alignment between *Saccharomyces cerevisiae* Sc POL3 (Boulet et al., 1989) and Sc POL2 (Morrison et al., 1990) and EBV DNA polymerase (Kouzarides et al., 1987) for sites ExoI and II (Bernad et al., 1989). Site ExoI' would correspond to POL3 or POL2 ExoI site deduced from that proposed by Bernad et al. (1989) for EBV DNA polymerase. Arrows designate the POL3 mutagenized loci ; for instance, 1DV means that the conserved aspartate residue (D) is changed into a valine (V) residue at locus 1. Numbers preceding the sequences correspond to amino acid rank starting from the first amino acid.

activities (Boulet et al., 1989). Second, three amino acid domains considered as the sites of the 3'-5' exonuclease activity in prokaryotic DNA polymerases (named ExoI, II and III) (Bernad et al., 1989) are present in the N-terminal portion of DNA polymerase δ (POL3).

We have mutagenized the ExoI and ExoII sites of DNA polymerase δ and asked whether the spontaneous mutation rate of two genes is increased in strains harboring such mutations.

a) Mutagenesis in ExoI and ExoII exonuclease sites

By oligonucleotide-mediated mutagenesis we obtained 4 mutant plasmids, each containing a single change : three located in the ExoI site (1DV, 1DA and 2EA) and one in the ExoII site (4DA) (Figure 1). Similar alterations at the analogous positions severely decrease the 3'-5' exonuclease activity of the ϕ29 DNA polymerase (Bernad et al., 1989). We introduced one change (3EA) in a site, named ExoI' in this work, corresponding to the domain of the EBV DNA polymerase proposed as ExoI by Bernad et al. (1989). We have also put together the 1DA and 4DA changes by ligating together the appropriate restriction fragments.

b) Reversion rate to lysine prototrophy and mutability from Can^S to Can^R

Each of the five mutant copies of POL3 (and the wild type copy as a control) cloned in a $LEU2^+$ centromeric *E. coli*-yeast vector was brought by transformation into a yeast strain carrying a chromosomal disrupted copy of POL3 complemented by a wild-type copy borne on an $URA3^+$ centromeric plasmid. Then the complementing $URA3^+$ plasmid was eliminated from the $LEU2^+$ transformants obtained. None of the mutants appeared to be either cold or temperature sensitive.

The reversion rate from lys^- to lysine prototrophy was determined by the fluctuation test developped by von Borstel et al. (1971). The results are shown in Table I. Four out of the six mutations confer a significantly increased mutation rate, compared with the wild-type, varying from 11-fold for the mutants 1DA and 2EA in the ExoI site to 15-fold for the double mutant 1DA-4DA in the ExoI and ExoII sites or a maximum of 18-fold for the mutant 4DA in the ExoII site. At the locus 1 of the ExoI site, the mutator effect is smaller when a valine (V) is put in place of the conserved acidic aspartate (D) than

TABLE I.

Site	**POL3**	**A**	**B**
	wt	1	1
ExoI	1DV	4	89
ExoI	1DA	12	372
ExoI	2EA	11	370
ExoI'	3EA	2	3
ExoII	4DA	18	536
ExoI-ExoII	1DA-4DA	15	323

A : Relative increase of reversion rates to lysine prototrophy (mutant/wt)
B : Relative increase of mutability rates from Can^S to Can^R (mutant/wt)

when it is an alanine (A). Mutant 3EA in ExoI' shows the weakest effect. It seems therefore that the veritable POL3 ExoI site is in fact that located some 60 amino-acids upstream from the ExoI' site predicted from the consensus proposed by Bernad et al. (1989) for EBV DNA polymerase.

The mutation rate from Can^S/Can^R was measured by the method of the median (Magni and von Borstel, 1962). The results are shown in Table I. They follow those observed for lys^- reversion though the rates are about 20 to 30 times higher.

c) Conclusion

Our results indicate that a single aminoacid change of the highly conserved aspartate or glutamate residues of the ExoI or ExoII domains of POL3 into non polar aminoacid residues gives rise to a mutator effect in *S. cerevisiae*. Similar changes in the ϕ29 DNA polymerase inhibit specifically its associated 3'-5' exonuclease activity (Bernad et al., 1989). When the aspartate residue of the ExoI site is changed into a valine instead of an alanine, the exonuclease seems to better recognize its substrate since the observed mutation rate is lower (Cf. Table I). Locus 3EA (in ExoI') appears not to belong to a catalytic domain, mutation 3EA giving rise to the weakest mutator effect.

Taken together the results presented herein strongly suggest that the 3'-5' exonuclease activity of the DNA polymerase III complex is located in the N-terminal portion of the POL3 polypeptide and is required for the accuracy of the DNA replication in yeast. All point mutations in the ExoI and ExoII sites studied here still complement a lethal chromosomal deletion of POL3. Furthermore since none of the resulting mutant strains is either temperature or cold sensitive and since each of them grows practically as fast and as far as wild-type strains, then DNA polymerase δ must retain most of its polymerizing activity.

2. POLIII and induced gene conversion

In the models of recombination (Meselson and Radding, 1975 ; Szostak et al., 1983), a DNA polymerase is needed for the propagation of strand-deplacement and/or for repair synthesis in the remaining gaps. DNA synthesis is also required for mismatch repair on heteroduplex DNA, leading eventually to gene conversion.

We have asked whether DNA polymerase δ is playing a part in induced mitotic gene conversion. The availability of different non-leaky thermo-sensitive pol3 mutations (cdc2) (Pringle and Hartwell, 1981) allowed us to construct diploids heteroallelic for POL3 and to test whether these cells are able to perform intragenic recombination (gene conversion) under the restrictive conditions.

The rationale of the experiment is the following. When cells arrested in the Go state are reincubated in a fresh medium, they progress through the G1 phase before entering the S phase. It is known that irradiations of Go or G1 diploid cells induce gene conversion events which occur before the cells reach the S phase (Wildenberg, 1970 ; Roman and Fabre, 1983 ; Fabre, 1978). pol3 (cdc2) thermosensitive mutants are blocked in the S phase at non permissive temperature because the replication is not complete. If POL3 is not involved in recombination, the irradiation of Go diploids with heteroallelic (cdc2) mutations in this gene should result in the formation of a wild-type POL3 gene, the transcription and translation of which will relieve the thermosensitive phenotype. On the contrary if the gene product is involved in recombination, no POL3 recombinant gene will be formed.

The induction of POL3 convertants under restrictive or permissive temperatures was studied as follows. First, we considered the effect of the restrictive temperature on cell viability. cdc2 heteroallelic cells, in the stationary phase (Go) were transplanted in a fresh YPD medium at 36°C. After 3,5 hours of incubation, the cells had formed a bud and they accumulate as doublets. The kinetics of lethality as a function of transient incubation times at 36°C, indicate that the cells remain viable for about 4 hours after the bud appearance. The lethality should therefore not interfere with the detection of POL3 convertants.

To study the induction of these convertants, stationary cdc2 heteroallelic cells were either UV or γ-ray-treated before being plated and incubated immediately at 36°C, or for six hours at 24°C and then at 36°C. The number of viable cells was determined by plating diluted aliquots of cell suspensions, followed by incubation at 24°C. The γ-ray doses were not lethal, nor were the UV doses up to 30 J/m^2. Survivals were 85 and 75 % after 40 and 50 J/m^2 respectively. The kinetics of viability as a function of time at 36°C was not significantly affected by the irradiations.

After γ-rays and immediate incubation at 36°C (Figure 2), a low but significant induction of POL3 convertants ($30/10^6$ cells) was already observed for the lowest dose (20 Gy). No further induction was obtained with higher doses. In the other part of the same experiment the plates were first incubated for 6 hours at 24°C. This was to allow conversion to occur under the

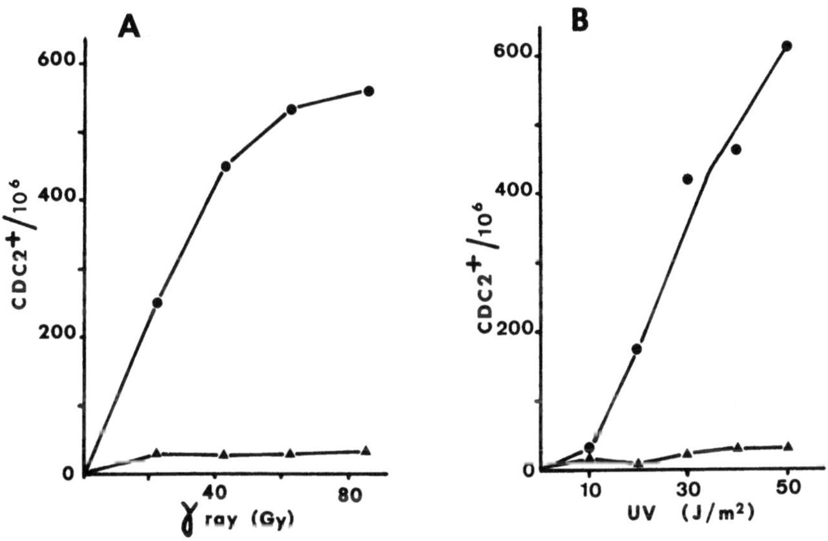

Figure 2. (A) Induction by γ-rays of POL3 convertants. (B) Induction by UV-light of POL3 convertants. The cells were incubated at 36°C immediately after irradiation and plating (▲) or for six hours at 24°C before incubation at 36°C (●). The background values (dose 0) were substracted.

permissive conditions, before applying the $POL3^+$ selection by incubation at 36°C. During these 6 hours at 24°C, the cells had divided at least once. As shown in Figure 2, dose-dependent induction of POL3 cells were obtained. The dose of 20 Gy induced 260 POL3 convertants per 10^6 cells. Between 20 and 80 Gy, the induction rose from 260 to $550/10^6$ (whereas no increase was seen if incubation was immediately at 36°C). Thus in this dose range, the induction appeared to be totally dependent on an active DNA polymerase δ.

The results were comparable after UV irradiation (Figure 2). A dose of of 50 J/m^2 induced $30/10^6$ revertants under the restrictive temperature, compared to $615/10^6$ if the cells were transiently incubated at 24°C.

To confirm that the revertants were indeed due to intragenic recombination, homoallelic cdc2-1 or cdc2-2 diploids were constructed and treated in the same conditions as those described above. No revertants were induced, even in the permissive conditions. The simplest interpretation of our results is that DNA polymerase δ is involved in induced intragenic recombination. The very low but significant level of induction of POL3 convertants, under the restrictive conditions may be due either to a residual activity of POL3 or to the substitution of POL3 by another polymerase.

3. Search for suppressors of thermosensitive POL3 mutants

We have mentioned above that DNA polymerase III forms a complex. Furthermore, several proteins might interact with it to drive or regulate its activity. For instance POL3 becomes highly processive when it is associated with PCNA (Bauer and Burgers, 1990). To bring some light about the particular function of POL3 in DNA replication (and eventually in DNA recombination) we are intending to identify some of the putative components which might permanently or transiently be associated with POL3.

For this purpose, we prepared a large collection of thermosensitive POL3 mutants. In a second stage we searched extragenic suppressors (dominant or recessive) of these mutants. Thermosensitive mutants were isolated using a plasmid exchange method. The N- or C-terminal part of gene POL3 were mutagenized in vitro with hydroxylamine. The mutated fragments were inserted into a LEU2 centromeric vector, then the ligation products were used to directly transform a strain bearing a chromosomal disrupted copy of POL3, complemented by a wild-type copy of POL3 carried on a URA3 vector (cf. Figure 3).

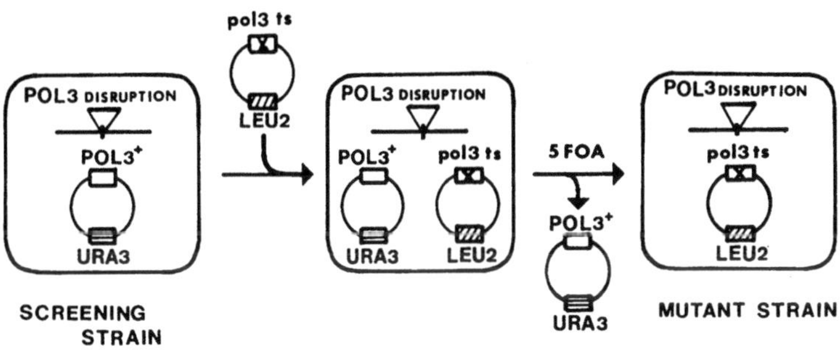

Figure 3. Introduction of POL3 mutations by plasmid exchange. The screening strain carries a POL3 chromosomal deletion complemented by a wild-type copy borne on a URA3 resident vector. It is transformed by a collection of mutagenized copies of POL3, carried by a LEU2 vector. The loss of the URA3 vector is selected with 5-fluoro-orotic acid (5FOA).

$LEU2^+$ transformants were selected at 24°C on plates containing no leucine but uracil in order to start removing the URA3 vector. They were then plated onto plates containing 5 fluoro-orotic acid and incubated at 24°C to select ura- subclones (Boeke et al., 1984) containing the LEU2+ vector as a unique copy of POL3 complementing the lethal disruption. Several ts mutants

```
NNAAGCTTTCCCAATTGTGGAAAAGTTGACTTTGTCCGGCTTCAATGAATAGCTGTTTTGTGTCAGTCATAGTTGCAGTTCGATCTGGTGTTCTCCTTTATATCTGCGTTTCTTTGCAGC  120

GTTCTTGCTTGACGGTTGATCTTGGTTTTGTGTGGTAAAAGTATGCCTTAACGTAATAGGCAATTTTTCGTATTACGCGTAACTTTTTATTCTATAAAATGTTCAATGAGGACATCTGCT  240

ATTCGCTTATGAAGAACAAACACTCAGTACTACTGATCTAAGGCAATTTTCAAGGATAAAGGAAAATAGATATTGAGCACTTGCTATTAAGCATTAATCTTTATACATATACGCACAGCA  360
                                                                                                   *

M  S  E  K  R  S  L  P  M  V  D  V  K  I  D  D  E  D  T  P  Q  L  E  K  K  I  K  R  Q  S  I  D  H  G  V  G  S  E  P  V
ATGAGTGAAAAAAGATCCCTTCCCATGGTTGATGTGAAGATCGATGACGAGGATACTCCCCAGTTGGAAAAGAAAATCAAACGGCAATCAATAGATCATGGTGTTGGAAGTGAACCTGTT  480

S  T  I  E  I  I  P  S  D  S  F  R  K  Y  N  S  Q  G  F  K  A  K  D  T  D  L  M  G  T  Q  L  E  S  T  F  E  Q  D  V  S
TCAACAATAGAGATTATTCCGAGTGATTCTTTTCGAAAATATAATAGTCAAGGCTTCAAAGCAAAGGATACAGATTTAATGGGTACGCAATTAGAGTCTACTTTTGAACAAGACGTATCG  600

Q  M  E  H  D  M  A  D  Q  E  E  H  D  L  S  S  F  E  R  K  K  L  P  T  D  F  D  P  S  L  Y  D  I  S  F  Q  Q  I  D  A
CAAATGGAACATGATATGGCCGACCAAGAAGAGCATGACCTGTCATCATTCGAGCGTAAGAAACTTCCAACCGATTTTGACCCAAGTTTGTATGATATTTCTTTCCAACAAATTGATGCG  720

                                                                                          T
E  Q  S  V  L  N  G  I  K  D  E  N  T  S  T  V  V  R  F  F  G  V  T  S  E  G  H  S  V  L  C  N  V  T  G  F  K  N  Y  L
GAACAGAGCGTACTGAATGGTATCAAAGATGAAAATACATCTACCGTGGTAAGGTTTTTTGGTGTCACTAGTGAAGGACACTCTGTACTTTGTAATGTTACAGGGTTCAAGAACTATCTT  840 M2
                                                                                           A
                                                                                    +
Y  V  P  A  P  N  S  S  D  A  N  D  Q  E  Q  I  N  K  F  V  H  Y  L  N  E  T  F  D  H  A  I  D  S  I  E  V  V  S  K  Q
TACGTCCCAGCGCCCAATTCTTCCGACGCTAACGATCAGGAGCAAATCAACAAGTTTGTGCACTATTTAAACGAAACATTTGACCACGCTATTGATTCGATTGAAGTTGTATCTAAACAG  960
                                                                                     *
         +
S  I  W  G  Y  S  G  D  T  K  L  P  F  W  K  I  Y  V  T  Y  P  H  M  V  N  K  L  R  T  A  F  E  R  G  H  L  S  F  N  S
TCTATCTGGGGTTATTCCGGAGATACCAAATTACCATTCTGGAAAATATACGTCACCTATCCGCATATGGTCAACAAACTGCGTACTGCGTTTGAAAGAGGTCATCTTTCATTCAACTCG 1080
          *                                                                                          *
                                                                                 F
W  F  S  N  G  T  T  T  Y  D  N  I  A  Y  T  L  R  L  M  V  D  C  G  I  V  G  M  S  W  I  T  L  P  K  G  K  Y  S  M  I
TGGTTTTCTAACGGCACGACTACTTATGATAACATTGCCTACACTTTAAGGTTAATGGTAGATTGTGGAATTGTCGGTATGTCCTGGATAACATTACCAAAAGGAAAGTATTCGATGATT 1200 M8
                                                                                  T

E  P  N  N  R  V  S  S  C  Q  L  E  V  S  I  N  Y  R  N  L  I  A  H  P  A  E  G  D  W  S  H  T  A  P  L  R  I  M  S  F
GAGCCTAATAACAGAGTTTCCTCTTGTCAGTTGGAAGTTTCAATTAATTATCGTAACCTAATAGCACATCCTGCTGAGGGTGATTGGTCTCATACAGCTCCATTGCGTATCATGTCCTTT 1320

                                                                              +  +  +  +  +  +  +  +  +  +  +  +  +  +
D  I  E  C  A  G  R  I  G  V  F  P  E  P  E  Y  D  P  V  I  Q  I  A  N  V  V  S  I  R  G  A  K  K  P  F  I  R  N  V  F
GATATCGAGTGTGCTGGTAGGATTGGCGTCTTTCCGGAACCTGAATACGATCCCGTCATCCAAATTGCCAACGTTGTGAGTATTCGTGGCGCTAAGAAACCATTCATTCGTAATGTGTTT 1440
                                                                              **   * *                                *
+  +  +                                                                                                  N
T  L  N  T  C  S  P  I  T  G  S  M  I  F  S  H  A  T  E  E  E  M  L  S  N  W  R  N  F  I  I  K  V  D  P  D  V  I  I  G
ACTCTGAATACATGCTCACCCATAACAGGTTCAATGATTTTTTCCCACGCCACTGAAGAGGAAATGTTGAGCAATTGGCGTAACTTTATCATCAAAGTTGATCCTGATGTTATCATTGGT 1560 M6
      *                                                                                                  A

Y  N  T  T  N  F  D  I  P  Y  L  L  N  R  A  K  A  L  K  V  N  D  F  P  Y  F  G  R  L  K  T  V  K  Q  E  I  K  E  S  V
TATAATACTACAAATTTTGATATCCCTTATCTTTTAAACCGTGCAAAGGCGCTAAAGGTGAATGATTTCCCATATTTTGGAAGGTTAAAAACCGTTAAGCAAGAAATTAAAGAGTCTGTG 1680

F  S  S  K  A  Y  G  T  R  E  T  K  N  V  N  I  D  G  R  L  Q  L  D  L  L  Q  F  I  Q  R  E  Y  K  L  R  S  Y  T  L  N
TTCTCTTCGAAGGCTTATGGTACAAGAGAAACCAAAAATGTCAATATTGACGGCCGATTACAGTTGGATCTTTTGCAATTTATTCAGCGTGAGTATAAACTAAGATCCTACACGTTGAAT 1800

A  V  S  A  H  F  L  G  E  Q  K  E  D  V  H  Y  S  I  I  S  D  L  Q  N  G  D  S  E  T  R  R  R  L  A  V  Y  C  L  K  D
GCAGTCTCTGCGCACTTTTTAGGTGAACAGAAGGAGGATGTACATTATAGCATCATTTCTGATCTACAAAATGGCGATAGTGAAACAAGAAGAAGGTTGGCCGTTTACTGTTTGAAAGAC 1920

                                                               C
A  Y  L  P  L  R  L  M  E  K  L  M  A  L  V  N  Y  T  E  M  A  R  V  T  G  V  P  F  S  Y  L  L  A  R  G  Q  Q  I  K  V
GCCTACCTGCCTTTAAGGCTTATGGAAAAACTAATGGCGTTAGTTAACTATACAGAAATGGCTCGTGTTACAGGTGTGCCATTTTCATATTTACTAGCTCGTGGTCAACAAATTAAAGTT 2040 M5
                                                     *         T

V  S  Q  L  F  R  K  C  L  E  I  D  T  V  I  P  N  M  Q  S  Q  A  S  D  D  Q  Y  E  G  A  T  V  I  E  P  I  R  G  Y  Y
GTTTCTCAACTATTTCGAAAGTGCCTGGAGATTGATACTGTGATACCTAACATGCAATCTCAGGCCTCTGATGACCAATATGAGGGTGCCACTGTTATTGAGCCTATTCGTGGTTATTAC 2160

D  V  P  I  A  T  L  D  F  N  S  L  Y  P  S  I  M  M  A  H  N  L  C  Y  T  T  L  C  N  K  A  T  V  E  R  L  N  L  K  I
GATGTACCGATTGCAACTTTGGATTTCAATTCTTTATATCCAAGTATTATGATGGCGCACAACCTATGTTATACAACACTTTGTAACAAAGCTACTGTAGAGAGATTGAATCTTAAAATT 2280

                  +  +  +
D  E  D  Y  V  I  T  P  N  G  D  Y  F  V  T  T  K  R  R  R  G  I  L  P  I  I  L  D  E  L  I  S  A  R  K  R  A  K  K  D
GACGAAGACTACGTCATAACACCTAATGGAGATTATTTTGTTACCACAAAAAGAAGGCGTGGTATATTACCAATTATTCTGGATGAATTAATAAGTGCTAGAAAACGCGCTAAAAAAGAT 2400
                  *   *  *

L  R  D  E  K  D  P  F  K  R  D  V  L  N  G  R  Q  L  A  L  K  I  S  A  N  S  V  Y  G  F  T  G  A  T  V  G  K  L  P  C
CTGAGAGATGAGAAGGATCCATTCAAAAGAGATGTTTTAAATGGTAGACAATTGGCTTTGAAGATTTCAGCTAACTCTGTCTATGGTTTTACAGGAGCGACGGTGGGTAAATTGCCATGT 2520

L  A  I  S  S  S  V  T  A  Y  G  R  T  M  I  L  K  T  K  T  A  V  Q  E  K  Y  C  I  K  N  G  Y  K  H  D  A  V  V  V  Y
TTAGCCATTTCTTCATCTGTTACTGCTTATGGTCGTACCATGATTTTAAAAACTAAAACCGCAGTCCAAGAAAAATATTGTATAAAGAATGGTTATAAGCACGATGCCGTTGTGGTTTAC 2640

G  D  T  D  S  V  M  V  K  F  G  T  T  D  L  K  E  A  M  D  L  G  T  E  A  A  K  Y  V  S  T  L  F  K  H  P  I  N  L  E
GGTGACACTGATTCCGTTATGGTAAAGTTTGGTACAACAGATTTAAAGGAAGCTATGGATCTTGGTACCGAAGCTGCCAAATATGTCTCCACTCTATTCAAACATCCGATTAACTTAGAA 2760

F  E  K  A  Y  F  P  Y  L  L  I  N  K  K  R  Y  A  G  L  F  W  T  N  P  D  K  F  D  K  L  D  Q  K  G  L  A  S  V  R  R
TTTGAAAAAGCATACTTCCCTTACCTTTTGATAAATAAAAAGCGTTATGCAGGTTTATTCTGGACTAATCCTGACAAGTTTGACAAGTTGGACCAAAAAGGCCTTGCTTCTGTCCGTCGT 2880

                                                                                       +
D  S  C  S  L  V  S  I  V  M  N  K  V  L  K  K  I  L  I  E  R  N  V  D  G  A  L  A  F  V  R  E  T  I  N  D  I  L  H  N
GATTCCTGTTCCTTGGTTTCTATTGTTATGAATAAAGTTTTAAAGAAAATTTTAATTGAAAGAAATGTAGATGGTGCTTTAGCTTTTGTCAGAGAAACTATCAATGATATTCTGCATAAT 3000
                                                                                        *

R  V  D  I  S  K  L  I  I  S  K  T  L  A  P  N  Y  T  N  P  Q  P  H  A  V  L  A  E  R  M  K  R  R  E  G  V  G  P  N  V
AGAGTAGATATTTCAAAGTTGATTATATCAAAGACGTTAGCCCCAAATTACACAAATCCACAGCCGCACGCCGTTTTGGCTGAACGTATGAAGAGGAGAGAGGGCGTTGGTCCAAATGTT 3120

G  D  R  V  D  Y  V  I  I  G  G  N  D  K  L  Y  N  R  A  E  D  P  L  F  V  L  E  N  N  I  Q  V  D  S  R  Y  Y  L  T  N
GGTGATCGTGTGGACTATGTCATTATCGGTGGTAATGATAAACTTTACAATAGAGCAGAAGATCCATTATTTGTACTAGAAAACAATATTCAAGTGGATTCGCGCTATTATTTAACTAAT 3240
```

```
                                    +
Q  L  Q  N  P  I  I  S  I  V  A  P  I  I  G  D  K  Q  A  N  G  M  F  V  V  K  S  I  K  I  N  T  G  S  Q  K  G  G  L  M
CAATTACAAAATCCAATCATTAGTATTGTTGCACCTATTATTGGCGACAAACAGGCGAACGGTATGTTCGTTGTGAAATCCATTAAAATTAACACAGGCTCTCAAAAAGGAGGCTTGATG 3360
                                   ***

S  F  I  K  K  V  E  A  C  K  S  C  K  G  P  L  R  K  G  E  G  P  L  C  S  N  C  L  A  R  S  G  E  L  Y  I  K  A  L  Y
AGCTTTATTAAAAAAGTTGAGGCTTGTAAAAGTTGTAAAGGTCCGTTGAGGAAAGGTGAAGGCCCTCTTTGTTCAAACTGTCTAGCAAGGTCTGGAGAATTATACATAAAGGCATTATAC 3480

D  V  R  D  L  E  E  K  Y  S  R  L  W  T  Q  C  Q  R  C  A  G  N  L  H  S  E  V  L  C  S  N  K  N  C  D  I  F  Y  M  R
GATGTCAGAGATTTAGAGGAAAAATACTCAAGATTATGGACACAATGCCAAAGGTGCGCTGGTAACTTACATAGTGAAGTTTTGTGTTCAAATAAGAACTGTGACATTTTTTATATGCGG 3600

V  K  V  K  K  E  L  Q  E  K  V  E  Q  L  S  K  W
GTTAAGGTTAAAAAAGAGCTGCAGGAGAAAGTAGAACAATTAAGCAAATGGTAAAAAACGATAGGGTGGCACATCATATTAGGATTAAGAAAGGCTAACAACTTTTTGCATGTTGGTGGA 3720

TATATATGTATATATAAATAGATACTTGGAAAAATCCAGATTCAAACAATGTTTTTGAAATAATGCTTCTCATGTTTAGAGGCAAGATAATTCTGAGTATGTTTTGGGTATTTTATTGTC 3840

AGTAATTTGTAAACGCTGCAATTCTAATGAGACCGAGCTTTGTTGCTGTATGTCATTGAACCAAGGGTGGTTCAAAGCTTN
```

Figure 4. Corrected version of gene POL3. Several errors were found in the original version of gene POL3 (Boulet et al., 1989). They are marked by stars. Some of them cause amino acid changes, they are capped by +. The position of four thermosensitive mutations are indicated (M2, M5, M6, M8).

have been obtained. The mutation of four of them was localized by DNA sequencing (Figure 4).

During the search for suppressors we observed that it is preferable to satisfy the following conditions to avoid the isolation of pseudo-suppressors. (i) the chromosomal POL3 gene should be completely deleted to keep clear of intragenic recombinants ; (2) the thermosensitive mutants should be inserted in place of the chromosomal wild-type copy of POL3 to avoid suppression due to mutations increasing the copy number of the LEU2 vector.

Acknowledgments. We thank A. Besnard for excellent secretarial assistance. This work was supported by the CNRS, the Institut Curie and the Association pour la Recherche sur le Cancer (contrat n° 6087).

References

Bauer, G.A. and Burgers, P.M.J. (1990) Nucl. Acids Res. 18:261-265.

Bauer, G.A., Heller, H.M. and Burgers, P.M.J. (1988) J. Biol. Chem. 263:917-924.

Bernad, A., Blanco, L., Lazaro, J.M., Martin, G. and Salas, M. (1989) Cell 59:219-228.

Boeke, J.D., Lacroute, F. and Fink, G.R. (1984) Mol. Gen. Genet. 197:345-346.

Boulet, A., Simon, M., Faye, G., Bauer, G.A. and Burgers, P.M.J. (1989) EMBO J. 8:1849-1854.

Burgers, P.M.J., Bambara, R.A., Campbell, J.L., Chang, L.M.S., Downey, K.M., Hübscher, U., Lee, M.Y., So, A.G. and Spadori, S. (1990) Eur. J. Biochem. 191:617-618.

Fabre, F. (1978) Nature 272:795-798.

Kouzarides, T., Bankier, A.T., Satchwell, S.C., Weston, K., Tomlinson, P. and Barrell, B.G. (1987) J. Virol. 61:125-133.

Magni, G.E. and von Borstel, R.C. (1962) Genetics 47:1097-1108.

Meselson, M.S. and Radding, C.M. (1975) Proc. Natl Acad. Sci. USA 72 : 358-361.

Morrison, A., Araki, H., Clark, A., Hamatake, R. and Sugino, A. (1990) Cell 62:1143-1151.

Pizzagalli, A., Valsasnini, P., Plevani, P. and Lucchini, G. (1988) Proc. Natl Acad. Sci. USA 85:3772-3776.

Pringle, J.R. and Hartwell, L.H. (1981) In: The Molecular Biology of the Yeast Saccharomyces : Life Cycle and Inheritance. J.N. Strathern, E.W. Jones and J.R. Broach (eds). Cold Spring Harbor Laboratory, Cold Spring Harbor, N.Y.

Roman, H. and Fabre, F. (1983) Proc. Natl Acad. Sci. USA 80:6912-6916.

Szostak, J.W., Orr-Weaver, T.L., Rothstein, R., Stahl, F. (1983) Cell 33:25-35.

von Borstel, R.C., Cain, K.T. and Steinberg, C.M. (1971) Genetics 69:17-27.

Wildenberg, J. (1970) Genetics 66:2291-2304.

Genetic Control of the DNA Polymerase α-Primase complex in the Yeast *Saccharomyces cerevisiae*

P. Plevani, Foiani M., Francesconi S., Pizzagalli A., Santocanale C., Muzi Falconi M., Piatti S., Piseri A., Derossi D., Longhese M.P., Locati F., Tazzi R. and Lucchini G.
Dipartimento di Genetica e di Biologia dei Microrganismi
Università di Milano
Via Celoria 26
20133 Milano
Italy

Three major control points can be envisaged in the cell division cycle of eukaryotic organisms. The transition from a quiescent state to active cell proliferation, the initiation of DNA synthesis and the induction of mitosis. By combining genetic approaches in the yeast *Saccharomyces cerevisiae* and *Schizosaccharomyces pombe* with the biochemical characterization of proteins that induce the mitotic state in other eukaryotic cells, a molecular description of the series of events that regulate the onset of mitosis is beginning to emerge (Cross et al. 1989; Nurse 1990). The $p34^{CDC28/cdc2+}$ protein kinase plays a major role in the induction of mitosis by associating with a subset of oscillating proteins (cyclins) and by recognizing and modifying a variety of mitotic substrates (Lewin 1990). It has been proposed that $p34^{CDC28/cdc2+}$ would allow the phosphorylation of a different set of substrates involved in the G1 to S phase transition by associating with other cyclin-like proteins (Nurse 1990; Pines and Hunter 1990). Indeed, three polypeptides with a significant homology to mitotic cyclins, but performing an essential overlapping function required for G1 to S phase passage have been identified in *S. cerevisiae* as the products of the *CLN1*, *CLN2* and *CLN3* genes (Wittenberg et al. 1990). Moreover, it appears that some essential proteins directly involved in DNA replication are phosphorylated in a cell-cycle dependent manner (Wang 1991) and a specific replication factor identical or very closely related to $p34^{CDC28/cdc2+}$ protein kinase is required for the activation of DNA replication in human and Xenopous egg extracts (Blow and Nurse 1990; D'Urso et al. 1990).

A prerequisite to address questions related to the regulation of DNA replication in eukaryotic organisms rests in the comprehensive knowledge of the proteins involved in this process and in the isolation of the corresponding genes. Such studies are limited by the low abundance of eukaryotic replication proteins, proteolysis during purification and the lack of useful genetic approaches in higher eukaryotic cells. These problems have been partially circumvented by the development of cell-free reconstituted viral DNA replication systems, by the relevance of immunoaffinity purification strategies and by the application of the powerful tools provided by the sophisticated improvement of yeast molecular genetics (Newlon 1988; Stillman 1989; Wang 1991).

In the last few years our laboratory has been mainly involved in the purification of DNA replication proteins from cells of the yeast *S. cerevisiae* and in the cloning and characterization of their respective genes by using immunological probes and reverse genetics strategies. This manuscript will briefly summmarize our present knowledge on the structure and function of the DNA polymerase α-primase complex and on the regulation of the expression of the genes coding for these essential proteins. The DNA polymerase α - primase complex was purified from yeast cells by using a highly selective immunoaffinity purification procedure and contains four subunits of 180, 86, 58 and 48 kilodalton (kDa). DNA polymerase activity is associated with the p180 polypeptide, while the p58 and p48 protein species represent the two DNA primase subunits (reviewed in Plevani et al. 1988). The function of the p86 polypeptide, often found as a 74 kDa protein band due to partial proteolysis, is still unknown. It is quite remarkable that the structure and function of this bifunctional protein complex has been highly conserved in a wide range of phylogenetically distant species (Wang 1991). Antibodies against the four isolated subunits of the yeast complex were produced and these immunological reagents have been used to clone the corresponding genes from a yeast genomic library constructed in the λgt11 expression vector (Plevani et al. 1988; Hinkle, personal communication). The first important conclusions derived from cloning of the genes coding for DNA polymerase α (*POL1*), the p74-86 subunit (*POL12*, Hinkle personal communication), the p48 (*PRI1*) and p58 (*PRI2*) subunits of DNA primase were represented by the finding that these genes are unique in the haploid yeast genome, map on different chromosomes and code for functions which are essential for cell viability (reviewed in Plevani et al. 1988). These results demonstrate that the *in vivo* role of the DNA polymerase α - primase polypeptides cannot be vicariated by anyone of the other DNA polymerases identified in yeast cells or by the DNA primase activity that has been claimed to be present as an isolated enzymatic activity (Burgers 1989; Wang 1991). A role for the polymerase-primase complex in the initiation of DNA replication at an origin and in the synthesis of the lagging strand at a replication fork has been proposed on the basis of *in vitro* studies with highly purified protein components from human cells (Tsurimoto et al. 1990). This assumption has been further substantiated by complementary genetics and biochemical approaches mostly performed in the yeast and viral systems. Computer-assisted protein sequence comparison and production and characterization of temperature-sensitive (ts) mutants in the DNA polymerase α and primase genes have been particularly informative. In fact, it has been found that yeast DNA polymerase α, δ and ε belong to a large family of enzymes characterized by six regions of extensive homology (regions I to VI) present in the same spatial arrangement (Burgers 1989; Wang 1991). The identification of amino acid sequence homologies in related proteins is considered to be indicative of important evolutionarily conserved functional domains. The available techniques of mutagenizing a cloned gene and the possibility to replace the wild-type copy with the mutated alleles in yeast provides a powerful tool to perform structure-function relationship studies.

Four DNA polymerase α ts mutants (*pol1-1*, *hpr3*, *cdc17-1*, *cdc17-2*) have been characterized and sequenced in our laboratory (Pizzagalli et al. 1988; Lucchini et al. 1988; Lucchini et al.

1990). Growth rate analysis and measurement of DNA synthesis show that only the *cdc17-1* mutation causes a quick-stop phenotype after shift to non-permissive temperature. The mutant phenotype is caused by a single amino acid substitution within region VI, which therefore appears to contribute to the formation of an active catalytic site. The remaining three slow-stop mutants seem to identify functionally distinct domains. Strain *cdc17-2* has suffered a single amino acid substitution within region IV that is conserved also in yeast DNA polymerase δ, while the *pol1-1* and *hpr3* alleles carry a point mutation causing the substitution of the same glycine-493 to arginine and glutamic acid, respectively (Lucchini et al. 1990). These last mutations alter an amino acid residue in a region exhibiting near perfect homology with human DNA polymerase α in the NH_2-terminal half of the enzymes. Since the *pol1-1* mutant shows a decreased stability of the polymerase-primase complex, and yeast and human DNA polymerases α are the only sequenced enzymes associated with DNA primase, we proposed that this region, designed as region P, might contribute to the formation of a DNA primase interaction domain (Plevani et al. 1988). Both yeast and human DNA polymerases α have a cystein-rich region towards the carboxyl terminus, potentially able to form a Zn-finger motif which might be required for proper binding with the template DNA molecule. Most of the mutations conferring an altered sensitivity of herpes simplex virus DNA polymerase α to aphidicolin or nucleotide and pyrophosphate analogues fall in the consensus regions II and III which can be directly involved in dNTPs binding and/or pyrophosphate hydrolysis (Wang 1991). Site specific mutagenesis in region I of adenovirus and ϕ29 DNA polymerases yielded non functional enzymes (Wang 1991), suggesting an essential role for this region in the catalytic function of α-like DNA polymerases. Altogether these data indicate that the amino acid conserved sequences in DNA polymerase α are critical for the interactions defining the multiple functional domains necessary for the proper activity of this complex enzyme.

It has been recently found that the amino acid sequences of the two subunits of yeast and mouse DNA primase are also highly homologous (Prussak et al. 1989; Francesconi et al. 1991). To start establishing the physiological role of the two yeast DNA primase subunits, we produced and characterized conditional and lethal mutations in the cloned *PRI1* and *PRI2* genes (Francesconi et al. 1991). Most of the mutations causing a null phenotype were due to the formation of stop codons generating the synthesis of truncated non-functional polypeptides. The *pri1-3* lethal allele was particularly informative since it allowed to establish that the last thirty amino acids of the p48 polypeptide are essential for cell viability. All the missense mutations in *PRI1* or *PRI2* fall in regions highly conserved in the mouse DNA primase subunits supporting the functional importance of amino acid sequence conservation. An highly selective autocatalytic affinity labeling procedure indicated that both primase subunits might partecipate to the formation of the primase active center (Foiani et al. 1989). Transplacement of the conditional primase alleles in the corresponding *PRI1* and *PRI2* chromosomal loci and a preliminary physiological characterization of the derived isogenic mutant strains further supported the notion that both polypeptides play an essential role in DNA replication. In fact, an evident decrease in the rate of cell growth and *in vivo* DNA synthesis was observed in all *pri1* and *pri2* mutant cells after shift to non-permissive temperature, suggesting that the primary defect in these strains is an impairment in DNA replication. Furthermore, dumbbell-shaped cells, a terminal phenotype typical of yeast DNA replication mutants, accumulated after shift to restrictive temperature of the mutant cells. Finally, alkaline sucrose gradient analysis of the DNA synthesized *in vivo* at the restrictive temperature showed that *pri1* and *pri2* conditional mutants failed to synthesize normal levels of high molecular weight DNA products (Francesconi et al. 1991). These results represent the first *in vivo* demonstration of the role of DNA primase in eukaryotic DNA replication.

The high degree of homology between α-type DNA polymerases from a variety of eukaryotic organisms and between yeast and mouse DNA primase, together with the availability of ts *pol1*, *pri1* and *pri2* mutant strains allowed to verify whether the phenotype of these mutants could be complemented by the expression of heterologous proteins with analogous function. We have found that the *in vivo* defect of *pri1* and *pri2* conditional and lethal mutants was not complemented by the expression of the corresponding mouse DNA primase genes driven by strong constitutive or inducible yeast promoters (unpublished observations). Similar results were obtained when the human DNA polymerase α gene was expressed in the yeast *pol1-1* mutant strain (Francesconi and Wang, personal communication). These data indicate that the DNA polymerase α - primase complex might represent a species-specific enzymatic function, in agreement with the results obtained with polyoma and SV40 cell-free DNA replication systems (Stillman 1989).

YEAST GENES INVOLVED IN DNA SYNTHESIS PERIODICALLY EXPRESSED IN THE CELL CYCLE AT THE G1/S PHASE BOUNDARY

GENE	GENE PRODUCT/FUNCTION	REFERENCES
CDC8	Thymidilate kinase	White et al., 1987
CDC21	Thymidilate synthase	Storms et al., 1984
CDC9	DNA ligase	White et al., 1987
POL1 (CDC17)	DNA polymerase α	Johnston et al., 1987
POL12	DNA polymerase α accessory factor	D. Hinkle (personal communication)
PRI1	Small subunit DNA primase	Johnston et al., 1987
PRI2	Large subunit DNA primase	Foiani et al., 1989
POL3 (CDC2)	DNA polymerase δ	Bauer and Burgers, 1990
POL30	yPCNA	Bauer and Burgers, 1990
CDC6	DNA synthesis	Zhou and Jong, 1990
DBF4	DNA synthesis	Chapman and Johnston, 1989

Table 1. The listed yeast genes code for functions related to DNA synthesis and the level of their transcripts fluctuate with the same kinetics during mitotic cell cycle, reaching a peak at the G1/S phase boundary.

In the yeast *S. cerevisiae* the expression of several DNA synthesis related genes, including *POL1*, *PRI1* and *PRI2*, fluctuates during the mitotic cell cycle with the level of

their transcripts peaking around the G1 to S phase boundary (Table 1).
Therefore, it is tempting to speculate that the coordinated expression of all these genes might involve the action of common cis- and trans-acting regulatory elements. Interestingly, a nucleotide sequence survey of the 5' non-coding regions of all the DNA synthesis genes listed in Table I and those of DNA polymerase ε (*POL2*), DNA topoisomerase II (*TOP2*) and yeast replication protein A (*RPA1*) revealed the presence, between -200 and -100 from the initial ATG, of the sequence 5'ACGCGT3' or a similar hexanucleotide with a single base-pair mismatch (Figure 1).

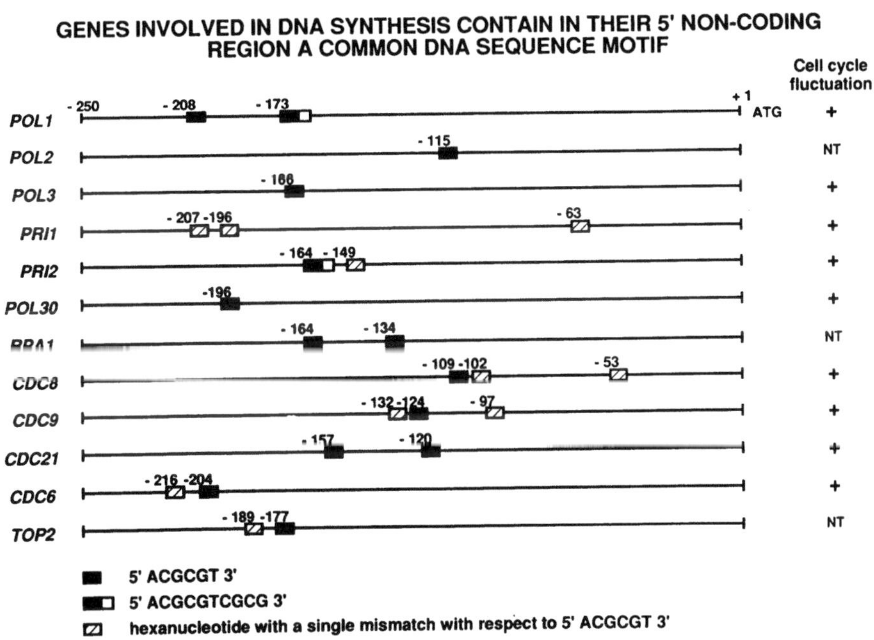

Figure 1. Identification and positioning of a common DNA sequence motif in the 5'non-coding region of DNA synthesis related genes whose transcripts fluctuate coordinately during the mitotic cell cycle (+). NT (not tested).

The analysis of *POL1-CYC1* hybrid promoters has further supported our initial observation that the fluctuation of *POL1* mRNA level during mitotic cell cycle involves a transcriptional control (Johnston et al. 1987). In fact, a 40 bp sequence from the upstream region of the *POL1* coding sequence containing the above hexanucleotide is sufficient to promote transcription of a *CYC1-lacZ* fusion gene deprived of its own UAS and the corresponding transcript fluctuates during the cell cycle with the same kinetics of *POL1* mRNA. Parallel deletion analysis of the *POL1* 5'non-coding region has confirmed that the same 40 bp sequence is necessary for proper control of the level of *POL1* mRNA during mitotic cell-cycle, while elements located downstream are necessary to allow initiation of *POL1* transcription at the correct positions (Pizzagalli et al., manuscript in preparation). These results, in conjunction with the search of mutants altered in trans-acting factors which regulate *POL1* gene expression, represent our initial effort to improve the knowledge of the mechanisms controlling the expression of DNA synthesis genes and possibly influencing the G1 to S phase transition.

ACKNOWLEDGEMENTS. Work in our laboratory has been partially supported by the CNR Target Project on Biotechnology and Bioinstrumentation. S.F., A.P. and S.P. were supported by fellowships from Fondazione Buzzati-Traverso. M.M.F. was supported by a fellowship from Istituto Mobiliare Italiano.

REFERENCES

Bauer GA, Burgers PMJ (1990) Molecular cloning, structure and expression of the yeast proliferating cell nuclear antigen gene. Nucleic Acids Res 18:261-265

Blow JJ, Nurse P (1990) A cdc2-like protein is involved in the initiation of DNA replication in Xenopous egg extracts. Cell 62:855-862

Burgers PMJ (1989) Eukaryotic DNA polymerases α and δ: conserved properties and interactions, from yeast to mammalian cells. Progress in Nucleic Acids Res and Molec Biol 37:235-282

Chapman JW, Johnston LH (1989) The yeast gene, *DBF4*, essential for entry into S phase is cell cycle regulated. Exp Cell Res 180:419-428

Cross F, Roberts J, Weintraub H (1989) Simple and complex cell cycles. Ann Rev Cell Biol 5:341-395

D'Urso G, Marracino RL, Marshak DR, Roberts JM (1990) Cell cycle control of DNA replication by a homologue from human cells of the p34^{cdc2} protein kinase. Science 250:786-791

Foiani M, Lindner AG, Hartmann GR, Lucchini G, Plevani P (1989) Affinity labeling of the active center and ribonucleoside triphosphate binding site of yeast DNA primase. J Biol Chem 264:2189-2194

Foiani m, Santocanale C, Plevani P, Lucchini G (1989) A single essential gene, *PRI2*, encodes the large subunit of DNA primase in *Saccharomyces cerevisiae*. Mol Cell Biol 9:3081-3087

Francesconi S, Longhese MP, Piseri A, Santocanale C, Lucchini G, Plevani P (1991) Mutations in conserved yeast DNA primase domains impair *in vivo* DNA replication. Proc Natl Acad Sci USA (in press)

Johnston LH, White JHM, Johnson AL, Lucchini G, Plevani P (1987) The yeast DNA polymerase I transcript is regulated in both the mitotic cell cycle and in meiosis and is also induced after DNA damage. Nucleic Acids Res 15:5017-5030

Lewin B (1990) Driving the cell cycle: M phase kinase, its partners and substrates. Cell 61:743-752

Lucchini G, Mazza C, Scacheri E, Plevani P (1988) Genetic mapping of the *S. cerevisiae* DNA polymerase I gene and characterization of a *pol1* temperature-sensitive mutant altered in DNA primase-polymerase complex stability. Mol Gen Genet 212:459-465

Lucchini G, Muzi Falconi M, Pizzagalli A, Aguilera A, Klein HL, Plevani P (1990) Nucleotide sequence and characterization of temperature-sensitive *pol1* mutants of *Saccharomyces cerevisiae*. Gene 90:99-104

Newlon C (1988) Yeast chromosome replication and segregation. Microbiological Rev 52:568-601

Nurse P (1990) Universal control mechanism regulating onset of M phase. Nature 344:503-508

Pines J, Hunter T (1990) p34^{cdc2}: the S and M kinase? The New Biologist 2:389-401

Pizzagalli A, Valsasnini P, Plevani P, Lucchini G (1988) DNA polymerase I gene of *Saccharomyces cerevisiae*: nucleotide sequence mapping of a temperature-sensitive mutation and protein homology with other DNA polymerases. Proc Natl Acad Sci USA 85:3772-3776

Plevani P, Foiani M, Muzi Falconi M, Pizzagalli A, Santocanale C, Francesconi S, Valsasnini P, Comedini A, Piatti S, Lucchini G (1988) The yeast DNA polymerase-primase complex: genes and proteins. Biochim Biophys Acta 951:268-273

Prussak CE, Almazan MT, Tseng BY (1989) Mouse primase p49 subunit molecular cloning indicates conserved and divergent regions. J Biol Chem 264:4957-4963

Stillman B (1989) Initiation of eukaryotic DNA replication in vitro. Ann Rev Cell Biol 5:197-245

Storms RK, Ord RW, Greenwood MT, Mirdamadi B, Chu FH, Belfort M (1984) Cell cycle-dependent expression of thymidilate synthase in *Saccharomyces cerevisiae*. Mol Cell Biol 4:2858-2864

Tsurimoto T, Melendy T, Stillman B (1990) Sequential initiation of lagging and leading strand synthesis by two different

polymerase complexes at the SV40 DNA replication origin. Nature 346:534-539
Wang TSF (1991) Eukaryotic DNA polymerases. Ann Rev Biochem (in press)
White JHM, Green SR, Barker DG, Dumas LB, Johnston LH (1987) The *CDC8* transcript is cell cycle regulated in yeast and is expressed coordinately with *CDC9* and *CDC21* at a point preceding histone transcription. Exp Cell Res 171:223-231
Wittenberg C, Sugimoto K, Reed SI (1990) G1-specific cyclins of *S. cerevisiae*: cell cycle periodicity, regulation by mating pherormone, and association with $p34^{cdc28}$ protein kinase. Cell 62:225-237
Zhon C, Jong A (1990) *CDC6* mRNA fluctuates periodically in the yeast cell cycle. J Biol Chem 265:19904-19909

Protein-primed Replication of Bacteriophage ∅29 DNA

M. Salas, M. Serrano, C. Gutiérrez, J.A. Esteban, M.A. Blasco, A. Bernad, J. Méndez, J. M. Lázaro, C. Garmendia, R. Freire, J.M. Sogo[1], J.M. Hermoso and L. Blanco.
Centro de Biología Molecular (CSIC-UAM)
Universidad Autónoma, Canto Blanco
28049 Madrid, Spain.

ABSTRACT Analysis by electron microscopy of the replicative intermediates synthesized in vitro in the ϕ29 replication system indicated that initiation of replication can occur from both DNA ends in the same template molecule. Type II molecules (DNA partially double-stranded and partially single-stranded) are produced by separation of the two displacement forks of double-stranded DNA molecules with two single-stranched branches when they meet.

In addition to the natural template ϕ29 DNA-terminal protein, the ϕ29 DNA polymerase is able to replicate single-stranded DNA with the sequences corresponding to the right 3' end of ϕ29 DNA by a protein-priming mechanism. In both cases, there was a strong preference for the formation of a covalent complex between the terminal protein p3 and dAMP. However, when poly dC was used as template, p3-dGMP was preferentially formed, indicating that the template specificity for the initiation reaction is provided by the 3' end of the template strand. In the absence of DNA, the ϕ29 DNA polymerase was able to deoxynucleotidylylate the terminal protein with any of the four dNTPs. In these conditions, the efficiency of the reaction is low, due to a strong decrease in affinity for the dNTPs. The ϕ29 DNA polymerase also catalyzed the +1 nucleotide addition on double-stranded DNA. The different dNMPs could act as template in this +1 addition reaction, as well as the 3' end of single-stranded DNA. The 3'→5' exonuclease activity of the ϕ29 DNA polymerase contributes to the fidelity of DNA synthesis by a factor of, at least, 100-fold. Six conserved

[1]ETH, Institut für Zellbiologie, Hönggerberg, CH-8093 Zürich, Switzerland

regions of amino acid homology have been found at the carboxyl-part and in the same linear arrangement in 27 DNA-dependent DNA polymerases belonging to the two main superfamilies: pol I-like and α-like DNA polymerases. According to the three-dimensional model of the Klenow fragment these regions are located in the proposed polymerization domain. Site-directed mutagenesis in three of these regions in the φ29 DNA polymerase supports this model.

The nature of the sequences recognized by the viral protein p6 at the φ29 DNA ends, and the structure of the protein p6-DNA complex have been further studied. The fact that stimulation of the initiation reaction by protein p6 is higher at conditions that stabilize the double-helix suggests that the nucleoprotein structure formed at the φ29 DNA ends facilitates unwinding at the replication origin allowing initiation to occur. The φ29 SSB protein p5 stimulates in vitro φ29 DNA replication by allowing reinitiations on new templates to take place.

INTRODUCTION

The Bacillus subtilis phage φ29 has a linear double-stranded DNA 19,285 bp long (Vlcek and Paces, 1986) with a short inverted terminal repeat of 6 bp (Escarmis and Salas, 1981; Yoshikawa et al., 1981) and the viral protein p3 (terminal protein) covalently linked at the two 5' ends (Salas et al., 1978) by a phosphoester bond between the OH group of Ser 232 and dAMP (Hermoso et al., 1985). Replication of φ29 starts at either DNA end by a protein-priming mechanism and proceeds by strand-displacement (Inciarte et al., 1980). An in vitro system with purified proteins has been developed that is able to initiate and elongate using terminal protein-φ29 DNA (p3-DNA) as template. A free molecule of the terminal protein p3 forms a complex with the viral DNA polymerase p2 (Blanco et al., 1987) and, in the presence of origin-containing templates and dATP, the DNA polymerase catalyzes the covalent linkage of dAMP to the terminal protein (Blanco and Salas, 1984; Watabe et al., 1984). The protein p3-dAMP initiation complex is elongated by the φ29 DNA polymerase to produce full-length φ29 DNA in a very processive way (Blanco and Salas, 1985; Blanco et al., 1989). In addition, the φ29 DNA polymerase has 3' → 5' exonuclease activity, with properties that enable it to act in a proofreading mechanism (C. Garmendia, A. Bernad, J. A. Esteban, L. Blanco and M. Salas, submitted). A pyrophosphorolytic activity has been recently characterized in the ϕ29 DNA polymerase (Blasco et al., 1991). Structural and functional studies have shown the presence of three regions of amino acid homology at the carboxyl part of α-like DNA polymerases, including the ϕ29 DNA polymerase (Bernad et al., 1987). By site-directed mutagenesis we have shown that these regions are involved in the synthetic activities of the ϕ29 DNA polymerase, i.e., initiation and polymerization, and do not affect the 3' → 5' exonuclease activity (Bernad et al., 1990; Blanco et al., 1991). In addition, three conserved segments of amino acid homology are present at the amino terminal region of α-like and E. coli DNA polymerase I (pol I)-like DNA polymerases (Bernad et al., 1989), where the 3' → 5' exonuclease activity of pol I lies (Derbyshire et al., 1988). By site-directed mutagenesis we have shown that, at least two of these regions, are involved in the 3' → 5' exonuclease activity of the ϕ29 DNA polymerase, but not in the synthetic activities (Bernad et al., 1989). By site-directed and deletion mutagenesis three functional domains have been characterized in the terminal protein, two of them for the interaction with the DNA polymerase and DNA, and another for the priming activity (Zaballos and Salas, 1989; Garmendia et al., 1988, 1990).

Two other viral proteins, p5 and p6, required for ϕ29 DNA replication in vivo (Mellado et al., 1980), stimulate the in vitro replication of p3-DNA. Protein p6 binds preferentially to the ϕ29 DNA ends (Prieto et al., 1988; Serrano et al., 1989), and stimulates the formation of the p3-dAMP initiation complex and its transition to the first elongation products (Pastrana et al., 1985; Blanco et al., 1988). Protein p5 has been shown to protect single-stranded DNA against nuclease degradation and to stimulate p3-DNA replication at incubation times where the replication in the absence of p5 levels off (Martín et al., 1989). Protein p5 binds to the single-stranded DNA portions of replicating molecules acting as a SSB protein that stimulates DNA replication by relieving from the inhibitory effect of single-stranded DNA (Gutiérrez et al., 1991).

In this paper we report on the further characterization of template and protein requirements of p3-DNA replication.

RESULTS AND DISCUSSION

a) Mechanism of replication of the two DNA strands of ϕ29 DNA.

To study whether or not the appearance of ϕ29 DNA type II replicative intermediates (linear DNA partially double - and partially single-stranded) requires the complete synthesis of one strand and the displacement of the complementary single-stranded DNA molecule, replication of p3-DNA, carried out in the in vitro purified system, was analyzed by electron microscopy. A significant amount of type II replicative intermediates (about 15% of the total replicative intermediates) were found after 2 min of incubation at 30ºC, when no synthesis of full-length ϕ29 DNA was detected. Moreover, when ϕ29 DNA molecules containing terminal protein at only one DNA end were used as template in the in vitro system, type II molecules were not seen, and single-stranded DNA molecules accumulated (C. Gutierrez, J.M. Sogo and M. Salas, submitted). These results support a model in which initiation of replication occurs from both DNA ends, and type II molecules are produced by separation of the two displacement forks when they meet.

b) Template-dependence for protein-primed initiation of replication

The formation of the p3-dAMP initiation complex using p3-DNA as template is stimulated about 50-fold when Mn^{2+} ions are used instead of Mg^{2+} (J.A. Esteban, A. Bernad, M. Salas and L. Blanco, submitted). Figure 1 shows the initiation reaction when each of the four dNTPs are used. With p3-DNA as template a strong preference for the formation of the p3-dAMP complex is observed, in agreement with the fact that a T is present at the 3' end. As also shown in Fig.1, a single-stranded DNA of 48 nucleotides, corresponding to the sequence of the 3' right terminus of ϕ29 DNA, gave rise to the formation of p3-dAMP with an efficiency about 15% of that obtained with the natural template p3-DNA. Some initiation reaction with dGTP and, to a lesser extent, with dTTP, was also observed. Being TTTCAT the sequence at the 3' end of this oligonucleotide these results suggest that, with single-stranded DNA templates, the initiation reaction may not always occur at the very 3' DNA end. Protein-primed initiation also occured with unspecific single-stranded DNA such as poly dC; the initiation complex formed was p3-dGMP with an efficiency about 2% with respect to the p3-dAMP complex formed with p3-DNA (Fig. 1). Using Mn^{2+} as activating metal ion, some initiation reaction could be

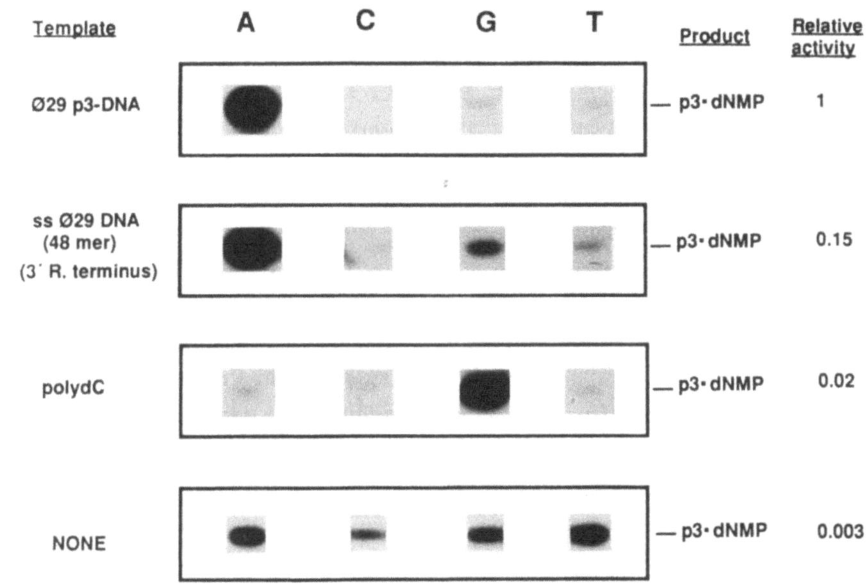

Figure 1. Template and nucleotide specificity of the initiation reaction catalyzed by the ϕ29 DNA polymerase using the ϕ29 terminal protein as primer. The initiation reaction was carried out essentially as described (Bernad et al., 1989) in the presence of 1mM $MnCl_2$, 20mM ammonium sulphate, 125 ng of purified terminal protein, 125 ng of purified ϕ29 DNA polymerase exonuclease deficient mutant, 0.25 μM either [α-^{32}P] dATP, [α-^{32}P] dCTP [α-^{32}P] dGTP or [α-^{32}P] dTTP, and either p3-DNA (0.5 μg), single-stranded oligonucleotide (48 mer) with the sequence of the 3' right terminus of ϕ29 DNA (130 ng), poly dC (0.5 μg), or without template. Incubation was carried out at 30ºC and then the samples were subjected to SDS-polyacrylamide gel electrophoresis. Autoradiography was performed for different times. The relative activity was calculated taking into account the incubation and exposure times.

detected in the absence of template, although in this case, the initiation complex could be formed between protein p3 and any of the nucleotides (see Fig. 1). The low level of terminal protein deoxynucleotidylylation catalyzed by the ϕ29 DNA polymerase in the absence of DNA is mainly due to a strong decrease in affinity for the dNTPs . All the above results indicate that the nucleotide specificity for the initiation reaction is provided by the 3' end at the template strand.

The template-independent deoxynucleotidylylation of the terminal protein described above was further characterized. The linkage was shown to be a phosphoester bond between the OH group of a serine residue and dAMP, as in the reaction in the presence of template. Moreover, the fact that the p3 mutant in which Ser^{232} was changed to Thr, was inactive in the template-independent initiation reaction indicates that Ser^{232} is also used in this reaction. In addition, the ϕ29 DNA polymerase mutants $Thr^{457} \rightarrow$ Pro and $Asp^{458} \rightarrow$ Gly, that had no initiation activity with p3-DNA as template, were also inactive in the template-independent reaction (L. Blanco, A. Bernad, J.A. Esteban and M. Salas, in preparation). In spite of the fact that deoxynucleotidylylation of the terminal protein by the ϕ29 DNA polymerase occurs in the absence of DNA, this reaction resembles the +1 nucleotide addition on double-stranded DNA catalyzed by several DNA polymerases (Clark, 1988), that has been also shown to be carried out by the ϕ29 DNA polymerase (see Fig. 2). The efficiency of this reaction depends on the nucleotide, being dATP>>dGTP>>dTTP>dCTP; the Km value for the dNTP is 10^5 to10^6-fold higher than in a template-dependent nucleotide addition. The different dNMPs can act as template in this +1 addition (see Fig. 2), decreasing the Km value for the complementary nucleotide. As also schematized in Fig. 2 the 3' end of single-stranded DNA can act as template in the +1 addition (unpublished results). Fig. 2 also shows an structural model for the initiation reaction catalyzed by the ϕ29 DNA polymerase when using the terminal protein as primer and single-stranded DNA as template, which resembles the incorporation of the first nucleotide when a template-primer is used.

c) Fidelity of replication by the ϕ29 DNA polymerase.

The 3' $\rightarrow$ 5' exonuclease activity of the ϕ29 DNA polymerase acts mainly on single-stranded DNA, showing a marked preference for excision of a mismatched versus a correctly paired 3' terminus (Garmendia et al., submitted). These characteristics enable the ϕ29 DNA polymerase to act in a proofreading mechanism. In fact, the 3' $\rightarrow$ 5' exonuclease activity of the ϕ29 DNA polymerase contributes to fidelity by a factor of, at least, 100-fold.

The insertion fidelity of the wild-type ϕ29 DNA polymerase is about 10^5, and the discrimination between the correct and the incorrect nucleotide is carried out mainly by differences in affinity. The fidelity decreases when Mn^{2+} ions are used instead of Mg^{2+}, due to an increase in the affinity for the incorrect nucleotide and in the capacity to elongate mismatches.

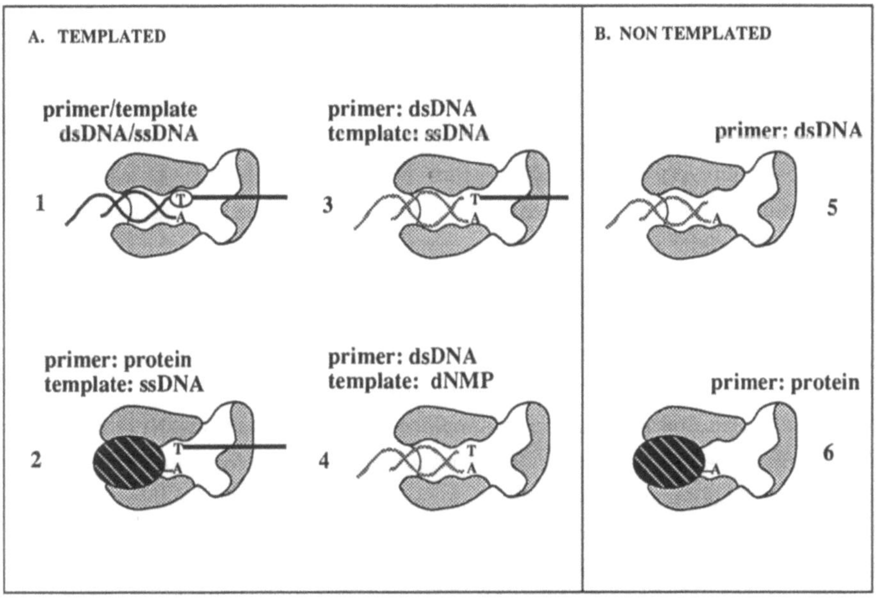

Figure 2. Structural model for the templated and non-templated polymerization reactions catalyzed by the ϕ29 DNA polymerase.

d) Functional domains in the ϕ29 DNA polymerase.

In addition to the 3' → 5' exonuclease domain already described (Bernad et al., 1989), significant amino acid sequence similarity has been found in the carboxy-terminal portion of 27 DNA-dependent DNA polymerases belonging to the two main superfamilies: pol I-like prokaryotic DNA polymerases and α-like prokaryotic and eukaryotic DNA polymerases (Blanco et al., 1991). The six most conserved carboxy-terminal regions are located in the same linear arrangement and contain highly conserved motifs and critical residues involved in the polymerization function. According to the three-dimensional model of the Klenow fragment of E. coli DNA polymerase I, these six conserved regions are located in the proposed polymerization domain, forming the metal and dNTP binding sites and the cleft for holding the DNA template. Site-directed mutagenesis carried out in three of these regions in the ϕ29 DNA polymerase, support some of these structural predictions (Blanco et al., 1991).

e) Function of protein p6 in ϕ29 DNA replication.

Protein p6 acts most probably as a dimer (Pastrana et al., 1985) and binds to double-stranded DNA cooperatively (Prieto et al., 1988). DNase I and hydroxyl radical footprinting analysis of protein p6-DNA complexes has revealed a regular pattern of repeating units of 24 bp, each one containing two contact sites. In addition, protein p6 restrains positive supercoils when bound to circular DNA. These results led us to propose a model in which protein p6 binds to DNA through the minor groove with a periodicity of 12 bp, bending the DNA and generating a right-handed superhelix in which the DNA wraps a multimeric protein p6 core (Serrano et al., 1990).
The main protein p6 recognition regions are located at positions 40 to 125, and 46 to 68 from the right and left ϕ29 DNA ends, respectively (Serrano et al., 1989). Not even a degenerated repeated sequence has been found in these regions, that would suggest a sequence-specific recognition. Thus, we favour the idea that sequence bendability plays a major role in protein p6 binding. This implies that bendability properties of these sequences should show also a periodicity of 12 bp, as it seems to be the case (Serrano et al., 1989). Furthermore, we have constructed concatemers of direct repeats of a 24 bp (dimer binding unit) sequence present in the mapped region from the left genome terminus. DNase I footprinting showed that protein p6 was bound to this sequence in the same position as in ϕ29 DNA. In addition, a higher binding affinity to this sequence was observed (to be published elsewhere).
To define more accurately the DNA structural features required for protein p6 binding, we have also studied binding to terminal fragments of the ϕ29-related phages Nf and GA-1. We have compiled the sequences from these phages recognized by p6 as well as the 200 bp terminal ϕ29 sequences (right and left) and arranged them in two sets of 7 bp, named *contact sites* and *DNase I hypersensitive sites*, in which those sites are centred. We have analysed for each set the relative frequency of each overlapping dinucleotide, finding that they are not random. The highest frequency sequence obtained was CAAT(A/C)(A/T)(A/T) for contact sites and ATAGACA for hypersensitive sites. A concatemer containing these sequences will be constructed to test the binding affinity of protein p6.
To obtain further information on the structure of the protein p6-DNA complex we have calculated the change in linking number induced by a dimer of protein p6 interacting with a binding unit (24 bp). Changes in superhelicity after protein p6 binding observed in pUC19-derived plasmids containing different number of binding units (9, 10 and 12) were measured. Results

obtained showed an average value of ΔLk/p6 dimer=0.13± 0.02. We cannot rule out a change in twist of the DNA bound to protein p6, however assuming that the DNA periodicity is 10.5 bp/turn, one turn of the superhelix would involve 184.6 bp.

It has been shown that protein p6 binding is required for the stimulation of the initiation reaction (Serrano et al., 1989). An interesting possibility to explain the mechanism of protein p6 activation is that the nucleoprotein structure would facilitate unwinding at the replication origin allowing the replication machinery to initiate DNA synthesis. This is in agreement with the higher stimulation by p6 at conditions that unfavour DNA breathing (low temperature, 0º-10ºC; high salt, 150 mM NaCl).

f) Function of protein p5 in p3-DNA replication.

Protein p5 is a single-stranded DNA binding protein (SSB) and stimulates *in vitro* DNA replication by the ϕ29 DNA polymerase with either p3-DNA or M13 single-stranded DNA as template (Martín et al., 1989; Gutiérrez et al., 1991). Other SSB proteins, such as *E. coli* SSB, T4 gp32, adenovirus DBP, human RF-A (Gutiérrez et al., 1991), and phage PRD1 protein p12 can functionally substitute for protein p5 (Pakula et al., 1990). Protein p5 readily binds to the single-stranded portions of ϕ29 replicating molecules; the stimulation of p3-DNA replication by protein p5 mainly occurs at incubation times when replication in the absence of p5 levels off, by allowing new initiations to occur (Martín et al., 1989).

To further study the above p5 effect, replication of p3-DNA was allowed to take place for 20 min at 30ºC in the absence of p5, conditions at which replication stops. Then, α-32P-dATP was added with and without protein p5 and, 30 seconds later, samples were taken, the DNA was digested with Hind III, and the fragments were separated by agarose gel electrophoresis. Densitometric analysis of the label in the different DNA fragments indicated that, in the absence of protein p5, label was distributed over the entire DNA length, indicating that no preferred stop or pause sites exist. However, a significant increase in the relative amount of label was found at both ϕ29 DNA ends when the pulse was carried out in the presence of protein p5. These findings are consistent with the idea that, after termination of the first replication round, protein p5 stimulates reinitiation on new templates.

g) Concluding remarks.

The fact that a linear single-stranded DNA can be replicated by a protein-priming mechanism of initiation suggests that the linear ends of the double-stranded ϕ29 DNA must be opened to function as replication origin. The special protein-priming mechanism of ϕ29 DNA replication require several virally-coded proteins, playing important roles in this process: 1) the terminal protein, which acts as a primer for initiation of replication of each DNA strand; 2) the DNA polymerase, specialized to use the terminal protein as a primer, replicating both strands with high processivity and by a strand-displacement mechanism. The strong 3' → 5' exonuclease activity of this enzyme acts coordinately during polymerization to ensure the accuracy of DNA replication; 3) protein p6, by forming a nucleoprotein structure at both ϕ29 DNA ends, is proposed to play an important role for opening both replication origins; 4) protein p5, a single-stranded DNA binding protein, is thought to play a dual role during ϕ29 DNA replication, to protect the long strectches of single-stranded DNA produced as replicative intermediates against nuclease degradation and to avoid the loss of replication factors with high affinity for single-stranded DNA, as the ϕ29 DNA polymerase.

ACKNOWLEDGEMENTS

This work has been aided by research grant 5R01 GM27242-12 from the National Institutes of Health, by grant nº PB87-0323 from Dirección General de Investigación Científica y Técnica and by an Institutional Grant from Fundación Ramón Areces. The technical assistance of L. Villar is gratefully acknowledged.

REFERENCES

- Bernad A, Blanco L, Lázaro JM, Martín G, Salas M (1989). A conserved 3'→ 5' exonuclease active site in prokaryotic and eukaryotic DNA polymerases. Cell 59: 219-228
- Bernad A, Lázaro JM, Salas M, Blanco L (1990). The highly conserved amino acid sequence motif Tyr-Gly-Asp-Thr-Asp-Ser in α-like DNA polymerases is required by phage ϕ29 DNA polymerase for protein-primed initiation and polymerization. Proc. Natl. Acad. Sci. USA 87: 4610-4614

- Bernad A, Zaballos A, Salas M, Blanco L (1987). Structural and functional relationships between prokaryotic and eukaryotic DNA polymerases. EMBO J. 6: 4221-4225.
- Blanco L, Bernad A, Blanco MA, Salas M (1991). A general structure for DNA-dependent DNA polymerases. Gene, in press
- Blanco L, Bernad A, Lázaro JM, Martín G, Garmendia C, Salas M (1989). Highly efficient DNA systhesis by the phage φ29 DNA polymerase. Symmetrical mode of DNA replication. J. Biol. Chem. 264: 8935-8940
- Blanco L, Bernad A, Salas M (1988). Transition from initiation to elongation in protein-primed φ29 DNA replication: salt-dependent stimulation by the viral protein p6. J. Virol. 62: 4167-4172
- Blanco L, Prieto I, Gutiérrez J, Bernad A, Lázaro JM, Hermoso JM, Salas M (1987). Effect of NH4+ ions on φ29 DNA-protein p3 replication: Formation of a complex between the terminal protein and the DNA polymerase. J. Virol. 61: 3983-3991
- Blanco L, Salas M (1984). Characterization and purification of a phage φ29-encoded DNA polymerase required for the initiation of replication. Proc. Natl. Acad. Sci. USA 81: 5325-5329
- Blanco L, Salas M (1985). Replication of phage φ29 DNA with purified terminal protein and DNA polymerase: Synthesis of full-length φ29 DNA. Proc. Natl. Acad. Sci. USA 82: 6404-6408
- Blasco MA, Bernad A, Blanco L, Salas M (1991). Characterization and mapping of the pyrophosphorolytic activity of the phage φ29 DNA polymerase. Involvement of amino acid motifs highly conserved in α-like DNA polymerases. J. Biol. Chem., in press
- Clark JM (1988). Novel non-templated nucleotide addition reactions catalyzed by prokaryotic and eukaryotic DNA polymerases. Nucl. Acids. Res 16: 9677-9686.
- Derbyshire V, Freemont PS, Sanderson MR, Beese L, Friedman JM, Joyce CM, Steitz TA (1988). Genetic and crystallographic studies of the 3' → 5' exonucleolytic site of DNA polymerase I. Science 240: 199-201
- Escarmis C, Salas M (1981). Nucleotide sequence at the termini of the DNA of Bacillus subtilis phage φ29. Proc. Natl. Acad. Sci. USA 78: 1446-1450
- Garmendia C, Hermoso JM, Salas M (1990). Functional domain for priming activity in the phage φ29 terminal protein. Gene 88: 73-79
- Garmendia C, Salas M, Hermoso JM (1988). Site-directed mutagenesis in the DNA linking site of bacteriophage φ29 terminal protein: isolation and characterization of a $Ser_{232} \rightarrow Thr$ mutant. Nucl. Acids Res. 16: 5727-5740
- Gutiérrez C, Martín G, Sogo JM, Salas M (1991). Mechanism of stimulation of DNA replication by bacteriophage φ29 SSB protein p5. J. Biol. Chem. 266: 2104-2111.
- Hermoso JM, Méndez E, Soriano F, Salas M (1985). Location of the serine residue involved in the linkage between the terminal protein and the DNA of φ29. Nucl. Acids. Res. 13: 7715-7728
- Inciarte MR, Salas M, Sogo JM (1980). Structure of replicating DNA molecules of Bacillus subtilis bacteriophage φ29. J. Virol. 34: 187-199
- Martín G, Lázaro JM, Méndez E, Salas M (1989). Characterization of the phage φ29 protein p5 as a single-stranded DNA binding protein. Function in φ29 DNA replication. Nucl. Acids. Res. 17: 3663-3672
- Mellado RP, Peñalva MA Inciarte MR, Salas M (1980). The protein covalently linked to the 5' termini of the DNA of Bacillus subtilis phage φ29 is involved in the initiation of DNA replication. Virology 104: 84-96

- Pakula T, Caldentey J, Serrano M, Gutiérrez C, Hermoso JM, Salas M, Bamford D (1990). Characterization of a DNA binding protein of bacteriophage PRD1 involved in DNA replication. Nucl. Acids. Res. 18: 6553-6557
- Pastrana R, Lázaro JM, Blanco L, García JA, Méndez E, Salas M (1985). Overproduction and purification of protein p6 of Bacillus subtilis phage φ29: Role in the initiation of DNA replication. Nucl. Acids Res. 13: 3083-3100
- Prieto I, Serrano M, Lázaro JM, Salas M Hermoso JM (1988). Interaction of the bacteriophage φ29 protein p6 with double-stranded DNA. Proc. Natl. Acad. Sci. USA 85: 314-318
- Salas M, Mellado RP, Viñuela E, Sogo JM (1978). Characterization of a protein covalently linked to the 5' termini of the DNA of Bacillus subtilis phage φ29. J. Mol. Biol. 119: 269-291
- Serrano M, Gutiérrez J, Prieto I, Hermoso JM, Salas M (1989). Signals at the bacteriophage φ29 DNA replication origins required for protein p6 binding and activity. EMBO J. 8: 1879-1885
- Serrano M, Salas M, Hermoso JM (1990). A novel nucleoprotein complex at a replication origin. Science 248: 1012-1016
- Vlcek V, Paces V (1986). Nucleotide sequence of the late region of Bacillus phage φ29 completes the 19285-bp sequence of φ29 genome. Comparison with the homologous sequence of phage PZA. Gene 46: 215-225
- Watabe K, Leusch M, Ito J (1984). Replication of bacteriophage φ29 DNA in vitro: The roles of terminal protein and DNA polymerase. Proc. Natl. Acad. Sci. USA 81: 5374-5378
- Yoshikawa H, Friedmann T, Ito J (1981). Nucleotide sequences at the termini of φ29 DNA. Proc. Natl. Acad. Sci. USA 78: 1336-1340
- Zaballos A, Salas M (1989). Functional domains in the bacteriophage φ29 terminal protein for interaction with the φ29 DNA polymerase and with DNA. Nucl. Acids. Res. 17: 10353-10366

Molecular Interactions at the Adenovirus Origin of DNA Replication

R.T. Hay, J. Bosher, A. Dawson and S.M. Temperley
Department of Biochemistry and Microbiology
University of St. Andrews
St. Andrews, Fife KY16 9AL
Scotland

ABSTRACT

The expression and purification of the proteins required for Ad2 DNA replication has resulted in the development of a completely defined system that can be used to study the molecular interactions that take place at the origin of DNA replication. Specific interactions between the replication proteins and the origin DNA as well as direct protein-protein interactions lead to formation of a large nucleoprotein complex that precedes the initiation reaction. Ad4 initiates DNA replication by the same mechanism utilised by Ad2 but has dispensed with the requirement for the sequence specific DNA binding proteins NFI and NFIII and as a result has a replication origin which consists of only 18bp. Although less well developed the remarkably simple requirements for Ad4 DNA replication make it an attractive system in which to study the basic processes of adenovirus DNA replication.

INTRODUCTION

The adenovirus genome is a linear double stranded DNA molecule of 35-36kb with inverted terminal repeats (ITR's) of about 100bp, the exact size depending upon serotype. DNA synthesis is initiated at either of the termini by transfer of dCMP, the terminal nucleotide, onto an 80kd precursor of terminal protein (pTP) in a template dependent reaction. The 3'OH of the pTP-dCMP complex serves as a primer for synthesis of the nascent strand by the viral DNA polymerase. Concomitant displacement of the non-template strand generates a single stranded molecule which then acts as a template for a second round of DNA synthesis (reviewed in Hay and Russell, 1991).

Adenovirus origins of replication are located at the molecular ends of the genome within the inverted terminal repeats. Covalently attached to each 5' end of the DNA is a terminal protein (TP) which is likely to be a *cis* acting protein component of the replication origin. While removal of the terminal protein reduces the efficiency of Ad2 and Ad4 DNA replication *in vitro* it does not abolish replication and plasmid templates, provided that the origin has been exposed by restriction enzyme cleavage, have 25% of the activity of protein linked genomes (Tamanoi and Stillman, 1982; Harris and Hay, 1988). Using plasmid templates, extensive mutational analysis has revealed that 4 regions within the terminal 51bp. of the Ad2 genome influence origin activity *in vitro* and *in vivo* (reviewed in Hay and Russell, 1989). The origin of Ad2 DNA replication (figure 1) consists of a core domain comprising the terminal 18bp of the genome which alone is only capable of supporting a low level of initiation and an auxiliary region encompassing nucleotides 25-50 which contains recognition sequences for the sequence specific DNA binding proteins nuclear factor I (NFI) and nuclear factor III (NFIII) (Nagata et al., 1983; Pruijn et al., 1986). Separating the core and the NFI binding site is a region of DNA where sequence changes are tolerated, but insertions or deletions are not (Adyha et al., 1986; Bosher et al., 1990; Wides et al., 1987).

Transfection assays carried out with Ad4 showed that in contrast to Ad2 only the terminal 18bp of the genome, which are identical in both serotypes and in Ad2 constitute the core origin, were required for efficient DNA replication *in vivo* (Hay 1985b). This is also the case *in vitro* where it was demonstrated that linearised plasmid containing only the terminal 18bp of the ITR could support initiation of DNA replication as effectively as a template containing a complete Ad4 ITR (Harris and Hay, 1988). The protein requirement for DNA replication differs markedly between Ad2 and Ad4 in that Ad4 appears to have circumvented the need for the host factors NFI/CTF and NFIII/Oct-1. It does not posses a NFI/CTF recognition site and whilst it does have a binding site for NFIII/Oct-1 neither factor is required for DNA replication *in vivo* or is capable of stimulating DNA replication *in vitro* (Hay *et al.*, 1988).

MATERIALS AND METHODS

Templates

Plasmid pHR18 contains the terminal 18bp of the Ad2 ITR which is identical in adenovirus types 2 and 4, while p4A2 contains the adenovirus type 4 ITR (Hay 1985a). pUDl67 pm9 to 16 (pM9 to pM16) all contain the terminal 67bp of the Ad2/5 ITR carrying a single base transition in each base pair from position 9 to 16 from the terminus and were obtained by C.R. Burrow and T.J. Kelly, Baltimore, U.S.A.. pMDC10 pm17 and 18 (pM17 and pM18) contain the terminal 69bp of the Ad2/5 ITR and carry single base transitions at positions 17 and 18 (Rawlins et al 1984). Plasmid DNA was prepared by two rounds of CsCl/ethidium bromide centrifugation. The ethidium bromide was removed by repeated extraction with caesium chloride saturated iso-butanol and the DNA was desalted by ultrafiltration in a centricon-30.

Purification of Ad4 replication proteins

HeLa S3 cells were infected with 100 plaque forming units/cell of adenovirus type 4 and after 90min medium containing 2% calf serum plus 10mM hydroxyurea was added. Cells were incubated for 22 hours at 37°C and extract was prepared as described (Challberg and Kelly, 1979). The NaCl concentration was adjusted to 50mM and the extract loaded on a DEAE-Sepharose column equilibrated in buffer containing 20mM HEPES (pH8.0), 5mM KCl, 0.5 mM $MgCl_2$, 0.5 mM DTT and 50mM NaCl. After washing with two volumes of buffer containing 50mM NaCl the column was eluted with two volumes of buffer containing 0.2 M NaCl. *In vitro* DNA replication activity was recovered in the 0.2M eluate and was applied to a denatured calf thymus DNA-sepharose column equilibrated in buffer containing 0.2M NaCl. The column was washed with two volumes of buffer containing 0.2M NaCl and developed with a gradient of 0.2M to 2M NaCl. Fractions active in DNA replication were dialysed against 20mM HEPES (pH8.0), 5mM KCl, 0.5mM $MgCl_2$, 5mM DTT and 1mM PMSF and stored at -70°C in small volumes.

DNA replication *in vitro*

Initiation of DNA replication was determined by the transfer of dCMP to pTP as described previously (Harris and Hay, 1988; Cleat and Hay, 1989). To assay for elongation of DNA initiated *in vitro* restriction enzyme cleaved template DNA was pre-incubated in a total volume of 7µl for 30 minutes at 30°C with 1-3µl of purified viral replication proteins in a buffer containing 25mM Hepes-KOH (pH 7.5), 4mM $MgCl_2$, 1mM DTT and 0.1 mg/ml BSA. 7µl of a buffer containing 80µM each of dTTP, dATP, and dGTP, 5µM dCTP, 0.3µCi [a^{32}P] dCTP (specific activity 3000Ci/mM), 4mM ATP, 10mM creatine phosphate, 10mg/ml creatine phosphokinase, 0.1mg/ml BSA, 25mMHepes-KOH, 4mM $MgCl_2$ and 1mMDTT was then added and the reaction incubated at 30°C for 90 minutes. Reaction products were resolved on a 2% agarose gel containing 0.1% SDS and labelled species quantitated by scintillation counting after excision from the gel.

Expression of replication proteins in insect cells

Growth of *Spodoptera frugiperda* cells, propagation of recombinant baculoviruses and DNA polymerase assays have been described previously (Watson and Hay, 1990).

HPLC Analysis

Proteins in 10mM potassium phosphate buffer pH 7.0, 0.4M KCl were applied to a Zorbax G-250 hydrophilic gel filtration column installed in a Waters HPLC system and the column effluent monitored by absorbance at 280nm. Fractions were collected and assayed for DNA polymerase activity. Sequence specific DNA binding activity was determined using a gel electrophoresis DNA binding assay as described (Clark et al., 1990).

RESULTS

Protein and DNA requirements for Ad2 and Ad4 DNA replication

The Ad2 origin of DNA replication is contained within the terminal 51bp of the genome and contains four recognisable domains: a conserved core; a spacer region; and the binding sites for NFI and NFIII (figure 1). Replication of Ad2 DNA requires three viral proteins: the 80,000 molecular weight preterminal protein (pTP); the 140,000 molecular weight DNA polymerase (Ad pol) and the 59,000 molecular weight DNA binding protein (DBP). The genes for each of these proteins have been inserted into recombinant baculoviruses and the proteins expressed at high levels after infection of *Spodoptera frugiperda* (Sf9) insect cells. In each case active protein has been purified by a combination of ion-exchange and DNA affinity chromatography (table 1). In addition to these viral proteins the two host proteins, NFI (Nagata et al., 1983) and NFIII (Pruijn et al., 1986) are also required for optimal levels of DNA synthesis. Both of these proteins are sequence specific DNA binding proteins which on binding to their cognate sites in the Ad2 origin of DNA replication can stimulate initiation up to 100 fold (Mul et al., 1990). cDNAs for NFI (also known as CTF, CAAT Transcription Factor) and NFIII (also known as oct 1 and OTF 1, Octamer Transcription Factor) have been isolated (Santoro et al., 1988; O'Neill et al., 1988) and in each case it has been demonstrated that the DNA binding domains alone are sufficient to stimulate adenovirus DNA replication (Santoro et al., 1988; Mermod et al., 1989; Gounari et al., 1990; Verrijzer et al., 1990). While these proteins have also been expressed in insect cells infected with recombinant baculoviruses it has been more convenient to express the small DNA binding domains of these proteins in bacteria. For this purpose the DNA binding domains of NFI and NFIII were fused to the gene for glutathione S-transferase (GST) and the recombinant plasmids transformed into E.coli. In these constructs the GST gene fusions are under control of the lac promoter and were therefore induced by the addition of IPTG. Expressed protein can be purified by affinity chromatography on glutathione agarose and the DNA binding domains of NFI and NFIII released from the fusion protein by treatment with thrombin, which cleaves at the C-terminus of the GST portion of the fusion. Addition of the five purified protein components in the presence of Mg^{2+}, ATP and the four dNTPs to protein linked Ad2 genomes or plasmids linearised to expose the Ad2 origin of DNA replication results in efficient initiation of DNA replication and extensive elongation of the initiated products (table 1).

In contrast to Ad2 only the terminal 18bp of the Ad4 genome (identical in both viruses) is required to constitute a fully functional origin of DNA replication. To examine the proteins involved in Ad4 DNA replication crude Ad4 infected extract which was capable of *in vitro* DNA replication was applied to DEAE-Sepharose equilibrated with 50mM NaCl and bound proteins eluted with 0.2M NaCl. Individual fractions were tested for their ability to initiate DNA replication *in vitro* by assaying the efficiency with which they could catalyse the transfer of dCMP onto pTP using EcoRI cleaved p4A2 as template. All DNA replication activity eluted with 0.2M NaCl and the efficiency of initiation was not increased by addition of other fractions. Fractions containing replication activity were combined, applied to denatured calf thymus DNA-Sepharose and bound proteins eluted with a linear gradient of 0.2-2M NaCl. One peak of protein eluted between 0.5-0.8M NaCl and a second at 1.0-1.2M NaCl (figure 2), but DNA replication activity, assayed as described above, was only detected in the fractions eluted with 1.0-1.2M NaCl (figure 2).

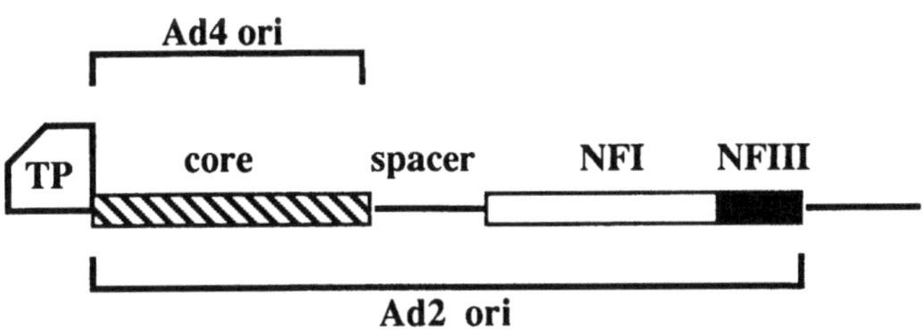

Figure 1. Structure of the Ad2 and Ad4 origins of DNA replication indicating the location of the minimal origin or core 1-18 sequence, the spacer region and the binding sites for the cellular transcription factors NFI and NFIII. The terminal protein (TP) is also induicated.

To examine the proteins present at each stage of the purification fractions were analysed by SDS polyacrylamide gel electrophoresis followed by silver staining. The fraction eluted with 1.0-1.2M NaCl, which contained all the DNA replication activity contained one predominant species of apparent molecular weight 65,000 and three less abundant species of apparent molecular weights 70,000 85,000 and 95,000. On the basis of its relatively high abundance and elution characteristics the predominant species was thought to be viral DBP. Western blotting with polyclonal antisera raised against Ad2 pTP and DBP confirmed the presence of the corresponding Ad4 proteins in the active fraction, while the expected presence of the Ad4 pol was indicated by the detection of an aphidicolin resistant DNA polymerase activity. In the presence of a linearised plasmid containing the Ad4 origin of DNA replication this protein fraction not only initiates DNA replication but efficiently elongates the nascent strands.

Effect of point mutations on the activity of the adenovirus type 4 origin of DNA

Within the terminal 18bp of the ITR which are identical in Ad2 and 4 the region between bp's 9 and 18 is perfectly conserved in all human serotypes so far examined suggesting that the integrity of this sequence is important for DNA replication. To examine the role of this sequence in Ad4 DNA replication we determined the consequence of single base changes from positions 9 to 18 on DNA replication *in vitro*. The relative efficiency of initiation was assayed

Adenovirus type 2		Adenovirus type 4	
51bp origin		18bp origin	
DBP pol pTP	Complete proteins purified from insect cells infected with recombinant baculovirus	DBP pol pTP X	Obtained in a single fraction from Ad4 infected cells
NFI NFIII	DNA binding domains purified from E.coli		

Table 1. Protein and DNA sequence requirements for Ad2 and Ad4 DNA replication in vitro. The polypeptide designated X is of unknown function.

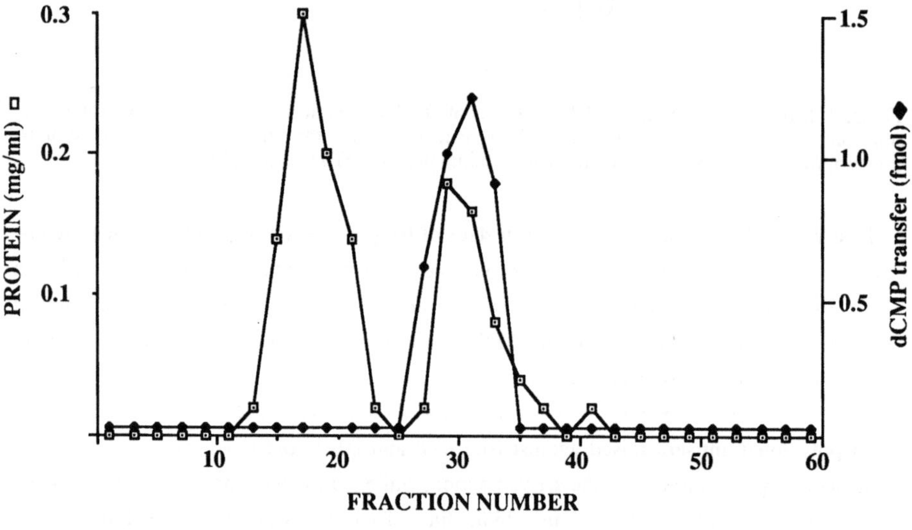

Figure 2. Purification of a fraction from Ad4 infected cells capable of replicating Ad4 DNA *in vitro*. The DEAE fraction from Ad4 infected HeLa cells was applied to denatured DNA Sepharose and bound proteins eluted with a gradient of 0.2M to 2M NaCl . Plamid p4A2 (50ng) containing a copy of the Ad4 ITR was cleaved with EcoRI and incubated with 8ml of each fraction (dialysed against 0.1M NaCl buffer) in a standard assay for initiation of DNA replication (see methods). After incubation at 32°C for 90 minutes and micrococcal nuclease digestion polypeptides were fractionated by SDS polyacrylamide gel electrophoresis, labelled species were excised from the gel and the radioactivity determined by liquid scintillation counting. Protein was determined by the method of Bradford (1976).

quantitatively by measuring the transfer of [^{32}P] dCMP to pTP, this being the first synthetic step of DNA replication. A series of ten plasmids, pM9 to pM18, each of which contains an Ad5 ITR with a single base change in positions 9 to 18 respectively where cleaved with EcoRI to expose the ITR sequence at a molecular end. Incubation of the template with the purified fraction from Ad4 infected HeLa cells, [α^{32}P] dCTP and optimal levels of $MgCl_2$ and ATP results in the transfer of dCMP to pTP. Reaction products were resolved by electrophoresis on an SDS containing polyacrylamide gel and the radioactivity in labelled species determined by liquid scintillation counting.

Plasmid p4A2 which contains one complete copy of an Ad4 ITR and pHR18 which contains the terminal 18bp of an Ad2/5 ITR, both supported initiation with approximately equal efficiency consistent with previous reports establishing that Ad4 requires only the terminal 18bp for efficient initiation (table 2). Similarly, single base substitutions from positions 12 to 15 from

ori plasmid	relative efficiency (%) initiation	elongation
p4A2	115	109
pHR18	85	92
pM9	51	60
pM10	30	26
pM11	53	33
pM12	113	102
pM13	101	78
pM14	116	118
pM15	87	81
pM16	88	48
pM17	19	56
pM18	16	28

Table 2. Effect of point mutations in the conserved 9-18 region on the efficiency of Ad4 DNA replication in vitro. Each template was linearised with Eco RI to expose the origin and incubated with the fraction purified from Ad4 infected cells under conditions described in the text. Initiation and elongation assays were quantitated as described in the text and the values compared to the activity of the wild type template.

the terminus had no deleterious effect on the efficiency of initiation. However, alterations at positions 9, 10, 11, 17 and 18 all resulted in a reduction in efficiency of initiation by between 50% and 90% (table 2).

To further examine the effect of the single point mutations we determined the relative efficiency with which wild type templates and plasmids pM9 to pM18 could support elongation from initiated products *in vitro*. Templates were cleaved with EcoRI and AvaII to give a 1.6kb fragment with the origin sequence at the EcoRI terminus and a 0.9kb fragment containing only plasmid sequences and incubated with the purified protein fraction, dNTP's (including [α^{32}P] dCTP), optimal $MgCl_2$ and an ATP regeneration system. Reaction products were fractionated by electrophoresis in an agarose gel containing SDS and replicated templates which had incorporated [^{32}P] dCMP were detected by autoradiography Correctly initiated products have pTP linked at the 5' end resulting in a lower electrophoretic mobility than input template molecules which incorporated ^{32}P at a low level as a result of repair synthesis. p4A2 and pHR18 both supported elongation with equal efficiency as did pM12 to pM16. However pM9, 10, 11, 17 and 18 all replicated at a reduced efficiency corresponding to the observed reduction for initiation. (table 2). These data indicate that as predicted, the conserved sequence from bp's 9 to 18 in the adenovirus ITR plays a vital functional role in viral DNA replication even though

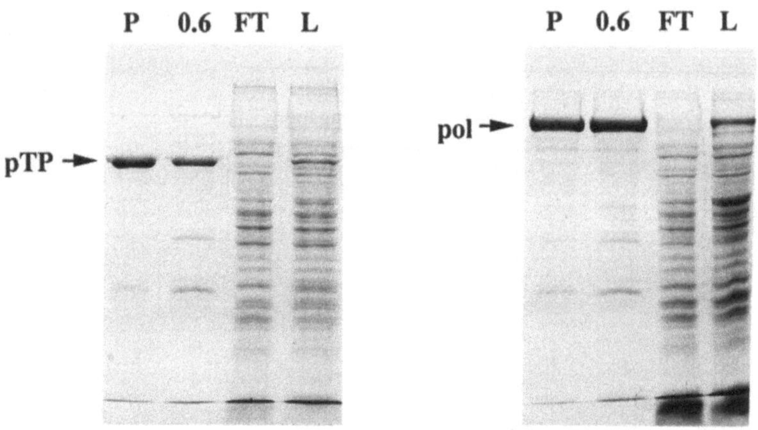

Figure 3. Binding of pTP and pol to denatured DNA. Extracts (L) from Sf9 cells infected with baculoviruses containing either the Ad2 pol or pTP genes were applied to denatured DNA Sepharose and unbound protein (FT) removed by washing with 0.2M NaCl. Bound proteins were eluted with 0.6M NaCl and samples analysed by SDS polyacrylamide gel electrophoresis and Coomassie staining.

certain limited sequence changes appear to be tolerable for *in vitro* replication . The corresponding decrease in the efficiency of elongation which is observed for mutants exhibiting impaired initiation suggests that it is only initiation of DNA replication that is being affected by changes in the origin sequence and that elongation can proceed unimpaired from those molecules which have initiated.

DNA binding properties of the adenovirus type 2 preterminal protein and DNA polymerase

To determine the individual DNA binding properties of pTP and pol, and the heterodimer formed between the two proteins, use has been made of recombinant baculoviruses expressing pol and pTP in insect cells. The ability of the proteins to bind non-specifically to

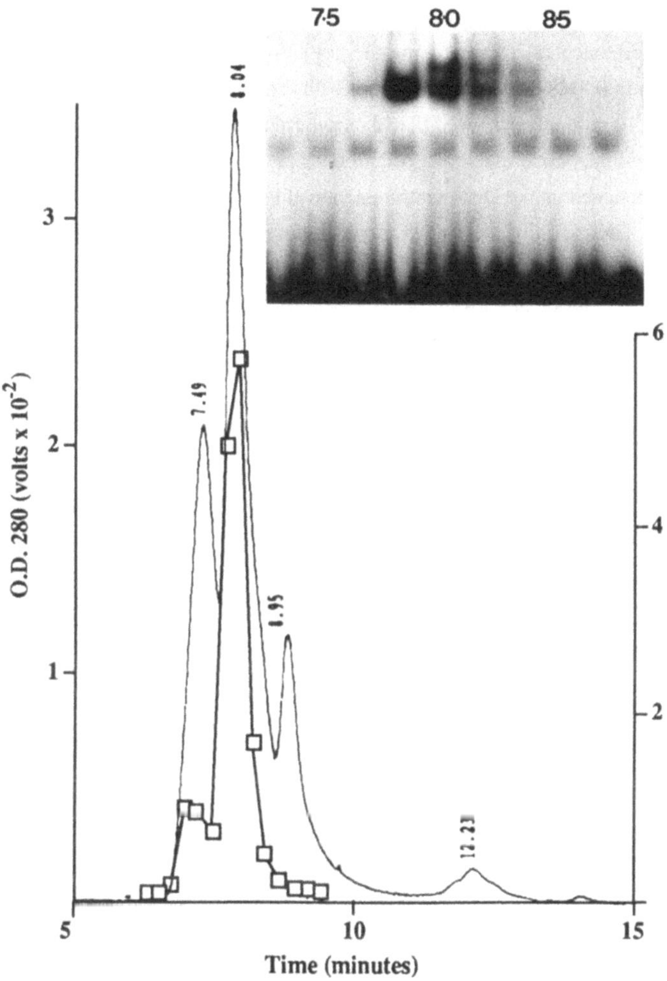

Figure 4. Sequence specific DNA binding by the Ad2 pTP/pol heterodimer. pol and pTP purified by denatured DNA Sepharose and hydroxylapatite chromatography were mixed and applied to a Zorbax G-250 gel filtration column installed in a Waters HPLC system. The column was developed isocratically in 10mM potassium phosphate buffer pH 7.0 containing 0.4M KCl and the column eluant monitored by absorbance at 280nm. Fractions were tested for DNA polymerase activity or for specific binding to a double stranded oligonucleotide containing the 1-18 core sequence, as described in the text.

DNA was examined by chromatography on denatured DNA Sepharose. Extracts from insect cells infected with baculovirus containing the pol and pTP genes were applied to denatured DNA sepharose in 0.2M NaCl. After washing with the same buffer bound protein was eluted with 0.6M NaCl and analysed by SDS polyacrylamide gel electrophoresis and Coomassie staining. Whereas the bulk of the insect cell proteins pass through the column both pol and pTP are completely retained on the DNA matrix and are eluted with 0.6M NaCl in a highly purified form (figure 3). DNA affinity chromatography therefore serves to demonstrate that pTP and pol bind non-specifically to single stranded DNA and also provides a useful first step in purification. Further purification was obtained by chromatography on hydroxylapatite (data not shown).

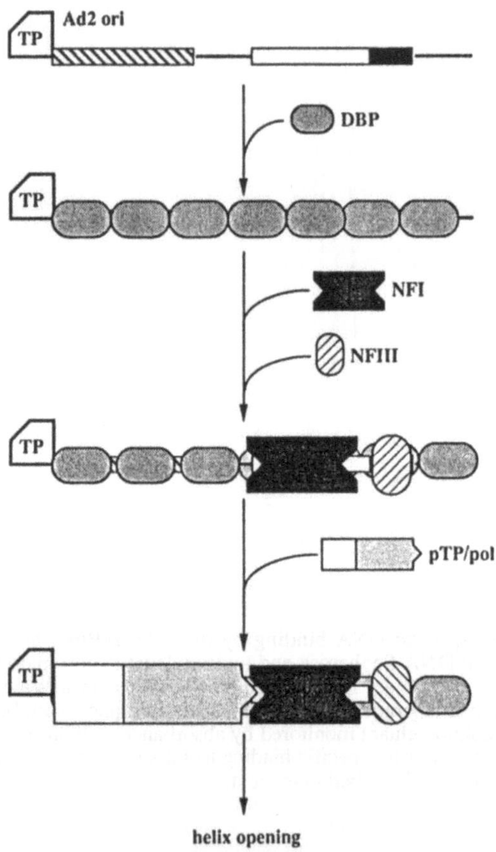

Figure 5. Model for the formation of a preinitiation complex at the Ad2 origin of DNA replication. Details are described in the text.

To investigate the sequence specific DNA binding properties of the pTP pol heterodimer the proteins were allowed to interact and analysed by high resolution gel filtration. The column eluate was monitored by absorbance at 280nm and fractions were collected for DNA polymerase and DNA binding assays. DNA polymerase activity of pTP/pol heterodimer co-eluted with sequence specific DNA binding activity as determined in a gel electrophoresis DNA binding assay using a double stranded synthetic oligonucleotide containing the Ad2 core (1-18bp) sequence (figure 4). Subsequent competition analysis confirmed the specificity of this interaction (data not shown).

DISCUSSION

Much of the information which has led to our current understanding of the mechanism of adenovirus DNA replication has been derived from work carried out using adenovirus types 2 and 5. Ad2 relies on the interaction of host factors NFI/CTF and NFIII/Oct-1 with their respective binding sites at the origin of replication for efficient DNA replication. Ad4 has apparently dispensed with the need to employ these host DNA binding proteins in the replication of its DNA and as a result has a remarkably simple origin construction which is reflected in the relatively simple complement of protein factors that are required, not only for initiation of DNA replication, but also for elongation. Within the minimal origin sequence two critical regions separated by a region in which single base changes could be tolerated were defined. The sequence between bp's 9 and 18 may therefore play an important part in recognition by and positioning of the replication proteins at the origin prior to initiation of DNA replication. Recent evidence demonstrating a direct interaction between pol and NFI suggests that Ad2 utilises the binding of NFI/CTF adjacent to the origin to stabilise a pre-initiation complex between the viral replication proteins and the template prior to initiation (Bosher et al., 1990; Chen et al., 1990; Mul et al., 1990). Insertion of more than 2bp between the NFI site and the 1-18 sequence reduces the efficiency of initiation *in vitro* (Adyha et al., 1986; Wides et al., 1987) and *in vivo* (Bosher et al., 1990) and stresses the importance of precise spacing between these functionally important sequences. The affinity of NFI for its binding site is in turn further increased by the presence of viral DBP although it has not been demonstrated that there is any direct protein-protein interaction involved in this cooperation (Cleat and Hay, 1989; Stuiver and van der Vliet, 1990). The three domain structure of the 9-18 region suggests it may be part of a protein recognition site since the boundaries of the two domains in which mutations have the profoundest effect on DNA replication lie approximately one turn of the helix apart and thus could be contact points in a DNA/protein interaction on one side of the DNA helix. Evidence presented here supports this contention and demonstrates that Ad2 pTP/pol heterodimer recognises and binds to a sequence in the terminal 18bp (figure 4).

Based on the available data it is therefore possible to formulate a model for the formation of a preinitiation complex at the adenovirus origin of DNA replication (figure 5). Viral genomes coated with DBP are bound by NFI which via interactions with the Ad2 DNA polymerase directs the pTP/pol heterodimer to the 1 - 18 minimal origin sequence. This complex is further stabilised by interactions between the pTP/pol heterodimer and specific DNA sequences within the origin region. Unwinding of the DNA in this complex would then be required to expose the single stranded DNA template for the initiation reaction. The observed ability of Ad4 to dispense with NFI suggests that the Ad4 replication proteins may have a higher intrinsic affinity for the minimal origin sequence.

This work was supported by the MRC, CRC and the SERC.

LITERATURE CITED

Adhya, S., P.S. Schneidman and J. Hurwitz (1986). Reconstruction of adenovirus replication origins with a human nuclear factor I binding site. J. Biol. Chem. 261: 3339-3346.
Bosher, J., E.C. Robinson and R.T. Hay (1990). Interactions between the adenovirus type 2 polymerase and the DNA binding domain of nuclear factor I. New Biologist. 2: 1083-1090.
Bradford, M.M. (1976). A rapid and sensitive method for the quantitation of microgram quantities of protein utilizing the principle of protein-dye binding. Anal. Biochem. 72: 248-254.
Challberg, M.D. and T.J. Kelly (1979). Adenovirus DNA replication in vitro. Proc. Natl. Acad. Sci. (USA). 76: 655-659.
Chen, M., N. Mermod and M.S. Horwitz (1990). Protein-protein interactions between adenovirus polymerase and nuclear factor I mediate formation of the DNA replication pre-initiation complex. J. Biol.Chem. 265: 18634-18642.
Clark, L., J.R. Matthews and R.T. Hay (1990). Interaction of enhancer binding protein EBP1 (NF-kB) with the human immunodeficiency virus type 1 enhancer. J. Virol. 64: 1335-1344.
Cleat, P.H. and R.T. Hay (1989). Co-operative interactions between NFI and the adenovirus DNA binding protein at the adenovirus origin of DNA replication. EMBO J. 8: 1841-1848.
Gounari, F., R. De Francesco, J. Schmitt, P.C. van der Vliet, R. Cortese and H. Stunnenberg (1990). Amino terminal domain of NFI binds to DNA as a dimer and activates adenovirus DNA replication. EMBO. J. 9: 559-566.
Harris, M.P.G. and R.T. Hay (1988). DNA sequences required for the initiation of adenovirus type 4 DNA replication. J. Mol. Biol. 201: 57-67.
Hay, R.T. (1985a). The origin of adenovirus DNA replication: minimal DNA sequence requirement in vivo. EMBO J. 4: 421-426.
Hay, R.T. (1985b). The origin of adenovirus DNA replication: role of the nuclear factor I site in vivo. J. Mol. Biol. 186: 129-136.
Hay. R.T., L. Clark, P.H. Cleat, M.P.G. Harris, E.C. Robertson and C.J. Watson (1988). Requirements for the initiation of adenovirus types 2 and 4 DNA replication. Cancer Cells: 6: 71-75. CSH Laboratory, New York.
Hay. R.T. and W.C. Russell (1989). Recognition mechanisms in the synthesis of animal virus DNA's. J. Biochem. 258: 3-16.
Mermod, N., E.A. O'Neill, T.J. Kelly and R. Tjian (1989). The proline rich transcriptional activator of CTF/NFI is distinct from the replication and DNA binding domain. Cell. 58:

741-753.
Mul, Y.M., C.P. Verrijzer and P.C. van der Vliet (1990). Transcription factors NFI and NFIII/Oct-1 function independently, employing different mechanisms to enhance adenovirus DNA replication. J. Virol. 64: 5510-5518.
Nagata, K., R.A. Guggenheimer and J.Hurwitz (1983). Specific binding of a cellular DNA replication protein to the origin of replication of adenovirus DNA .Proc. Natl. Acad. Sci.(USA) 80: 6177-6181.
O'Neill, E.A., C. Fletcher, C.R. Burrow, N. Heintz, R.G. Roeder and T.J. Kelly (1988). The transcription factor OTF-1 is functionally identical to the adenovirus DNA replication factor NFIII. Science. 241: 1210-1213.
Pruijn, G.J.M., W.Van Driel and P.C. Van der Vliet (1986). A novel sequence specific DNA binding protein from HeLa cells stimulating adenovirus DNA replication.Nature 322: 656-659.
Rawlins, D.R., P.J. Rosenfeld, R.J.Wides, M.D. Challberg and T.J. Kelly (1984). Structure and function of the adenovirus origin of DNA replication. Cell 37: 309-319.
Santoro, C., N. Mermod, P.C. Andrews and R. Tjian (1988). A family of human CCAAT box binding proteins active in transcription and DNA replication: cloning and expression of multiple cDNAs. Nature. 334: 218- 224.
Stuiver, M.H., and van der Vliet, P.C. (1990). Adenovirus DNA binding protein forms a multimeric protein complex with double-stranded DNA and enhances binding of nuclear factor 1. J Virol 65:379-389.
Tamanoi, F.and Stillman, B.W. (1982). Function of the adenovirus terminal protein in the initiation of DNA replication. Proc Natl Acad Sci U. S. A. 79:2221-2225.
Verrijzer, C.P., Kal, A.J.and van der Vliet, P.C. (1990). The DNA binding domain (POU domain) of transcription factor oct-1 suffices for stimulation of DNA replication. EMBO J 9:1883-1888.
Watson, C.J.and Hay, R.T. (1990). Expression of adenovirus type 2 DNA polymerase in insect cells infected with a recombinant baculovirus. Nucl Acids Res 18:1167-1173.
Wides, R.J., Challberg, M.D., Rawlins, D.R.and Kelly, T.J. (1987). Adenovirus origin of replication: sequence requirements for replication in vitro. Mol Cell Biol 7:864-874.

Analysis *in vitro* of Mutations in the Cloned Precursor to the Terminal Protein (pTP) and the Adenovirus DNA Polymerase (Ad Pol) Genes

Jeffrey A. Engler, Insil Joung, Jeffrey N. Fredman, Steven C. Pettit[1], Carl Abraham[1], and Marshall S. Horwitz[1]
Department of Biochemistry
University of Alabama at Birmingham
Schools of Medicine and Dentistry
UAB Station
Birmingham, Alabama 35294
USA

INTRODUCTION

Human adenoviruses have been a good model system for the study of eukaryotic DNA replication (for a review, see Freifeld et al., 1984; Tamanoi and Stillman, 1984; Challberg and Kelly, 1989); this virus family shares a protein-priming mechanism with the *B. subtilis* bacteriophage ϕ29 and several other bacterial phages. In the case of adenovirus DNA replication, an in vitro system has been widely used to study the molecular requirements of the replication reactions (Challberg and Kelly, 1979). This system has been used to characterize the initiation process, to show how a deoxyribonucleotide triphosphate (dCTP) is covalently attached to the precursor to the terminal protein (pTP[2]) in a reaction that also requires an adenovirus-specific DNA polymerase (Ad Pol), and two nuclear factors (NFI and NFIII). Elongation then occurs by the formation of 5' to 3' phosphodiester bonds from this initiation complex and requires the action of Ad DBP and a type I topoisomerase (NFII).

---

[1]Address for these authors: (C.A. and M.S.H.) Department of Microbiology-Immunology, Albert Einstein College of Medicine, Bronx, New York 10461; current address for S.C.P.: Lineberger Cancer Research Center, Campus Box #7295, University of North Carolina, Chapel Hill, North Carolina 27599-7295.
This work was supported by Public Health Service grants AI20408 (to J.A.E.) and CA11512 and P30-CA13330 (to M.S.H.). J.N.F. and S.C.P. were predoctoral trainees supported by NIH training grants T32GM08111 and 5T32CA09467, respectively. Costs for DNA synthesis were supported in part by grant CA13148 to the UAB Comprehensive Cancer Center.
[2]Abbreviations used: Ad, adenovirus; Ad Pol, the adenovirus DNA polymerase; DBP, the 59K single-stranded DNA binding protein; ORF, open reading frame; pTP, the adenovirus precursor to the terminal protein; TP, terminal protein.

The adenovirus DNA replication proteins (Ad DBP, Ad Pol, and pTP) are all encoded in early region 2 on the leftwardly-transcribed strand of the Ad genome (Figure 1). The structure of the mRNAs that encode these proteins have been determined by electron microscopy (Stillman et al., 1981) and by S1 nuclease mapping (Shu et al., 1988). Both Ad Pol and pTP share an upstream exon near genome coordinate 39 that provides the initial methionine and two additional amino acids. RNA splicing then connects this upstream exon to the main ORFs that encode pTP and Ad Pol. These proteins share a high degree of nucleotide and predicted amino acid sequence homology in the different adenovirus serotypes.

We have sought to understand the molecular details of the interactions in in vitro adenovirus DNA replication by studying mutants in either the pTP or the Ad Pol proteins. For these initial studies, a transient expression system for pTP and Ad Pol has been employed (Shu et al., 1987; Pettit et al., 1988). This system utilized a vector, p91023, carrying different segments of adenovirus DNA (Kaufman, 1985); these plasmids were transfected into COS or into CMT-4 cells

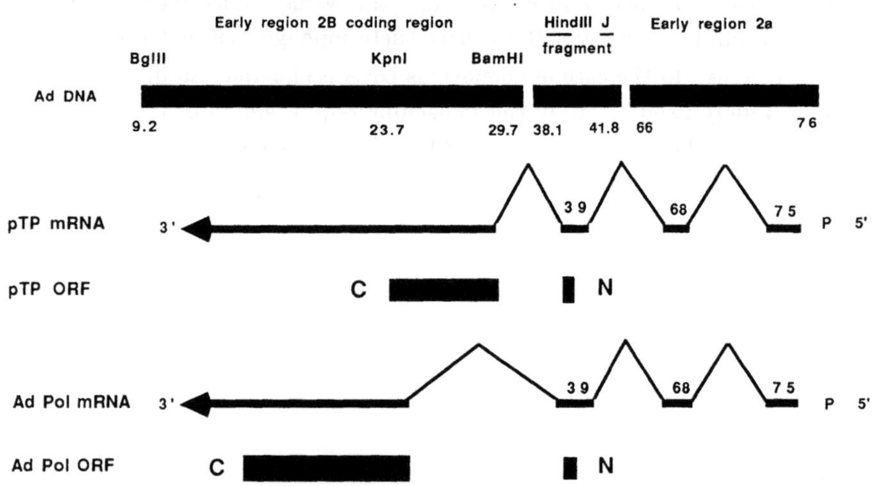

Figure 1: The structure and predicted coding regions of the mRNA from which functional pTP and Ad Pol are synthesized. Ad DNA segments from which Ad2 pTP and Ad Pol exons are derived are shown as solid rectangles on the top line; the approximate genome coordinate of each segment is shown below (numbered according to Roberts et al., 1986). Below the representation of these DNA molecules, the structures of the processed mRNAs that encode pTP or Ad Pol are shown as thick arrows interrupted by introns (thin diagonal lines) and pointing from 5' to 3'. These structures contain the exons mapped by Stillman et al. (1981). The promoter for transcription of early region 2 is designated as "P". The pTP and Ad Pol ORFs are shown below their respective mRNA structures. The filled boxes encoded in the HindIII J fragment show the upstream ORF that supplies the initiator AUG to both pTP and Ad Pol. Adapted from Pettit et al. (1988).

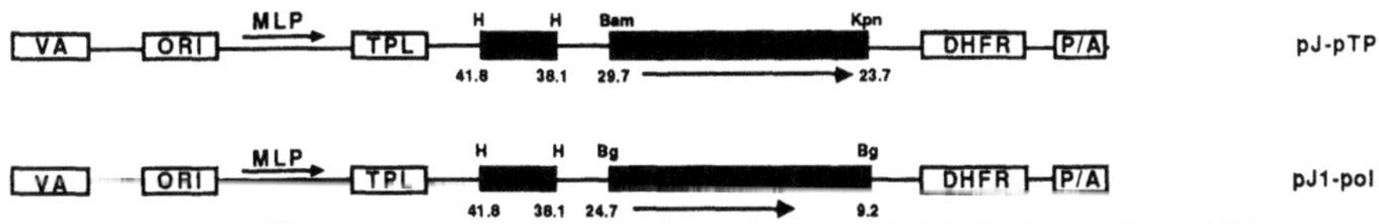

Figure 2: Plasmids for transient expression of pTP or Ad Pol. Adenovirus DNA segments cloned into expression vector 91023 are shown as filled rectangles, with restriction endonuclease cleavage sites and genome coordinates written above and below. Features of the vector that are important for expression include the Ad2 VA genes (VA) that enhance translation of mRNAs containing the tripartite leader (TPL), the SV40 origin of replication and enhancer sequences (ORI) that increase copy number in SV40 T-antigen expressing cells, the Ad2 major late promoter (MLP) for transcription, the dihydrofolate reductase gene (DHFR), and the SV40 polyadenylation site (P/A). Adapted from Shu et al. (1987) and Pettit et al. (1988).

(Gerard and Gluzman, 1985) by the modified calcium phosphate method (Chen and Okayama, 1987). A crude cytoplasmic or nuclear lysate that showed pTP or Ad Pol activity could be isolated from the transfected cells by Dounce homogenization. The amounts of pTP and Ad Pol in these extracts was quantitated by densitometry of immunoblots developed with antipeptide antibodies specific for the C-terminus of pTP and of Ad Pol; in each quantitation experiment, known quantities of wild type pTP or Ad Pol extracts was included in each blot. The vector system for expression of pTP and for Ad Pol is shown schematically in Figure 2.

A number of assays have been used to study these mutants in vitro. Both Ad Pol and pTP mutants can be tested in vitro in adenovirus-specific assays for initiation and elongation. Initation was measured by the covalent attachment of ^{32}P-dCMP to the 80 Kd pTP protein band on a SDS-polyacrylamide gel; densitometry of autoradiograms from each initiation assay was used to measure activity. Elongation was measured by the specific incorporation of α-^{32}P-dTTP into origin-containing end fragments in an SmaI digest of Ad35 DNA-pro; the densitometry values for incorporation into end fragments were corrected for non-specific labeling due to other processes occuring in the crude extracts, such as DNA repair reactions. The results of each assay was normalized for the amount of adenovirus-specific protein, quantitated from Western blots of the extracts.

Two additional assays were also used to characterize Ad Pol mutants. Incorporation of α-^{32}P-dCTP into nicked calf thymus DNA was used to measure the general level of DNA polymerase activity in each extract. When aphidicolin, a potent inhibitor of cellular DNA polymerase α but not Ad Pol, was added to each reaction, Ad Pol-specific activity could be measured. In addition, Field et al. (1984) described an assay for Ad Pol and Ad DBP dependent incorporation of α-^{32}P-dATP into poly(dT)/oligo(dA) templates; this assay was very sensitive and discriminated Ad Pol activity from other cellular DNA polymerases present in the crude extracts. As described above, the results of these assays were normalized, based on the relative levels of Ad Pol or its mutants in each extract.

LINKER INSERTION MUTANTS IN pTP

Both Freimuth and Ginsberg (1986) and Roovers et al. (1991) have studied the ability of linker insertion mutants of pTP to form virus plaques by recombination with adenovirus DNA fragments. These mutants were characterized as silent (producing an apparent wild type virus), lethal (unable to make a virus plaque), temperature sensitive (making a plaque at 32°C but not 39.5°C) or replication defective (lower than normal virus yield). We have studied many of the linker insertions described by Freimuth and Ginsberg (1986) by cloning each mutant into pJ-pTP (Fig. 2), expressing the resulting protein in transfected CMT-4 cells, and testing for activity in vitro.

Initial studies focused on the amino terminal end of pTP. When only the main ORF of pTP was expressed in vector p91023 (a net deletion of 18 amino acids from the amino terminus of the protein), a nonfunctional polypeptide was obtained. Deletion of six amino acids was sufficient to reduce pTP activity 10 to 20 fold (Pettit et al., 1989). Similar results have been observed in N-terminal deletions of the terminal protein of ϕ29 (Zaballos et al., 1988, 1989). Freimuth and Ginsberg (1986) reported two linker insertion mutants in this region, each of which had an unusual phenotype (see Table I). When tested for initiation and elongation activity in vitro at 30°C (permissive temperature) and at 38°C (nonpermissive temperature), mutant 423 showed the same temperature-sensitive phenotype as it did when studied in recombinant adenovirus.

	Phenotype	Initiation			Elongation		
Sample		30°C	38°C		30°C	38°C	
J-pTP	wild type	100	346	(3.5 x ⇑)	100	183	(1.8 x ⇑)
423-pTP	temp sens	66	89	(no change)	101	17	(6 x ⇓)
437-pTP	rep defective	80	189	(2.4 x ⇑)	126	185	(1.5 x ⇑)

Table I: Temperature sensitive and replication defective pTP mutants assayed in vitro. The data has been normalized so that J-pTP at 30°C has 100% activity. 423-pTP contains a 2 amino acid insertion after amino acid 11; this mutant gave a temperature sensitive phenotype when incorporated into virus. 437-pTP contains a 4 amino acid insertion after amino acid 5 and gave a lower yield of virus when studied in a recombinant adenovirion (Freimuth and Ginsberg, 1986). Table adapted from Pettit et al. (1989).

Further, this mutant showed a much more drastic reduction in elongation activity than in initiation activity, when activities at 30°C and 38°C are compared. Although mutant 437 was replication defective in virus, it appeared to have nearly wild type activity in the in vitro assays. Taken together, this data suggests either that the amino terminus of pTP must play an essential role in both the initiation and the elongation reactions or that it is responsible for directing the overall folding of the molecule. Studies to distinguish these two possibilities are currently underway.

REPLACEMENTS AT SERINE 580 IN pTP

Serine 580 in the pTP polypeptide is the amino acid to which dCTP is covalently attached during the initiation step of adenovirus DNA replication (Desiderio and Kelly, 1981). The amino acid sequences around this residue are highly conserved in many serotypes of human adenovirus, suggesting that this region of pTP encodes an essential domain for activity. In order to probe the importance of this region, serine 580 was replaced with both threonine (a conservative change that retains the hydroxyl group on the side chain) and alanine (a nonconservative change). Surprisingly, both substitutions were inactive (Pettit et al., 1989). This result suggests that there are severe constraints on the structure of this domain of the pTP molecule. Similar results were observed for a serine to threonine replacement in the precursor to the terminal protein of bacteriophage ϕ29 (Garmendia et al., 1988); however, Garmendia et al. (1990) showed that substitution of a cysteine for the serine required for covalent linkage to dAMP retained low (0.7%) activity.

MUTATIONS IN THE CONSERVED REGION I OF AD POL

Ad Pol is similar to other eukaryotic DNA polymerases of the polα-family. since it shares five of the six conserved amino acid sequences that are characteristic of these enzymes (Wong et al., 1988; Wang et al, 1989); these regions are numbered I through VI, based on their degree on homology (Figure 3). Region I is the most highly conserved and contains the sequence YGDTDSLF. A similar motif, YGDD and YMDD, is often found in RNA polymerases and reverse transcriptases (Kamer and Argos, 1984; Argos, 1988). Little is known about the function or structure of region I. Delarue et al. (1990)

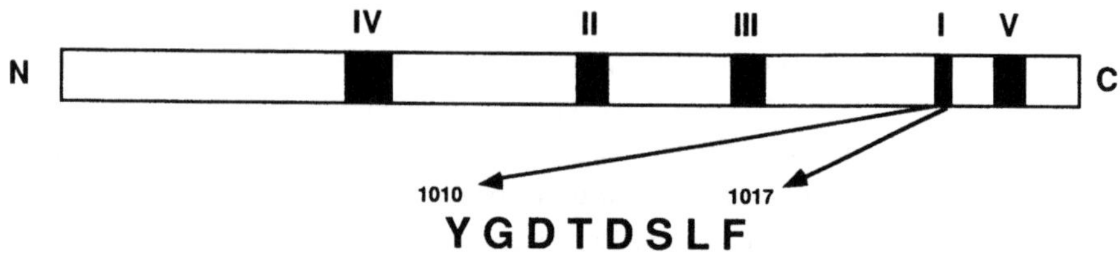

Figure 3: Conserved regions in Ad Pol. The homologous regions found in Ad Pol and other members of the polα-family are marked I through V (Wang et al., 1989).

have suggested that region I is located within a loop at the end of a β-hairpin and that one or both of the aspartic acid residues in region I coordinate with a metal ion, as in the large fragment of E. coli DNA Pol I (Ollis et al., 1985). Marcy et al. (1990) have suggested that region I is involved in substrate recognition, since several herpes pol mutants in region I increase sensitivity to aphidicolin and decrease sensitivity to acylovir and phosphonoacetic acid (PAA).

In order to understand the functions encoded within region I, we have constructed and studied a series of conservative and nonconservative amino acid replacements within region I of Ad Pol. Conservative mutations retained the same hydrophobicity or charge with a similar size, to avoid major alteration of secondary structure within region I. Nonconservative mutations generally changed charge or hydrophobicity of the side chain. In addition, a series of replacements at glycine 1011 were studied. Some of these changes were designed to mimic alterations of this amino acid in other DNA polymerases, such as G1011C, found in the ϕ29 DNA polymerase (Yoshikawa and Ito, 1982), or G1011S, found in the PGKL1 DNA polymerase. The replacement G1011M was constructed to resemble the YMDD motif found in several virus reverse transcriptases (Delarue et al., 1990).

DNA fragments containing each of the replacement mutations were cloned into pJ1-Pol (Figure 2) and expressed as mutant proteins in transfected CMT-4 cells; twenty-four hours after transfection, crude cytoplasmic extracts were prepared by Dounce homogenization (Challberg and Kelly, 1979). When the extracts were assayed in vitro, each of the conservative replacements showed 2 to 10 fold reduced activity on nicked calf thymus DNA and 5 to 10 fold reductions on poly(dT)/oligo(dA) templates (Figure 4). Similar reductions in activity were also observed for each mutant in the initiation and elongation reactions. As might be expected from these results, nonconservative replacements also showed dramatically lower activities in these assays.

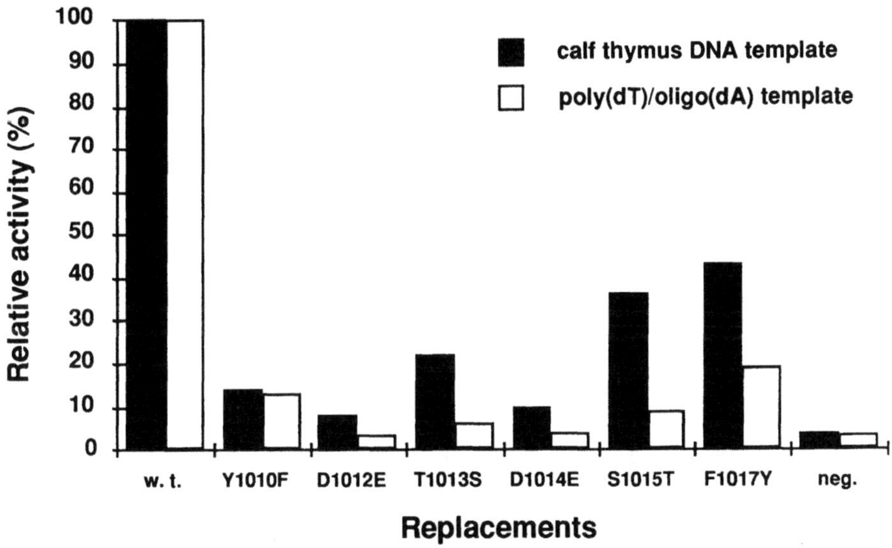

Figure 4: Assays for enzyme activity of conservative Ad Pol mutants in region I. The relative activities of different mutants were compared to that of wild type Ad Pol (w.t.), which was set at 100%. Filled rectangles show the activity of crude extracts on nicked calf thymus DNA (Pharmacia) in the presence of 120 μM aphidicolin; open rectangles show the activity on poly(dT)/oligo(dA) templates in the presence of Ad DBP. The data shown was normalized for the relative amounts of Ad Pol protein in each crude extract, as determined by quantitation of Ad Pol bands on immunoblots visualized with an Ad Pol-specific anti-peptide antibody.

The results obtained from analysis of the glycine replacement mutants showed activities on nicked calf thymus DNA and on poly(dT)/oligo(dA) templates that were generally higher than were replacements at other amino acids. These results were observed even with nonconservative amino acid changes such as gly to met or gly to ser (Figure 5). When assayed in vitro for adenovirus-specific initiation and elongation, changes that mimicked amino acids found in other DNA polymerases that use a protein priming mechanism showed activities similar to those observed for wild type (40 to 120%). However, the gly to ala changed reduced the observed activity ten fold in these in vitro assays.

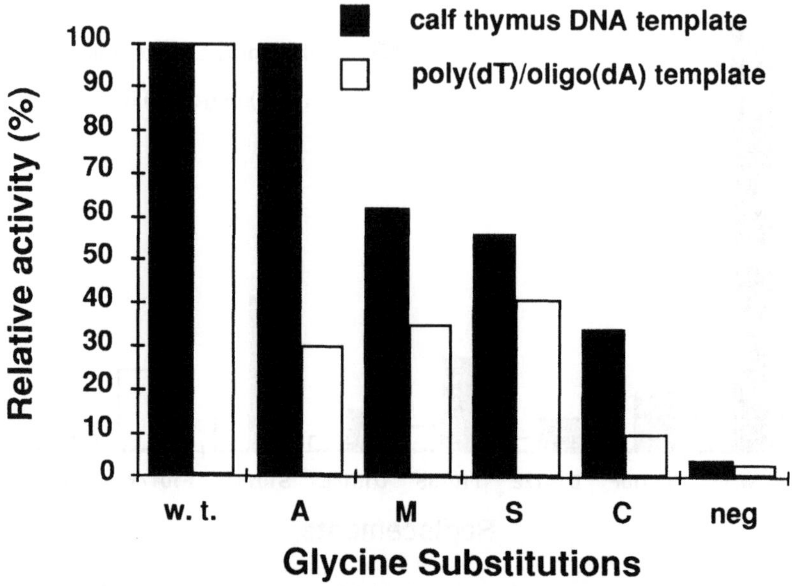

Figure 5: Replacements at glycine 1011 in Ad Pol. The relative activities of different mutants were compared to that of wild type Ad Pol (w.t.), which was set at 100%. Filled rectangles show the activity of crude extracts on nicked calf thymus DNA (Pharmacia) in the presence of 120 μM aphidicolin; open rectangles show the activity on poly(dT)/oligo(dA) templates in the presence of Ad DBP. The data shown was normalized for the relative amounts of Ad Pol protein in each crude extract, as determined by quantitation of Ad Pol bands on immunoblots visualized with an Ad Pol-specific anti-peptide antibody.

SUMMARY

A series of mutations in the pTP and Ad Pol molecules have been used to study the function of various domains within these polypeptides. These mutations have been characterized using a series of in vitro assays for DNA synthesis activity and for adenovirus-specific initiation and elongation reactions. Deletions and linker insertion mutants in the amino terminus of pTP suggest that this domain is essential for pTP function and that steps in the elongation process may be more severely impaired than the initiation reaction in these mutants. Similarly, even conservative changes at ser580 abolish the ability of pTP to be covalently attached to dCMP. Most replacements in the conserved region I of Ad Pol also abolish activity in vitro: even conservative changes had drastic effects. Some substitutions at glycine 1011 that mimicked changes found in DNA polymerases from other protein-priming systems retained partial or full activity, depending on the assay used.

REFERENCES

Argos, P. (1988) A sequence motif in many polymerases. Nucl. Acids Res. 16:9909-9916.
Challberg, M.D., and Kelly, T.J., Jr. (1979) Adenovirus DNA replication in vitro. Proc. Natl. Acad. Sci. USA 76:655-659.
Challberg, M.D., and Kelly, T.J., Jr. (1989) Animal virus DNA replication. Ann. Rev. Biochem. 58:671-717.
Chen, C., and Okayama, H. (1987) High-efficiency transformation of mammalian cells by plasmid DNA. Mol. Cell. Biol. 7:2745-2752.
Delarue, M., Poch, O., Tordo, N., Moras, D., and Argos, P. (1990) An attempt to unify the structure of polymerases. Prot. Engineering 3:461-467.
Desiderio, S.V. and Kelly, T.J., Jr. (1981). Structure of the linkage between adenovirus DNA and the 55,000 molecular weight terminal protein. J. Mol. Biol. 145:319-337.
Field, J. Gronostajski, R.M. and Hurwitz, J. (1984) Properties of the adenovirus DNA polymerase. J. Biol. Chem. 259:9487-9495.
Freimuth, P.I. and Ginsberg, H.S. (1986) Codon insertion mutants of the adenovirus terminal protein. Proc. Natl. Acad. Sci. USA 83:7816-7820.
Friefeld, B.R., Lichy, J.H., Field, J., Gronostajski, R.M., Gugenheimer, R.A., Krevolin, M.D., Nagata, K., Hurwitz, J., and Horwitz, M.S. (1984) The in vitro replication of adenovirus DNA. Curr. Top. Microbiol. Immunol. 110:221-255.
Garmendia, C., Salas, M., and Hermoso, J.M. (1988). Site-directed mutagenesis in the DNA linking site of bacteriophage ϕ29 terminal protein: isolation and characterization of a ser232$\rightarrow$thr mutant. Nucl. Acids Res. 16:5727-5740..
Garmendia, C., Hermoso, J.M., and Salas, M. (1990). Functional domain for priming activity in the phage ϕ29 terminal protein. Gene 88:73-79.
Gerard, R.D. and Gluzman, Y. (1985) New host cell system for regulated simian virus 40 DNA replication. Mol. Cell. Biol. 5:3231-3240.
Kamer, G. and Argos, P. (1984) Primary structural comparison of RNA-dependent polymerases from plant, animal and bacterial viruses. Nucl. Acid Res. 12:7269-7282.
Kaufman, R.J. (1985). Identification of the components necessary for adenovirus translation control and their utilization in cDNA expression. Proc. Natl. Acad. Sci. USA 82:689-693.
Marcy, A.I., Hwang, C.B.C., Ruffner, K.L., and Coen, D.M. (1990) Engineered Herpes simplex virus DNA polymerase point mutants: the most highly conserved region shared among α-like DNA polymerases is involved in substrate recognition. J. Virol. 64:5883-5890.
Ollis, D.L., Brick, P., Hamlin, R., Xuong, N.G., and Steitz, T.A. (1985) Structure of large fragment of Escherichia coli DNA polymerase I complexed with dTMP. Nature 313:762-766.
Pettit, S.C., Horwitz, M.S., and Engler, J.A. (1988) Adenovirus preterminal protein synthesized in COS cells from cloned DNA is active in DNA replication in vitro. J. Virol. 62:496-500.
Pettit, S.C., Horwitz, M.S., and Engler, J.A. (1989) Mutations of the precursor to the terminal protein of adenovirus serotypes 2 and 5. J. Virol. 63:5244-5250.

Roberts, R.J., Akusjärvi, G., Aleström, P., Gelinas, R.E., Gingeras, T.R., Sciaky, D., and Pettersson, U. (1986) A consensus sequence for the adenovirus 2 genome, pp.1-51. In: W. Doerfler (ed.), Adenovirus DNA: The viral genome and its expression. Martinus Nijhoff Publishing, Boston.

Roovers, D.J., Overman, P.F., Chen, X., and Sussenbach, J.S. (1991) Linker mutation scanning of the genes encoding the adenovirus type 5 terminal protein precursor and DNA polymerase. Virol. 180:273-284.

Shu, L., Horwitz, M.S., and Engler, J.A. (1987) Expression of enzymatically active adenovirus DNA polymerase from cloned DNA requires sequences upstream of the main open reading frame. Virology 161:520-526.

Shu, L., Pettit, S.C., and Engler, J.A. (1988) The precise structure and coding capacity of mRNAs from early region 2B of human adenovirus serotype 2. Virol. 165:348-356.

Stillman, B.W., Lewis, J.B., Chow, L.T., Matthews, M.B., and Smart, J.E. (1981) Identification of the gene and mRNA for the adenovirus terminal protein precursor. Cell 23:497-508.

Tamanoi, F., and Stillman, B.W. (1984) The origin of adenovirus DNA replication. Curr. Top. Microbiol. Immunol. 109:75-87.

Wang, T.S.-F.,Wong, S.W., and Korn, D. (1989) Human DNA polymerase α: predicted functional domains and relationships with viral DNA polymerases. FASEB J. 3:14-21.

Wong, S.W., Wahl, A.F., Yuan, P.-M., Arai, N., Pearson, B.E., Arai, K.-i., Korn, D., Hunkapiller, M.W., and Wang, T.S.-F. (1988) Human DNA polymerase α gene expression is cell proliferation dependent and its primary structure in similar to both prokaryotic and eukaryotic replicate DNA polymerases. EMBO J. 7:37-47.

Yoshikawa, H. and Ito, J. (1982) Nucleotide sequence of the major early region of bacteriophage ϕ29. Gene 17:323-335.

Zaballos, A., Mellado, R.P., and Salas, M. (1988). Initiation of phage ϕ29 DNA replication by mutants with deletions at the amino end of the terminal protein. Gene 63:113-121.

Zaballos, A., Lázaro, J.M., Méndez, E., Mellado, R.P., and Salas, M. (1989). Effects of internal deletions on the priming activity of the phage ϕ29 terminal protein. Gene 83:187-195.

The Role of Transcription Factors in Adenovirus DNA Replication

P.C. van der Vliet, C.P. Verrijzer, Y.M. Mul, J.A.W.M. van Oosterhout and W. van Driel.

Laboratory for Physiological Chemistry
University of Utrecht
Vondellaan 24a
3521 GG Utrecht
The Netherlands

Abstract

Replication of adenovirus DNA initiates by a protein-priming mechanism followed by elongation via strand displacement. Minimal initiation in vitro requires the complex of the precursor terminal protein and the Ad DNA polymerase (pTP-pol) and a core origin of 18 basepairs. Initiation is enhanced over 100-fold by two cellular transcription factors, NFI/CTF and NFIII/oct-1 binding to the auxiliairy region between nucleotides 18 and 50. We purified

these two proteins after overexpression and studied their mechanism of action. For both proteins, the DNA binding domain suffices for enhancement. NFI/CTF recruits the pTP-pol complex and positions it correctly at the origin through site-specific binding. NFI/CTF also increases the stability of initiation complexes. NFIII/oct-1 induces a bend in the DNA, thereby presumably facilitating the formation of a multiprotein initiation complex. Bending is mediated by the POU-specific subdomain of NFIII/oct-1 which is essential for replication enhancement.

Introduction

Replication of the 36 kb long, double stranded adenovirus genome in HeLa cell nuclei requires the interaction between three viral replication proteins and at least three cellular proteins (see Table I). The viral proteins are all derived from the same transcription unit, E2. The most abundant protein is the 72 kDa DNA binding protein

Table I

PROTEINS REQUIRED FOR ADENOVIRUS DNA REPLICATION

VIRAL PROTEIN	REQUIRED FOR
DNA binding protein (DBP)	Elongation, initiation
Terminal Protein (TP)	Initiation
Precursor Terminal Protein (pTP)	Initiation
Ad DNA polymerase (pol)	Initiation, elongation
CELLULAR	
Nuclear Factor I (NFI/CTF)	Initiation
Nuclear Factor II (NFII)	Elongation
Nuclear Factor III (NFIII/oct-1)	Initiation

(DBP). This multifunctional protein binds to double-stranded DNA forming a dynamic, multiprotein-DNA complex which facilitates the subsequent binding of NFI (Cleat and Hay, 1989; Stuiver and Van der Vliet) DBP also enhances the processivity of the Ad DNA polymerase (Lindenbaum et al, 1986; Mul et al, 1989) and binds to the displaced single-stranded DNA, thereby forming a rigid, extended structure (Van Amerongen et al, 1987). The Ad DNA polymerase (pol) and the precursor terminal protein (pTP) form a strong heterodimer. This complex is able to recognize and bind to the core origin when part of the template strand is in a single-stranded form (Kenny and Horwitz, 1988). In the presence of ATP and dCTP, the DNA polymerase catalyzes the formation of a pTP-dCMP initiation-complex by covalent coupling of a dCMP residue to serine residue 580 in pTP. This reaction is considered to be the initiating step. Further elongation occurs by using the 3'-OH-group of dCMP as a primer (reviews, Challberg and Kelly, 1989; Stillman, 1989; Hay and Russell, 1989; Van der Vliet, 1990).

Shortly after the development of an in vitro system and the subsequent isolation of the three viral replication proteins it became clear that nuclear extracts from uninfected cells stimulate initiation considerably. Based upon this property two proteins were purified to homogeneity and further characterized. Surprisingly, both proteins appeared to be cellular transcription factors. Nuclear factor I (NFI) (Nagata et al, 1982) is a member of a family of transcription factors recognizing the CCAAT box in promoters of several cellular genes and has also been called CCAAT binding transcription factor (CTF) (Jones et al, 1987). The protein binds as a dimer to one side of the helix, mainly contacting the major groove (De Vries et al, 1987). Its recognition sequence in the Ad2 origin of replication is located between base pairs 25 and 40 (Fig. 1). NFI/CTF contains a distinct, conserved DNA binding domain as well as variable transcription activation domains which are rich in proline residues (Santoro et al, 1988; Mermod et al, 1989; Gounari et al, 1990).

Nuclear Factor III (NFIII, Pruijn et al, 1986) is a member of the POU-protein family (Herr et al, 1988) and is identical to Oct-1 (O'Neill et al, 1988; Pruijn et al, 1989). The protein recognizes a sequence adjacent to the NFI site (Fig. 1) by contacting both the major and the minor groove (Pruijn et al, 1988) and it stimulates DNA replication independently of NFI (Mul et al, 1990). NFIII/oct-1 is composed of a conserved DNA binding domain, the POU domain, as well as an N-terminal glu-rich region which is

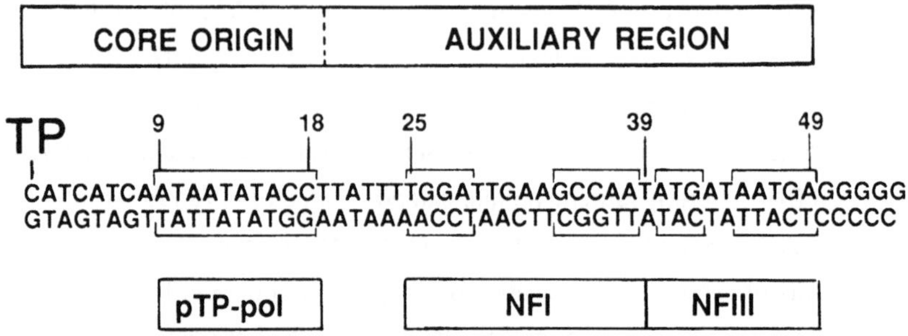

Fig. 1 The origin of replication of adenovirus type 2. The terminal protein (TP) is covalently bound to the 5'-dCMP residue of the parental strand. The regions conserved between most human serotypes are indicated by brackets. The pTP-pol binding site is only defined for a partially single stranded template (Chen et al, 1990).

involved in transcription activation (Tanaka and Herr, 1990).

Both for NFI/CTF and NFIII/Oct-1 it was shown that the DNA-binding domains (NFI-BD and the POU-domain, respectively) suffice for stimulation of Adenovirus DNA replication (Mermod et al, 1989; Gounari et al, 1990; Verrijzer et al, 1990a).

Here we describe experiments that are aimed to obtain a better understanding of the mechanisms by which these two transcription factors enhance DNA replication.

METHODS

Replication proteins

Recombinant vaccinia virus containing the N-terminal DNA-binding domain (NFI-BD) of rat NFI (amino acids 4-240) were used to infect Hela cells and the protein domain was purified to homogeneity from nuclear extracts by a three-step procedure as described in detail elsewhere (Mul et al, 1990). The construction of recombinant vaccinia viruses containing the Oct-1 POU domain or the homeodomain as well as their purification has been described (Verrijzer et al, 1990a, 1990b). The viral pTP-pol complex was purified from cells infected with a mixture of vaccinia viruses expressing pTP and pol, respectively (Mul et al, 1989). DBP was purified from Ad5 infected Hela cells (Stuiver and Van der Vliet, 1990).

Conditions for replication in vitro

As templates we employed either Ad2 or Ad5 DNA-TP complex predigested with XbaI. The optimal replication conditions were as described in Mul et al (1990). Preincubation of the template with pTP-pol, with or without NFI, was for various times at 10°C in a buffer containing 25mM Hepes (pH 7.5) - 1mM DTT - 4mM $MgCl_2$ - 2mM ATP - 50mM NaCl. After preincubation, dNTPs including [α-^{32}P] dCTP as well as the lacking replication components were added and replication was allowed to proceed for 15 minutes as 37°C. The products were analysed by SDS-agarose gel electrophoresis as described (Mul et al, 1990).

Detection of complexes between NFI and pTP-pol

NFI-BD and pTP-pol were mixed in a volume of 25 µl in a buffer containing 25mM Hepes (pH 7.5) - 1mM DTT - 5mM $MgCl_2$ - 4% glycerol - 60mM NaCl - 3µg BSA. After 30 min. at 30°C, the solution was mixed with protein A-Sepharose containing anti-pTP antibodies. After extensive washing the proteins binding to the beads were eluted by boiling in sample buffer, electrophoresed in a SDS-polyacrylamide gel and blotted onto nitrocellulose. NFI was detected employing a polyclonal rabbit antibody raised against NFI-BD. Details of the procedure will be described elsewhere (Mul et al, in preparation).

Analysis of bending

The oct-1 protein was bound to various radiolabelled 143 bp fragments containing a permutated octamer binding site present in the Ad4 origin. The fragments originated from pOctaBend 2 (Verrijzer et al, submitted) after digestion with 8 different restriction enzymes. After 30 minutes at room temperature bound and free DNA were resolved on a 4 % polyacrylamide gel run in 1xTBE at 25°C.

Results and Discussion

NFI and NFIII function independently

Several mechanisms can be envisaged by which NFI and NFIII enhance initiation. These proteins might change the structure of the core origin, they could initiate or stabilize the formation of a multiprotein initiationcomplex, or change the kinetics of the initiation reaction itself. These various mechanisms are not mutually exclusive and could operate simultaneously.

To study the effects of the two proteins separately or in combination we purified both transcription factors as well as their functional DNA binding domains to homogeneity. Their enhancing effect on initiation was studied employing a reconstituted replication

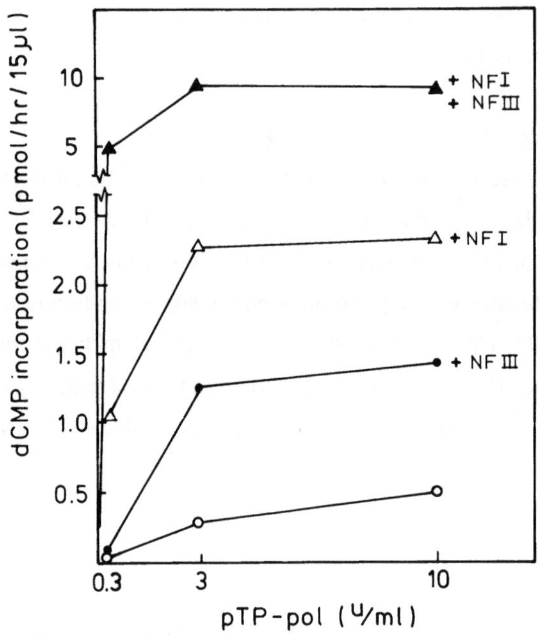

Fig. 2 Enhancement of initiation by NFI and NFIII is dependent on the pTP-pol concentration.
Replication was performed in a final volume of 15 µl containing 25 ng of XhoI-digested Ad5 DNA-TP, 0.6 µg DBP and pTP-pol as indicated. Replication was for 1 hour at 37°C either without transcription factors (0) or with 3.5 ng NFI-BD (Δ), 2.7 ng POU domain (•) or both (▴) and was expressed as pmol dCTP incorporated per hour.

system. As shown in Fig. 2, saturating amounts of NFI or NFIII stimulated DNA replication to a different extent. Saturation is defined as the level of protein that is required to give a 100 % bandshift in a gel retardation assay and corresponds to the maximum level of stimulation that can be obtained by each protein (Mul et al, 1990).

At 0.3 U/ml of the pTP-pol complex, NFI stimulated the reaction 46-fold and NFIII 5-fold. When these proteins were added together, both at saturating levels, the total enhancing effect was the product of both values, i.e. 230-fold. This could indicate that the two proteins act synergistically, suggesting that they enhance the final reaction along independent mechanisms but in a non-cooperative fashion. Interestingly, the binding of NFI and NFIII to the origin is also independent and non-cooperative despite the proximity of the recognition sites (Mul et al, 1990).

At increasing pTP-pol concentrations the final level of replication continues to increase slowly and at 3-10 U/ml pTP-pol a maximal level was obtained. Under these conditions, in the presence of NFI and NFIII, the input template could be more than tripled indicating the synthesis of multiple rounds of replication. This was supported by the appearance of labelled single-stranded DNA originating from displacement of a labelled daughter strand. Remarkably, at high pTP-pol concentrations the amount of stimulation by NFI was much less than at low pTP-pol levels. At 3.3 U/ml pTP-pol NFI stimulated only 8-fold and NFIII still stimulated 4.6-fold (Fig. 2). At very high pTP-pol concentrations (30 U/ml) NFI stimulated less than 2-fold (Mul et al, 1990), despite the presence of saturating amounts of the two transcription factors or their DNA binding domains. Thus, enhancement of stimulation, in particular by NFI, is dependent upon the pTP-pol concentration. This suggests that NFI acts by facilitating the pTP-pol complex to bind to the core origin or to form a stable initiation complex.

<u>NFI can bind to the pTP-pol complex through its DNA-binding-domain</u>

To investigate these possibilities further we used two independent approaches. First, we studied whether NFI could interact with the pTP-pol complex. Second, we investigated the half-life of an initiation complex as a function of the presence of NFI, using a competing template.

We mixed purified pTP-pol and NFI-BD and studied complex formation under variable conditions by immune precipitation using anti-pTP antibodies (Stunnenberg et al, 1988) followed by Western blotting to detect any NFI-BD bound to the pTP-pol complex. As shown in Fig. 3, NFI-BD appeared to be present in a stable complex with pTP-pol. Almost all NFI-BD could be bound in the absence of DNA. Controls show that without pTP-pol, hardly any NFI-BD could be detected (Fig 3, lanes 1-3).

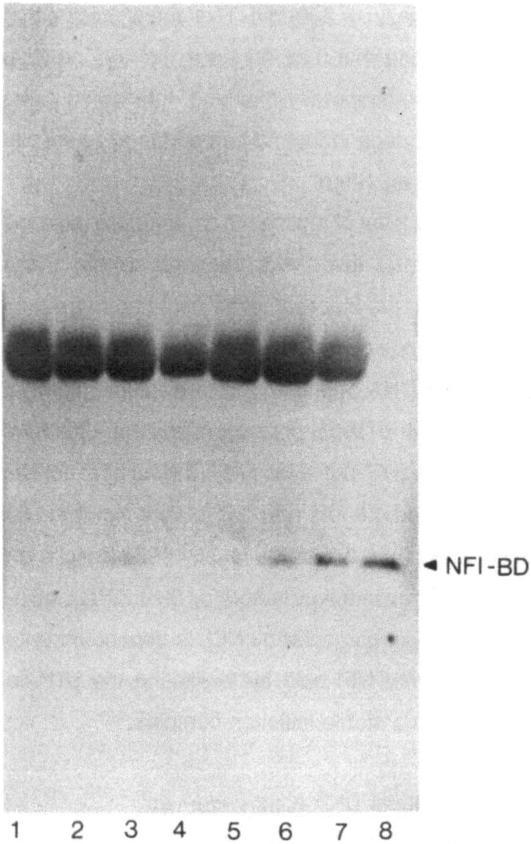

Similar studies were performed by Chen et al (1990) who could establish that NFI interacts specifically with the Ad DNA polymerase in the complex and that a region of NFI between amino acids 68-150 is involved. Thus, these data can be interpreted to mean that NFI is able to recruit, in solution, the pTP-pol complex and, by virtue of its specific binding site in the auxiliary region, may be able to position the pTP-pol complex in such a way to enable initiation. It should be stressed, however, that direct binding studies of pTP-pol to origin DNA have not been successful so far. Therefore it is not possible to establish directly that NFI enhances the binding of pTP-pol to the core origin. It may also function just to keep pTP-pol close to the core origin. We favour the first interpretation because this could better explain the highly conserved character of

Fig. 3 NFI co-precipitates with the pTP-pol complex.
NFI-BD was mixed at increasing concentrations in the presence of pTP-pol(500mU) (lanes 5-7) or in the absence (lanes 1-3) with protein A-sepharose to which anti-pTP antibodies were coupled. After washing the eluate was separated on a SDS-polyacrylamide gel and NFI-BD was detected by immunoblotting. The amounts of NFI-BD used were 30 ng (lanes 1,5), 90 ng (lanes 2,6) and 180 ng (lanes 3,7). Lane 4 contained only pTP-pol, lane 8 contained 180 ng untreated NFI. The broad band below NFI-BD consists of protein-A eluted from the sepharose beads.

the core origin. Moreover, at least on a partially single-stranded core origin binding of pTP-pol can be detected both by gel retardation and copper-phenantrolin footprinting (Chen et al, 1990).

NFI stabilizes a pre-initiation complex

The sequence of events that lead to the initiation reaction have not yet been clarified. We have tried to obtain more information aboute the various steps by studying the stability of a presumed pre-initiation-complex. For that purpose we used a competing template consisting of the same origin but producing a reaction product that can be distinguished by virtue of its different length. This method was used previously to study the stability of transcription complexes (Klein-Hitpass et al, 1990). We used Ad5 DNA-TP and Ad2 DNA-TP digested with XbaI giving rise to templates of 5508 bp and 4140 bp, respectively, in addition to fragments of identical lengths (1332 bp). A limiting amount of pTP-pol was preincubated with Ad5 template under different conditions. After 5 or 10 minutes the second template (Ad2) as well as the remaining replication components were added and replication was allowed for 15 min. As shown in Fig. 4, lanes 4-9, the ratio of Ad2 to Ad5 replication was almost 1:1 after addition to the competing template in the absence of NFI or pTP-pol indicating a short half-life of the complex. However, in the presence of both NFI and pTP-pol competitor resistance could be observed as shown by the increased Ad5 to Ad2 ratio indicating that the half-life of the complex was increased. This result was independent of the presence of ATP and occurred at several temperatures. We interprete this to mean that the presence of NFI stabilizes the preinitiation complex.

NFIII/oct 1 changes the origin structure

We performed similar assays employing NFIII/oct-1, or the purified POU-domain, but with negative results. Also stimulation of replication by the POU domain was much less dependent on the pTP-pol concentration (Mul et al, 1990). Therefore, we looked for other properties of the POU domain and investigated whether binding of NFIII/oct1 might change the origin structure. In particular we concentrated on DNA bending.
To study possible bending of DNA we employed several methods. First, we used a circular permutation assay in which the binding site is located at different positions in

the same fragment. Position-dependent electrophoretic mobilities of the protein-bound DNA indicates bending (Wu and Crothers, 1984). We employed the vector pBend2 (Kim et al, 1989) in which the Ad4 octamer site was cloned into a unique XbaI site leading to a 143 bp fragment upon digestion with several restriction enzymes. The results of such an assay with NFIII/oct-1 is shown in Fig.5. The protein-bound fragment is most retarded when the binding site is in the middle of the fragment which is explained by the reduced end-to-end distance caused by bending. Two other independent techniques confirmed that NFIII/oct-1 bends DNA, i.e. the measurement of relative bend direction employing prebent DNA and an enhanced ligase-mediated cyclization rate. Details of these assays will be published elsewhere (Verrijzer et al, submitted).

The POU domain consists of a conserved POU-specific domain of about 75 amino acids connected to a POU-homeodomain by a

Fig. 4 NFIII/oct1 bends the origin.
Purified NFIII/oct1 was bound to eight [^{32}P]-labeled 143 bp fragments (1-8) each with a uniquely positioned octamer motif. In lane 4 the octamer is positioned in the middle of the fragment, in lanes 1 and 8 at the extreme right and left sides, respectively. The complexes were resolved on a 4% polyacrylamide gel.

flexible linker (Herr et al, 1988). Detailed contact point analysis has shown that, although direct DNA binding of the POU-specific domain is very difficult to detect, this domain nevertheless contributes to the specificity and strength of DNA binding by the POU homeodomain by providing additional contacts (Verrijzer et al, 1990b). This indicates that the octamer is a bipartite binding site.

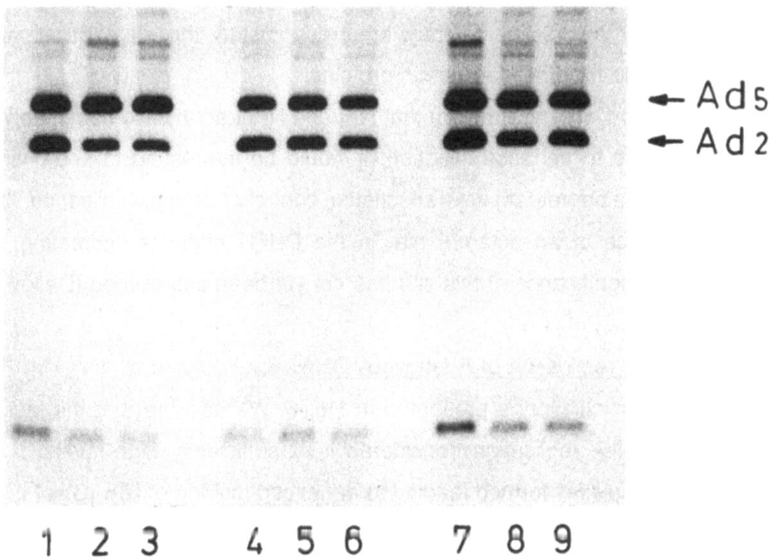

Fig. 5 Formation of a stable pre-initiation complex depends on the presence of NFI.
Ad5 DNA-TP was preincubated at 10°C with NFI alone (lanes 4-6), with pTP-pol alone (lanes 7-9), or with a combination of both (lanes 1-3). After 0 min. (lanes 1,4,7) 5 min. (lanes 2,5,8) or 10 min. (lanes 3,6,9) the second template (Ad2) as well as the other replication components were added followed by a 15 min. replication period.

From the absence of replication stimulation by the POU homeodomain we inferred that the POU-specific domain is essential for enhancement of initiation (Verrijzer et al, 1990a). In order to study a possible relation between DNA bending and stimulation of DNA replication we also studied bending by the POU homeodomain, but none was found (Verrijzer et al, submitted). This suggests that the POU-specific domain is responsible for DNA bending as well as stimulation of initiation. Further studies are required to study a possible causal relationship between these two properties.

How can bending contribute to initiation? Several viral origins as well as cellular origins contain bent DNA. Many origins are prebent and binding or proteins enhance bending (Zahn and Blattner, 1987; Cadlle et al, 1990). Bending may facilitate the formation of initiation complexes by making protein-protein contacts more easy. Alternatively, bending may help in unwinding origins at a distance by using elastic energy (Liu-Johnson et al, 1986), or may otherwise distort the DNA structure to make it recognizable for other replication proteins.

In conclusion, we have shown that NFI and NFIII act by two completely independent mechanisms to enhance initiation. It would be interesting to know whether similar mechanisms operate during transcription control or cellular replication. In this respect, the presence of an octamer site in the DHFR origin is interesting, although the functional significance of this site has not yet been established (Dailey et al, 1990).

A model for replication of Adenovirus DNA

A model for initiation is presented in Fig. 6. We assume that the level of DBP will influence DNA replication considerably. At sufficiently high levels of DBP a multi protein-complex is formed leading to enhanced binding of NFI (Cleat and Hay, 1989; Stuiver et al, 1990). NFI already recruits the pTP-pol complex in solution and the heterotrimer binds to the origin, thereby displacing DBP which is in dynamic equilibrium with the DNA. At the same time, NFIII/oct1 bends the DNA thereby presumably facilitating binding of pTP-pol to the core origin. This leads to a relatively stable, committed pre-initiation complex which has become resistant to competing templates (Fig.4). In an undefined step which may require ATP as well as structural changes in the pTP-pol complex a dCMP residue is bound to serine 580 in pTP. This may be followed by dissociation of pTP and pol after which the polymerase starts to synthesize the daughter strand employing the 3'OH group of dCMP as a primer . The start of polymerization and fork translocation could lead to partial unwinding of the recognition sites of NFI and NFIII and dissociation of these transcription factors from DNA. Note that we have not included a role for factor TP in this model and have suggested dissociation of pTP-pol after initiation, although evidence for such a dissociation is still lacking. Finally, elongation proceeds by displacement and DBP covers the displaced single strands. Due to the availability of purified replication proteins details of this process are now amenable to further study.

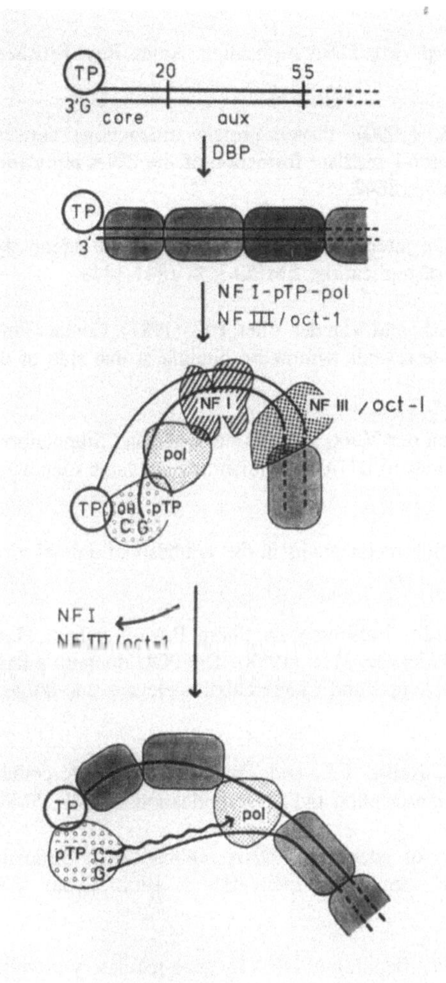

Fig. 6 A model for initiation of Adenovirus DNA replication. For details see text.

Acknowledgements

This work was supported in part by the Netherlands Foundation for Chemical Research (SON) with financial support from the Netherlands Organization for Scientific Research (NWO).

REFERENCES

Caddle, M.S., Dailey, L., and Heintz, N.H. (1990). RIP60, a mammalian origin-binding protein, enhances DNA bending near the Dihydrofolate reductase origin of replication. Mol. Cell. Biol. 10, 6236-6243.

Challberg, M.D., and Kelly, T.J. (1989). Animal virus DNA replication. Annu. Rev. Biochem. 58, 671-717.

Chen, M., Mermod, N., and Horwitz, M.S. (1990). Protein-protein interactions between Adenovirus DNA polymerase and Nuclear Factor-1 mediate formation of the DNA replication preinitiation complex. J. Biol. Chem. 265, 18634-18642.

Cleat, P.H., and Hay, R.T. (1989). Co-operative interactions between NFI and the adenovirus DNA binding protein at the adenovirus origin of replication. EMBO J. 8, 1841-1848.

De Vries, E., Van Driel, W., Van den Heuvel, S.J., and Van der Vliet, P.C. (1987). Contact Point analysis of the Nuclear Factor I recognition site reveals symmetric binding at one side of the DNA helix. EMBO J. 6, 161-168.

Gounari, F., De Francesco, R., Schmidt, J., Van der Vliet, P.C., Cortese, R., and Stunnenberg, H.G. (1990). Amino terminal domain of NFI binds to DNA as a dimer and activates adenovirus DNA replication. EMBO J. 9, 559-566.

Hay, R.T., and Russell, W.C. (1989). Recognition mechanisms in the synthesis of animal virus DNA. Biochem. J. 258, 3-16.

Herr, W., Sturm, R.A., Clerc, R.G., Corcoran, L.M., Baltimore, D., Sharp, P.A., Ingraham, H.A., Rosenfeld, M.G., Finney, M., Ruvkun, G., and Horvitz, H.R. (1988). The POU domain: a large conserved region in the mammalian pit-1, oct-1, oct-2 and Caenorhabditis elegans unc-86 gene products. Genes Dev. 2, 1513-1516.

Jones, K.A., Kadonaga, J.T., Rosenfeld, P.J., Kelly, T.J., and Tjian, R. (1987). A cellular DNA-binding protein that activates eukaryotic transcription and DNA replication. Cell 48, 79-89.

Kenny, M., and Hurwitz, J. (1988). Initiation of adenovirus DNA replication II, Structural requirements using synthetic oligonucleotide adenovirus templates. J. Biol. Chem. 263, 9809-9817.

Kim, J., Zwieb, C., Wu, C., and Adhya, S. (1989). Bending of DNA by gene-regulatory proteins: construction and use of a DNA bending vector. Gene 85, 15-23.

Klein-Hitpass, L., Tsai, S.Y., Weigel, N., Allan, G.F., Riley, D., Rodriguez, R., Schrader, W.T., Tsai, M.-J., and O'Malley, B.W. (1990). The progesterone receptor stimulates cell-free transcription by enhancing the formation of a stable preinitiation complex. Cell 60, 247-257.

Lindenbaum, J.D., Field, J., and Hurwitz, J. (1986). The adenovirus DNA-binding protein and adenovirus DNA polymerase interact to catalyze elongation of primed DNA templates. J. Biol. Chem. 261, 10218-10227.

Liu-Johnson, H.-N., Gartenberg, M.R., and Crothers, D.M. (1986). The DNA binding domain and bending angle of E.coli CAP protein. Cell 47, 995-1005.

Mermod, N., O'Neill, E.A., Kelly, T.J., and Tjian, R. (1989). The proline-rich transcriptional activator of CTF/NFI is distinct from the replication and DNA binding domain. Cell 58, 741-753.

Mul, Y.M., Van Miltenburg, R.T., De Clercq, E., and Van der Vliet, P.C. (1989). The mechanism of inhibition of adenovirus DNA replication by the acyclic nucleoside triphosphate analogue (S)-HPMPApp: Influence of the adenovirus DNA binding protein. Nucleic Acids Res. 17, 8917-8929.

Mul, Y.M., Verrijzer, C.P., and Van der Vliet, P.C. (1990). Transcription factors NFI and NFIII/oct-1 function independently, employing different mechanisms to enhance Adenovirus DNA replication. J. Virol. 64, 5510-5518.

Nagata, K., Guggenheimer, R.A., Enomoto, T., Lichy, J.H., and Hurwitz, J. (1982). Adenovirus DNA replication in vitro: Identification of a host factor that stimulates synthesis of the preterminal protein-dCMP complex. Proc. Natl. Acad. Sci. USA 79, 6438-6442.

O'Neill, E.A., Fletcher, C., Burrow, C.R., Heintz, N., Roeder, R.G., and Kelly, T.J. (1988). Transcriptional factor OTF-1 is functionally identical to the DNA replication factor NFIII. Science 241, 1210-1213.

Pettit, S.C., Horwitz, M.S., and Engler, J.A. (1988). Adenovirus preterminal protein synthesized in COS cells from cloned DNA is active in DNA replication in vitro. J. Virol. 62, 496-500.

Pruijn, G.J.M., Van Driel, W., and Van der Vliet, P.C. (1986). Nuclear Factor III, a novel sequence-specific DNA-binding protein from HeLa cells stimulating Adenovirus DNA replication. Nature 322, 656-659.

Pruijn, G.J.M., Van Miltenburg, R.T., Claessens, A.J., and Van der Vliet, P.C. (1988). Interaction between the octamer-binding protein Nuclear Factor III and the adenovirus origin of DNA replication. J. Virol. 62, 3092-3102.

Pruijn, G.J.M., Van der Vliet, P.C., Dathan, N.A., and Mattaj, I.W. (1989). Anti-OTF-1 antibodies inhibit NFIII stimulation of in vitro adenovirus DNA replication. Nucleic Acids Res. 17, 1845-1863.

Santoro, C., Mermod, N., Andrews, P.C., and Tjian, R. (1988). A family of human CAAT-box-binding proteins active in transcription and DNA replication: cloning and expression of multiple cDNAs. Nature 334, 218-224.

Stillman B.M. (1989). Initiation of eukaryotic DNA replication in vitro. Annu. Rev. Cell. Biol. 5, 197-245.

Stuiver, M.H., and Van der Vliet, P.C. (1990). The adenovirus DNA binding protein forms a multimeric protein complex with double-stranded DNA and enhances binding of nuclear factor I. J. Virol. 64, 379-386.

Stunnenberg, H.G., Lange, H., Philipson, L., Van Miltenburg, R.T., and Van der Vliet, P.C. (1988). High expression of functional adenovirus DNA polymerase and precursor terminal protein using recombinant vaccinia virus. Nucleic Acids Res. 16, 2431-2444.

Tanaka, H., and Herr, W. (1990). Differential transcriptional activation by oct-1 and oct-2: Interdependent activation domains induce oct-2 phosphorylation. Cell 60, 375-386.

Van Amerongen, H., Van Grondelle, R., and Van der Vliet, P.C. (1987). The interaction between the adenovirus DNA binding protein and single-stranded polynucleotides studied by circular dichroism and ultraviolet absorption. Biochem. 26, 4646-4652.

Van der Vliet, P.C. (1990). Adenovirus DNA replication in vitro, in The Eukaryotic Nucleus (Straus, P., Wilson S.H. eds). Telford Press, Caldwell NJ 1-32.

Verrijzer, C.P., Kal, A.J., and Van der Vliet, P.C. (1990a). The DNA binding domain (POU domain) of transcription factor oct-1 suffices for stimulation of DNA replication. EMBO J. 6, 1883-1888.

Verrijzer, C.P., Kal, A.J., and Van der Vliet, P.C. (1990b). The oct-1 homeo domain contacts only part of the octamer sequence and full oct-1 DNA binding activity requires the POU-specific domain. Genes Dev. 4, 1964-1974.

Wu, H.M., and Crothers, D.M. (1984). The locus of sequence-directed and protein-induced DNA bending. Nature 308, 509-513.

Zahn, K., and Blattner, F.R. (1987). Direct evidence for DNA bending at the lambda replication origin. Science 236, 416-422.

The Replication Initiator Protein of P1 Is Activated by two *E. coli* Heat Shock Proteins, DnaJ and DnaK

Sue Wickner, Joel Hoskins# and Keith McKenney#
Laboratory of Molecular Biology
National Cancer Institute
National Institutes of Health
Bethesda, MD 20892

SUMMARY

Two *E. coli* heat shock proteins, DnaK (the hsp70 homolog) and DnaJ activate the specific DNA binding function of the replication initiator protein, RepA, of plasmid P1 by about 100-fold (Wickner *et al.*, 1991). The activation is ATP-dependent and DNA-independent. The mechanism of activation is the conversion of RepA dimers to monomers and the monomer form binds with high affinity to P1 origin DNA. Treatment of RepA with reversible chemical denaturants also converts dimers to monomers and simultaneously activates P1 origin binding. Increasing protein concentration converts monomers to dimers and deactivates RepA.

INTRODUCTION

We have been interested in the function of *E. coli* heat shock proteins in the initiation and regulation of DNA replication of plasmid P1.

#Center for Advanced Research in Biotechnology, National Institute of Standards and Technology, Rockville, MD 20850

The P1 origin of replication is flanked on one end by two 9 bp DnaA binding sites and on the other end by five 19 bp RepA binding sites and three DnaA binding sites (Fig. 1). The P1 *repA* gene, which codes for the initiator protein, is beside the origin with its promoter in the RepA binding sites. A replication control region downstream from the *repA* gene contains nine RepA binding sites and is required for the plasmid to be maintained at a copy number of one or two. RepA, in addition to functioning in initiation, also regulates plasmid copy number by binding to the control locus and regulates transcription by binding to sites in the origin that overlap the *repA* promoter (Chattoraj *et al.*, 1985).

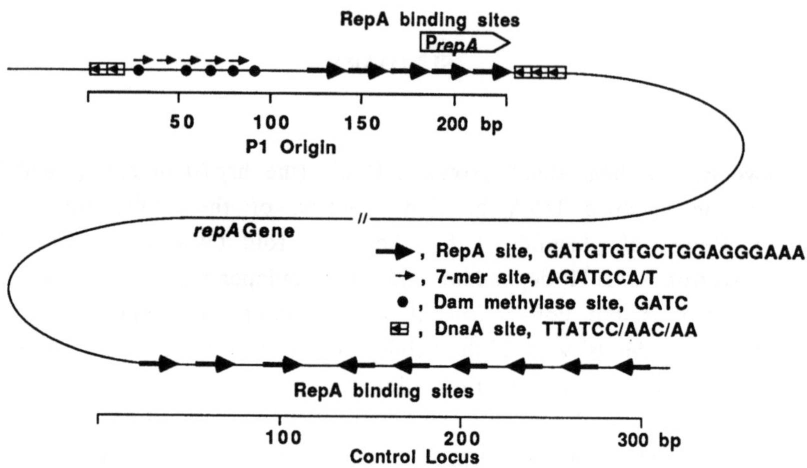

Fig. 1. Structure of the P1 plasmid replicon (Abeles *et al.*, 1984).

In vitro studies have shown that DNA replication of plasmid DNA carrying the P1 origin is catalyzed by crude protein fractions of *E. coli*. Replication requires, in addition to RepA, the *E. coli* DnaA initiator, DnaB helicase, DnaC, DnaG primase, DNA polymerase III holoenzyme, DNA gyrase, and RNA polymerase (Wickner and Chattoraj, 1987). Three heat shock proteins, DnaJ, DnaK and GrpE, are also involved in both *in vitro* replication (Wickner, 1990) and *in vivo* replication of P1 (Tilly and Yarmolinsky, 1989; Bukau and Walker, 1989). They are also required for replication of phage λ (Dodson *et al.*, 1986; Alfano and McMacken, 1989) and plasmid mini-F (Ezaki *et al.*, 1989; Kawasaki *et al.*, 1990).

The P1 RepA protein is a dimer in solution (Swack *et al.*, 1987; Wickner, 1990) and forms a stable complex with DnaJ, containing a dimer each of RepA and DnaJ (Wickner, 1990). The RepA-DnaJ tetramer can be purified from cell extracts or reconstituted with purified RepA and DnaJ.

RESULTS

We have examined DNA binding by RepA. Although RepA binds specifically to P1 origin DNA (Abeles, 1986), about a 100-fold molar excess of RepA dimers to binding sites (or about a 500-fold excess of RepA dimers to DNA fragments) was required for retention of *ori*P1 DNA on nitrocellulose filters (Fig. 2) and also for retardation of *ori*P1 DNA on

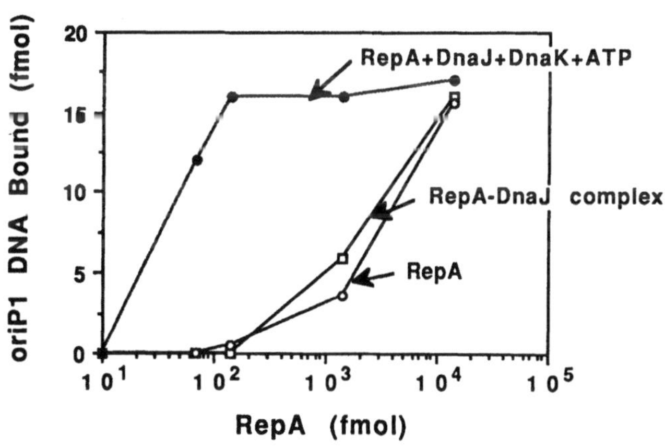

Fig. 2. *ori*P1 DNA binding by RepA alone, RepA-DnaJ complex and RepA in combination with DnaJ (30 ng), DnaK (100 ng) and ATP. Nitrocellulose filter binding reactions were carried out as described (Wickner *et al.*, 1991).

gel electrophoresis. These results suggested that the protein preparation was largely inactive or there was a low proportion of bound versus free protein at equilibrium in the reactions. We found that the RepA-DnaJ complex bound as poorly to DNA as RepA (Fig. 2). However, the combination of of DnaJ, DnaK, and ATP increased DNA binding by RepA by about 100-fold (Fig. 2). DnaJ, DnaK and ATP were required, but GrpE was not (Fig. 3).

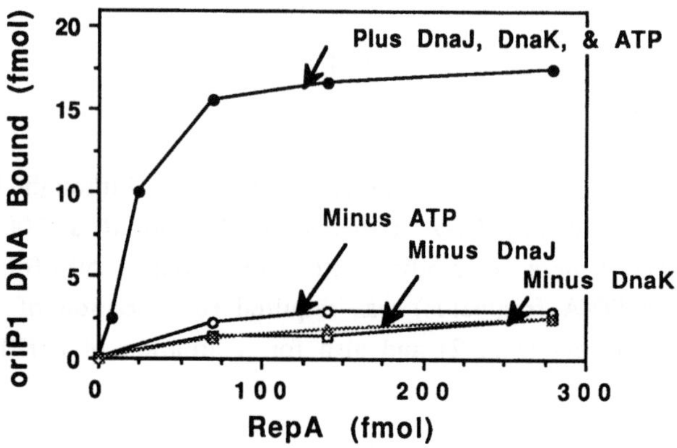

Fig. 3. Requirements for activation of *ori*P1 DNA binding by RepA. DNA binding reactions were carried out as described (Wickner *et al.*, 1991).

*ori*P1 DNA binding by RepA in the presence of DnaJ, DnaK, and ATP proceeded linearly for 20 minutes at 24°. There was very little DNA binding when the reactions were carried out at 0°. Using temperature shift experiments, we were able to separate the reaction into two steps. The first reaction was for 15 minutes at 24° and the second reaction was for 15 minutes at 0°. When DnaJ, DnaK, RepA, and ATP were incubated in the first reaction, DNA binding occurred upon the addition of *ori*P1 DNA during the second reaction at 0° (Table 1). A time course of each step revealed that the first reaction in the absence of DNA proceeded linearly for 20 minutes and the second reaction at 0° was complete by 30 seconds. These experiments show that DnaJ and DnaK, in an ATP-dependent reaction, activate RepA in the absence of DNA such that it is capable of rapidly and specifically binding to *ori*P1 DNA.

Table 1. Temperature requirement for activation of RepA.

Incubation at 24° (15 min)	Incubation at 0° (15 min)	*ori*P1 DNA bound (%)
Complete	No additions	26
No additions	Complete	<2
Minus ATP	+ ATP	<2
Minus *ori*P1 DNA	+ *ori*P1 DNA	32
Minus DnaJ	+ DnaJ	4
Minus DnaK	+ DnaK	4
Minus RepA	+ RepA	3

DNA binding was measured by retention on nitrocellulose filters as described (Wickner *et al.*, 1991). Reaction mixtures for the first incubation at 24° contained 20 ng of RepA, 30 ng of DnaJ, and 100 ng of DnaK where indicated. After 15 min, the reactions were chilled on ice, the omitted components were added, and the reactions were continued for 15 min at 0°.

Activation of the RepA-DnaJ complex also required ATP and incubation at room temperature for 20 minutes. We had previously observed that DnaJ and RepA complex formation occurred rapidly at 0°. Thus the rate limiting and temperature dependent step in the activation of RepA is the action of DnaK on the RepA-DnaJ complex.

We wanted to know which of the proteins involved in the reaction ultimately end up bound to *ori*P1 DNA. The protein-DNA complex was isolated from a large reaction mixture by gel filtration and analyzed by SDS-PAGE. The only protein detected was RepA, suggesting that DnaJ and DnaK function as true chaperones in this reaction (Fig. 4). In control reactions lacking DnaJ or ATP, no proteins were detected bound to the DNA. Thus DnaJ and DnaK activate RepA to bind to DNA but are not part of the protein-DNA complex. The electrophoretic mobility of the bound RepA was indistinguishable from that of the input 32 kDa RepA subunits, suggesting that RepA had not undergone proteolytic processing and that the change in RepA was subtle.

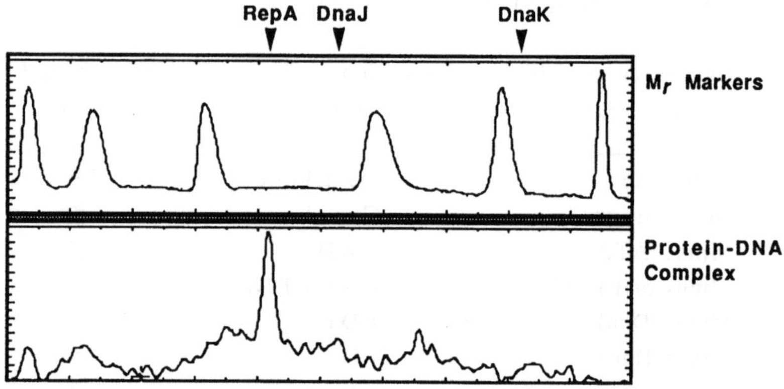

Fig. 4. Densitometer scan of coomassie blue stained SDS polyacrylamide gel of proteins associated with *ori*P1 DNA. M_r markers are BioRad low-range SDS-PAGE standards.

We sought to discover the change in RepA to understand the mechanism of action of DnaJ and DnaK. We isolated the activated form of RepA by first isolating the RepA-DNA complex by gel filtration and then dissociating RepA by treating the complex with 1 M NaCl. The dissociated RepA was separated from the P1 DNA by gel filtration in 0.4 M NaCl. The released RepA was in the activated form because (i) it bound to P1 DNA at 0°, (ii) DNA binding was not further stimulated by incubation with DnaJ, DnaK, and ATP, and (iii) its specific activity was similar to that of the activated form of RepA and 50-100 fold higher than that of the unactivated form.

DnaK and hsp70s have been implicated in the disaggregation of proteins, suggesting that DnaK might carry out that function in activating RepA. However, RepA does not aggregate under our reaction conditions. By analogy, it seemed possible that DnaK catalyzes the dissociation of dimers into monomers and that this is the form needed for origin binding. This possibility was also attractive because the 19 bp RepA binding site has no obvious 2-fold symmetry. To test this possibility, we analyzed the activated form of RepA by gel filtration using conditions that could cleanly separate RepA monomers from dimers (Fig. 5).

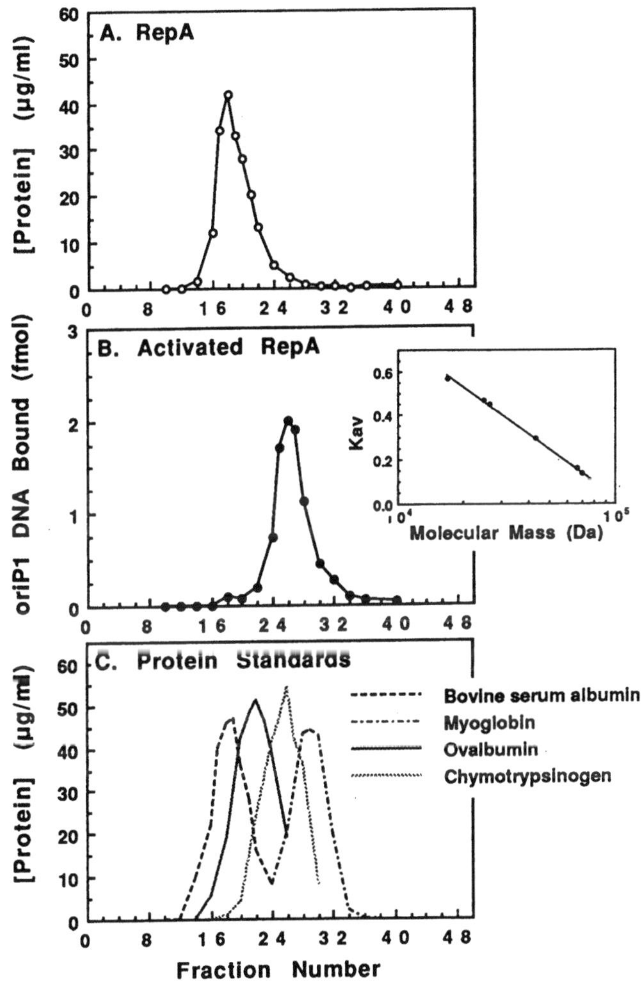

Fig. 5. G100 column chromatography of (A) RepA, (B) the activated form of RepA, and (C) Protein standards. B inset: estimation of the molecular weight of the activated form of RepA. In B, DNA binding was measured minus DnaJ, DnaK, and ATP.

Purified RepA chromatographed as a protein of 70 kDa, the dimer of 32 kDa monomers (Swack *et al.*, 1985; Wickner, 1990). The activated form chromatographed as a 27 kDa protein. This result in conjunction with the previous result that activated RepA was indistinguishable from the unactivated form by SDS-PAGE, indicates that the activated form is a monomer. Possibly RepA monomers are active because the DNA binding domain overlaps the dimerization domain.

The observation that DnaJ, DnaK, and ATP convert RepA dimers to monomers suggested that chemical denaturants might also dissociate RepA dimers to produce activated RepA. RepA was treated with various chemicals, dialyzed and then assayed for *ori*P1 DNA binding with and without DnaJ, DnaK, and ATP (Table 2). Treatment with NaCl did not activate RepA. In contrast, treatment with urea or guanidine resulted in the conversion of RepA to the activated form. Triton X-100 or SDS activated RepA but to a lesser extent.

Table 2. Activation of RepA by chemical denaturation.

	*ori*P1 DNA Bound (fmol)	
	- DnaJ, DnaK, ATP	+ DnaJ, DnaK, ATP
Untreated RepA	0.6	27.0
0.4 M NaCl-treated	1.2	37.4
8 M Urea-treated	26.4	41.7
4 M Guanidine-treated	18.4	29.5

RepA was diluted to 10 μg/ml in 5% glycerol, 20 mM Tris·HCl, pH 7.5, 0.1 mM EDTA, 1 mM dithiotreitol, 50 μg/ml bovine serum albumin and chemicals as indicated. After 30 min at 0°, samples were dialyzed. 20 ng of each sample was assayed for DNA binding activity +/- DnaJ, DnaK, and ATP by nitrocellulose filter binding.

Urea-treated RepA was analyzed by gel filtration. It migrated as a monomer. Untreated RepA, with the same conditions, migrated entirely as a dimer. These results show that the activation of RepA by DnaJ and DnaK can be mimicked by chemically converting RepA dimers to monomers. They suggest that DnaJ and DnaK do not modify RepA but simply dissociate it into monomers, which bind to DNA with high affinity.

The activation of RepA by urea treatment was reversible. Urea activated RepA bound *ori*P1 DNA at 0° and was stimulated about 2-fold by DnaJ and DnaK. Following a 10-fold concentration of protein, DNA binding was 10-fold lower but could be stimulated about 10-fold by DnaJ and DnaK. This reversibility demonstrates that activation and deactivation need not involve a covalent change in RepA.

DISCUSSION

DnaJ and DnaK also activate DNA binding by RepA *in vivo*. Mini-P1 plasmids can transform *dnaJ, dnaK*, and *grpE* mutant strains but are not stably maintained (Bukau and Walker, 1989; Tilly and Yarmolinsky, 1989). Moreover, repression of the autoregulated *repA* promoter by RepA is 2 to 10-fold less in *dnaJ, dnaK* and *grpE* mutant strains compared to wild-type strains (Tilly *et al.*, 1990). Both of these *in vivo* observations could be explained at the molecular level by our *in vitro* results that DnaJ and DnaK activate the DNA binding function of RepA. One remaining puzzle is the biochemical function of GrpE.

Purified RepA monomers are active in *in vitro ori*P1 DNA replication reactions. This result and the *in vivo* results show that the monomer form of RepA is the form which is active in DNA replication. Perhaps the dimer form is involved in regulating P1 DNA replication possibly by sequestering RepA and thus limiting DNA replication.

Our model of how DnaJ and DnaK might function is shown in Fig. 6. It is analogous to that proposed by Pelham (1986, 1990) for the heat shock response, but emphasizes the role of DnaJ and DnaK with native proteins in unstressed conditions. Pelham suggested that during heat shock damaged proteins are identified as targets for DnaK through the specific recognition of exposed hydrophobic regions. We suggest that during normal growth conditions native proteins are identified as targets by DnaK through the specific recognition of a protein tag. In our case DnaJ tags RepA for binding by DnaK, perhaps also by exposing hydrophobic regions. There may be two conformations of DnaK. The stable form in the absence of ATP does not bind to the target protein; the ATP-bound form does. When DnaK binds to the RepA-DnaJ complex, RepA or DnaJ acts as an allosteric effector of the DnaK ATPase. The hydrolysis of ATP leads to the return of DnaK to the ground state conformation and to its release from the complex. The binding of DnaK to RepA leads to dissociation of RepA dimers. DnaJ is also released because only the dimer form of RepA binds DnaJ. DnaK bound to ATP acts like urea, which binds to hydrophobic regions and promotes denaturation (Tanford, 1970). ATP hydrolysis and the accompanying conformational change of DnaK acts like the removal of urea by dialysis, allowing spontaneous refolding to the active form of the target.

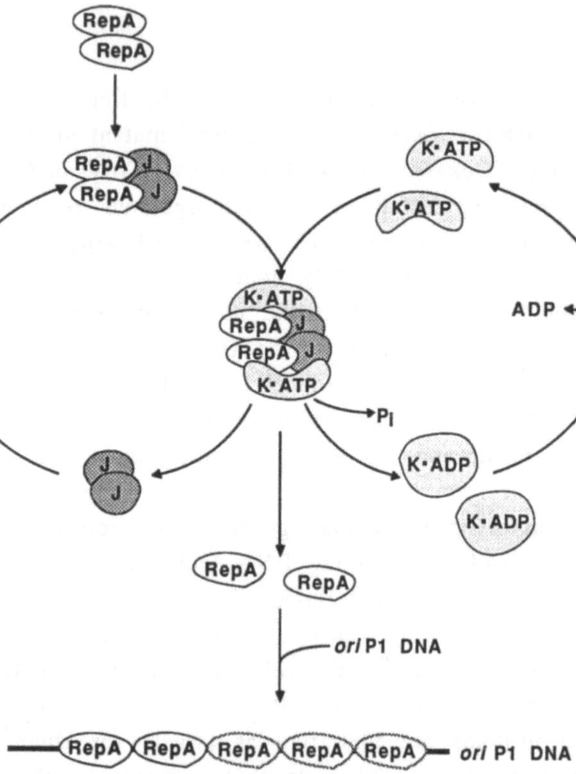

Fig. 6. Model of the mechanism of action of DnaJ and DnaK on native proteins.

REFERENCES

Abeles, A. L., Snyder, K. M., and Chattoraj, D. K. (1984) P1 plasmid replication: Replicon structure. *J. Mol. Biol.* 173, 307-324.

Abeles, A. L. (1986) P1 plasmid replication: Purification and DNA-binding activity of the replication protein, RepA. *J. Biol. Chem.* 261, 3548-3555.

Alfano, C. and McMacken, R. (1989) Heat shock protein-mediated disassembly of nucleoprotein structures is required for the initiation of bacteriophage DNA replication. *J. Biol. Chem.* 264, 10709-10718.

Bukau, B., and Walker, G. (1989) Δ*dnaK*52 mutants of *Escherichia coli* have defects in chromosome segregation and plasmid maintenance at normal growth temperatures. *J. Bacteriol.* 171, 6030-6038.

Chattoraj, D.K., Snyder, K.M., and Abeles, A.L., (1985) P1 plasmid replication: Multiple functions of RepA protein at the origin. *Proc. Natl. Acad. Sci. U. S. A.* 82, 2588-2592.

Dodson, M., Echols, H., Wickner, S., Alfano, C., Mensa-Wilmot, K., Gomes, B., LeBowitz, J., Roberts, J., and McMacken, R. (1986) Specialized nucleoprotein structures at the origin of replication of bacteriophage lambda: Localized unwinding of DNA by a six-protein reaction. *Proc. Natl. Acad. Sci. U. S. A.* 83, 7638-7642.

Ezaki, B., Ogura, T., Mori, H., Niki, H., and Hiraga, S. (1989) Involvement of DnaK protein in mini-F plasmid replication: Temperature-sensitive *seg* mutations are located in the *dnaK* gene. *Mol. Gen. Genet.* 218, 183-189

Kawasaki, Y., Wada, C., and Yura, T. (1990) Roles of *Escherichia coli* heat shock proteins DnaK, DnaJ and GrpE in mini-F replication. *Mol. Gen. Genet.* 220, 277-282

Pelham, H. B. R. (1986) Speculations on the functions of the major heat shock and glucose-regulated proteins. *Cell* 46, 959-961.

Pelham, H.B.R. (1990) Functions of the hsp70 protein family: An overview. *In*: Morimoto, R., Tissieres, A. and Georgopoulos, C. (eds.) Stress proteins in biology and medicine. Cold Spring Harbor Laboratory, Cold Spring Harbor, NY. pp. 287-300.

Swack, J. A., Pal, S. K., Mason, R. J., Abeles, A. L., and Chattoraj, D. K., (1987). P1 plasmid replication: measurement of initiator protein concentration *in vivo*. *J. Bacteriol.* 169, 3737-3742.

Tanford, C. (1970) Protein denaturation, part C. Theoretical models for the mechanism of denaturation. *Advances in Protein Chemistry* 24, 1-95.

Tilly, K., and Yarmolinsky, M., (1989) Participation of *Escherichia coli* heat shock proteins DnaJ, DnaK, and GrpE in P1 plasmid replication. *J. Bacteriol.* 171, 6025-6029.

Tilly, K., Sozhamannan, S., and Yarmolinsky, M., (1990) Participation of the *Escherichia coli* heat shock proteins DnaJ, DnaK, and GrpE in autorepression of the P1 plasmid *repA* promoter. *The New Biologist* 2, 1-6.

Wickner, S. (1990) Three *Escherichia coli* heat shock proteins are required for P1 plasmid DNA replication: Formation of an active complex between E.coli DnaJ protein and the P1 initiator protein. *Proc. Natl. Acad. Sci. U. S. A.* 87, 2690-2695.

Wickner, S., Hoskins, J., and McKenney, K. (1991) DnaJ and DnaK Function as Chaperones in Origin Specific DNA Binding by RepA. *Nature*, 350, 165-167.

Wickner, S., and Chattoraj, D. (1987) Replication of mini-P1 plasmid DNA *in vitro* requires two initiation proteins encoded by the *repA* gene of phage P1 and the *dnaA* gene of *Escherichia coli*. *Proc. Natl. Acad. Sci. U. S. A.* 84, 3668-3672.

Zylicz, M., Ang, D., Liberek, K., and Georgopoulos, C. (1989) Initiation of λ DNA replication with purified host- and bacteriophage-encoded proteins: The role of the dnaK, dnaJ and grpE heat shock proteins. *EMBO J.* 8, 1601-1608.

Bacteriophage λ DNA Replication and the Role of the Universally Conserved dnaK, dnaJ and grpE Heat Shock Proteins

K. Liberek[1]*, D. Skowyra*, J. Marszalek*, J. Osipiuk*, M. Zylicz*, D. Ang[1], A. Maddock[1], C. Johnson[1] and C. Georgopoulos[1]
*Division of Biophysics
Department of Molecular Biology
University of Gdansk
Kladki 24, 80-822 Gdansk
Poland

Introduction

There are many similarities in the replication mechanisms employed by bacteriophage λ and its host *E. coli* (for reviews, see Bramhill and Kornberg, 1988, Keppel et al., 1988, and McMacken et al., 1987). In both cases, the host-coded primase enzyme (the *dnaG* gene product) is mostly responsible for synthesizing the RNA primers for both the leading and lagging strand, to be extended into DNA by the DNA polymerase III holoenzyme. A major problem in both systems is that the primase enzyme is normally unable to make RNA primers by itself (except in the case of a few single-stranded bacteriophages). Instead, it must first make contact with the dnaB helicase, properly positioned on single-stranded DNA. The dnaB helicase unwinds DNA by traveling in the 5'→3' direction (LeBowitz and McMacken, 1986). To do so, the dnaB helicase must somehow "climb" onto DNA at a single-stranded region. Such a single-stranded DNA region is created at the *E. coli oriC* site by the binding of the dnaA•ATP protein form (Bramhill and Kornberg, 1988). To enter the single-stranded region at *oriC*, the dnaB helicase must be escorted by the dnaC initiation protein (Bramhill and Kornberg, 1988).

The λ DNA replication system

Genetic studies originally defined the *E. coli dnaB, dnaK, dnaJ* and *grpE* genes as essential for bacteriophage λ DNA replication. This was done through the isolation of *E. coli* mutants unable to propagate bacteriophage λ (Georgopoulos and Herskowitz, 1971; Sunshine et al., 1977; Saito and Uchida, 1977). Table 1 shows the plating properties of *E. coli* strains

[1]Department of Cellular, Viral and Molecular Biology
University of Utah School of Medicine
Salt Lake City, Utah 84132
U.S.A.

E. coli host	λ	λ*P*	λ*dnaB*$^+$	λ*dnaK*$^+$	λ*dnaJ*$^+$	λ*grpE*$^+$
wild type	+	+	+	+	+	+
*dnaB*558	-	+/-	+	-	-	-
*dnaK*756	-	+/-	-	+	-	-
*dnaJ*259	-	+/-	-	-	+	-
*grpE*280	-	+/-	-	-	-	+

Table 1. Plating properties of various bacteriophage λ derivatives on various *E. coli* strains. Some of the λ*P** mutations that enable λ to overcome the host block to λ growth are allele-specific, while others are not. However, none of the λ*P** mutations bypass the requirement for the host protein in question, i.e., λ*P** bacteriophage will not form plaques on *dnaB*$^{\Delta}$, *dnaK*$^{\Delta}$, *dnaJ*$^{\Delta}$, or *grpE*$^{\Delta}$ deleted mutant bacteria. The (+) symbol indicates good plaque size, the (+/-) symbol indicates small plaque size, and the (-) symbol indicates no phage growth.

with mutations in these genes. In all cases studied, compensatory mutations in bacteriophage λ could be isolated that overcome the host-imposed block to λ growth. All such mutations were shown to be localized in the λ replication gene *P*.

Additional genetic and biochemical studies identified other host-encoded components of the DNA replication machinery essential for λ DNA replication (summarized in Furth and Wickner, 1983; McMacken et al., 1987; Keppel et al., 1988). On the basis of these studies, our laboratories, as well as that of Dr. R. McMacken, developed *in vitro* λ DNA replication systems composed entirely of purified components (Zylicz et al., 1989; Mensa-Wilmot et al., 1989). The conclusions of these studies have been nicely complemented by electron microscopic work as well (Dodson et al., 1985, 1986, 1989).

Fig. 1 summarizes the steps involved in the *in vitro* λ DNA replication system. The first step involves the binding of dimers of the replicon-specific λO protein to the replication origin, *ori*λ, leading to a large protein•DNA structure, termed the "O-some" (Dodson et al., 1985). It has been shown that λO binding induces bending of DNA near the *ori*λ site (Zahn and Blattner, 1987). Such binding results in local unwinding of the DNA duplex within a 60-bp AT-rich region located immediately to the right of the λO-binding sites on the conventional λ map. This unwinding requires supercoiled DNA, but not ATP (Schnos et al., 1988). If the DNA is relaxed, λO protein can still bind to *ori*λ, but in this case the local unwinding is not observed. It has been proposed that such partial unwinding of DNA is necessary for the assembly of other replication proteins near the origin, in particular, positioning of the dnaB helcase on the two single strands of DNA created by the local unwinding (Bramhill and Kornberg, 1988). In λ DNA replication, the dnaB helicase must be complexed with another

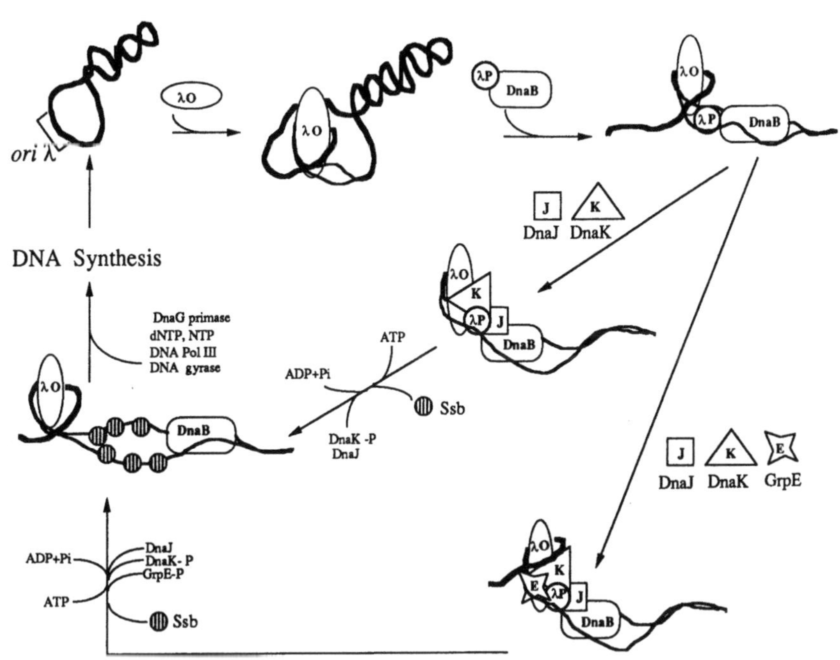

Fig. 1. Model of initiation of λ DNA replication in the grpE-dependent and -independent systems. See text for details.

bacteriophage λ replication protein, λP, to be properly positioned near *ori*λ (Furth and Wickner, 1983). Approximately 3 protomers of λP are complexed with a dnaB hexamer (Mallory et al., 1990). Formation of the *ori*λ•λO•λP•dnaB complex is mediated through the previously described weak λO•λP interaction (Zylicz et al., 1984; Wickner and Zahn, 1986). The addition of dnaK and dnaJ proteins leads to the formation of an even larger supramolecular protein-DNA structure at *ori*λ (Dodson et al., 1986).

At this point, although the dnaB helicase has been successfully captured and brought to *ori*λ by the λP protein, it is kept in an inactive state, since the λP•dnaB interaction is known to suppress all of the known activities of dnaB (Wickner, 1979; Mallory et al., 1990). Thus, additional steps are required to activate the dnaB helicase in λ as opposed to *ori*C DNA replication (Bramhill and Kornberg, 1988). These steps include (a) the liberation of the dnaB helicase from λP, and (b) the successful and continuous sequestering of the λP protein. This is so because once the dnaB helicase has been liberated and starts to unwind DNA, it is still subject to capture by the λP protein at the unwinding fork.

We have developed two *in vitro* systems that successfully replicate *ori*λ-containing plasmid DNA (Zylicz et al., 1989). The first system, called grpE-dependent, requires the presence of all three heat shock proteins, dnaK, dnaJ and grpE at approximately 60 protomers per *ori*λ site. The second system, called grpE-independent, relies on the presence of a 10-fold excess of dnaK protein (approximately 600 protomers per *ori*λ site), in addition to dnaJ protein (approximately 60 protomers per *ori*λ site). The two systems function equally efficiently *in vitro* to promote λ DNA replication in the presence of hydrolyzable ATP and deoxyribonucleotides (Zylicz et al., 1989). Why do the two systems possess different qualitative and quantitative protein requirements? We believe that the answer lies in the strength of the various protein-protein interactions among the five or six protein assembled at *ori*λ, and whether such interactions are sensitive to the presence of hydrolyzable ATP. Table 2 summarizes the known protein-protein interactions among the six proteins, λO, λP, dnaB, dnaK, dnaJ, and grpE. In most instances, both genetic and biochemical evidence exists demonstrating pairwise specific interactions among these proteins (Georgopoulos et al., 1990; Liberek et al., 1990).

	λO	λP	dnaB	dnaK	dnaJ	grpE
λO	+	+ *		+		+
λP		+	+ *	+ *	+ *	+ *
dnaB			+		+	
dnaK				+	+ *	+ *
dnaJ					+	
grpE						+

Table 2. Protein-protein interactions among the known bacteriophage λ- and host-coded proteins assembled at *ori*λ. The (+) symbol denotes an interaction demonstrated *in vitro* between the two purified proteins. The (*) symbol denotes an interaction demonstrated genetically *in vivo*, i.e., mutations in one gene partially or totally compensate for the defective phenotype of mutations in another gene. See Georgopoulos et al. (1990) for details.

In the grpE-independent system, we know that most of the λP, dnaK and dnaJ proteins (and some λO and dnaB proteins) are released from *ori*λ (Liberek et al., 1988; Alfano and McMacken, 1989a, 1989b; Dodson et al., 1989). We believe that an excess of dnaK protein is needed to sequester λP protein, resulting in the subsequent release of λP from *ori*λ. Because the dnaK•λP interaction is inherently weak and sensitive to the presence of ATP (Liberek et al., 1990), a relatively large number of dnaK protomers may be needed to successfully "bump" the λP protein and keep it from "recapturing" the dnaB helicase at the replication fork.

In the case of the grpE-dependent system, we believe that the demonstrated grpE•λP interaction, being ATP-resistant, is biologically important in sequestering the released λP protein (Ang, 1988; Georgopoulos et al., 1990). Thus, fewer dnaK molecules are needed because the λP protein can be efficiently sequestered following its release from *ori*λ. As discussed below, the simultaneous presence of both dnaJ and grpE is needed to stimulate dnaK's weak ATPase activity. Such stimulation may lead to faster dnaK protein action, thus lowering the requirement for high levels of dnaK protein. The grpE-dependent system is the biologically relevant one, since we have constructed *grpE*-deleted *E. coli* strains (Ang and Georgopoulos, 1989) and shown that bacteriophage λ will not replicate on such hosts. Furthermore, no λ mutants capable of growth on *grpE*-deleted hosts can be isolated, even after heavy mutagenesis.

Activation of dnaK's ATPase by dnaJ and grpE

A great deal of biochemical evidence exists demonstrating that dnaK, a prototype of a chaperone protein, is capable of binding to various "unfolded" or other proteins, including dnaJ and grpE (Table 2) (reviewed in Georgopoulos et al., 1990; Morimoto et al., 1990; Rothman, 1989). Various experimental facts suggested to us that one of the biological roles of the dnaJ and grpE proteins could be to help dnaK carry out its chaperone functions, perhaps by stimulating its weak ATPase activity (Zylicz et al., 1983).

Fig. 2 shows that, indeed, in the **joint** presence of dnaJ and grpE, a substantial stimulation of dnaK's ATPase activity can be observed. Only wild type grpE protein was capable of stimulating dnaK's ATPase activity. The mutant grpE280 protein (Ang, 1988), in the presence of dnaJ, did not appreciably stimulate dnaK's ATPase activity. Titration of grpE or dnaJ protein in the dnaK ATPase assay showed that the ratio of 1 mol of dnaK : 1-2 mol of

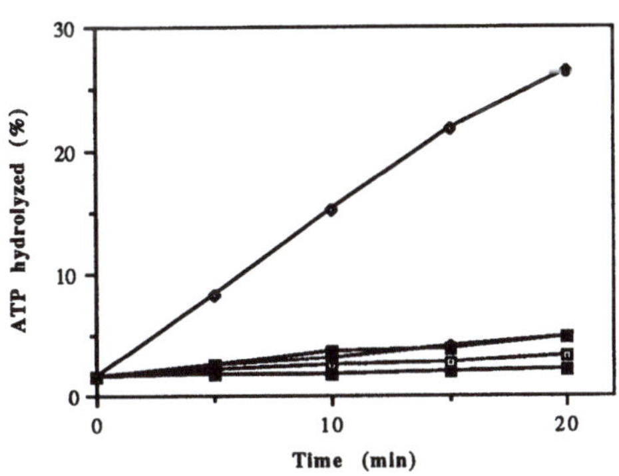

Fig. 2. Kinetics of ATP hydrolysis. The reactions were carried out as described by Liberek et al. (1991). The 25-μl reaction mixtures contained 0.46 μM dnaK, 0.8 μM dnaJ, and 0.25 μM grpE. The concentration of ATP was 100 μM. (◽) dnaK alone; (■) dnaK and grpE; (◆) dnaK and dnaJ; (■) dnaJ and grpE; (◇) dnaK, dnaJ and grpE.

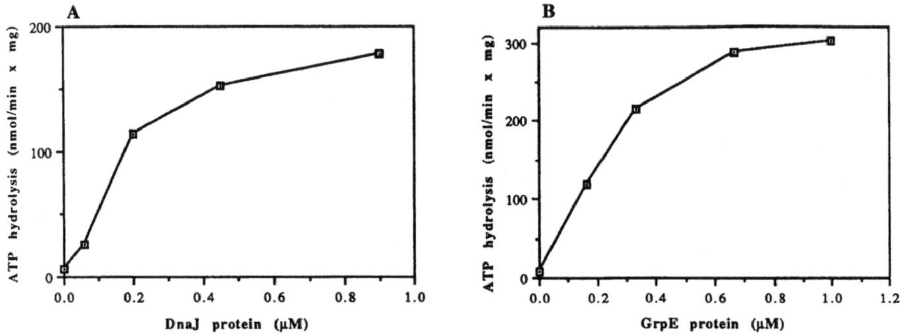

Fig. 3. Titration of dnaJ and grpE in the dnaK ATPase assay. (A) A reaction mixture of 0.57 μM dnaK and 0.83 μM grpE was incubated with various amounts of dnaJ protein. (B) A reaction mixture of 0.57 μM dnaK and 0.7 μM dnaJ was incubated with various amounts of grpE protein. In all assays, the concentration of ATP was 100 mM. The reactions were carried out as described by Liberek et al. (1991).

dnaJ : 1-2 mol of grpE substantially stimulated the activity (Fig. 3). Since all the proteins used in our experiments belong to the family of heat shock proteins, which are involved in cellular metabolism under stress conditions (Morimoto et al., 1990), such as high temperature, and since we had previously shown that dnaK protein possesses an unusually stable ATPase activity (Zylicz et al., 1983), we determined the initial velocity of dnaK's ATPase activity as a function of temperature. We found that (a) the maximal initial velocity of ATP hydrolyzed by dnaK alone was within the 50°-55° range (reaching a four-fold higher value at 50°C than at 30°C), and (b) the ratio of dnaK's ATP activity in the absence or in the simultaneous presence of dnaJ and grpE remained the same, up to 52°C (Liberek et al., 1991).

In an attempt to pinpoint the molecular mechanism of action of the stimulation of dnaK's ATPase activity, we carried out the following experiments (Fig. 4). We incubated dnaK with [α-^{32}P]-ATP for a short period, separated the protein-nucleotide complex by size chromatography and determined the form of the nucleotide bound to dnaK by thin layer chromatography. In a typical experiment (Fig. 4A-D, lane a), 60% of the bound nucleotide was ADP and the rest was ATP. Gel filtration of this complex showed that although the ADP was slowly and preferentially released from dnaK, neither ATP nor ADP were released efficiently (Fig. 4A, fractions 4-8). Instead, the majority of ATP and ADP remained bound to dnaK (Fig. 4A, fractions 2, 3).

The addition of either dnaJ or grpE protein alone resulted in dramatically different effects on the dnaK-bound nucleotides. The addition of dnaJ for 1 minute at 30°C resulted in the quantitative conversion of ATP into ADP (Fig. 4B, lane b); however, the ADP product was still bound to dnaK (Fig. 4B, fractions 4-8). The addition of grpE alone for 1 minute at 30°C resulted in an almost opposite result, i.e., no stimulation of hydrolysis of ATP to ADP was observed; rather, both bound ATP and ADP were quantitatively released from dnaK (Fig. 4C, fractions 4-8). [As a control, we showed that purified grpE280 mutant protein did not stimulate either the hydrolysis of bound ATP to ADP or the release of dnaK-bound nucleotide.] As expected, the presence of both grpE and dnaJ led to the efficient hydrolysis of ATP to ADP and the efficient release of all nucleotides bound to dnaK (Fig. 4D).

We have shown that dnaK binds tightly to denatured, but not native, bovine pancreatic trypsin inhibitor (BPTI). Furthermore, the addition of hydrolyzable ATP is needed for the efficient release of BPTI from its complex with dnaK (unpublished data). We tested to see whether unfolded BPTI will influence dnaK's ATPase reaction. We found that unfolded BPTI had no discernible effect on either the hydrolysis of bound ATP to ADP or the release of the dnaK-bound nucleotides.

Conclusions

We have shown that the dnaK, dnaJ and grpE heat shock proteins of *E. coli* are required for the successful disassembly of the *oriλ*•λO•λP•dnaB complex. We believe that dnaK is the main protagonist in this system because of its ability to bind to various "unfolded" or other polypeptides and release them following ATP hydrolysis. The dnaJ and grpE protein, through their demonstrated ability to accelerate dnaK's ATPase activity, help dnaK recycle efficiently in the *oriλ* replication system, as well as *in vivo*. In the absence of either dnaJ or grpE, we would expect to find a "shortage" of dnaK activity in the cell, consistent with the identical phenotypes exhibited by mutations in either the *dnaK*, *dnaJ* or *grpE* genes (Georgopoulos et al., 1990; Gross et al., 1990). The λ DNA replication system has also allowed us to uncover an ATP-resistant grpE•λP protein-protein interaction (Ang, 1988), which we believe is important in "sequestering" the λP protein once it is released from its complex at *oriλ*. Thus, the presence of grpE protein not only allows dnaK protein to work more efficiently (through the acceleration of dnaK's ATPase activity), but also effectively prevents λP protein from inactivating the dnaB protein once λP is released from the "O-some" complex.

The exact role that the dnaJ protein plays in λ DNA replication is not certain. It can add alone onto the *oriλ*•λO•λP•dnaB complex, perhaps aided by its demonstrable interactions with both the λP and dnaB proteins (Liberek et al., 1990). Previously, we had argued that perhaps the dnaJ protein "targets" certain λP molecules to be acted upon by dnaK (Zylicz et al., 1989; Liberek et al., 1990). Our recent unpublished results indicate that the dnaJ protein helps either the initial formation or stabilization of a dnaK•λP complex in the presence of ATP. Thus, *in vivo* dnaJ (and grpE, as well) may play an additional role(s) besides simply acting as dnaK's "cohort," stimulating dnaK's ATPase activity.

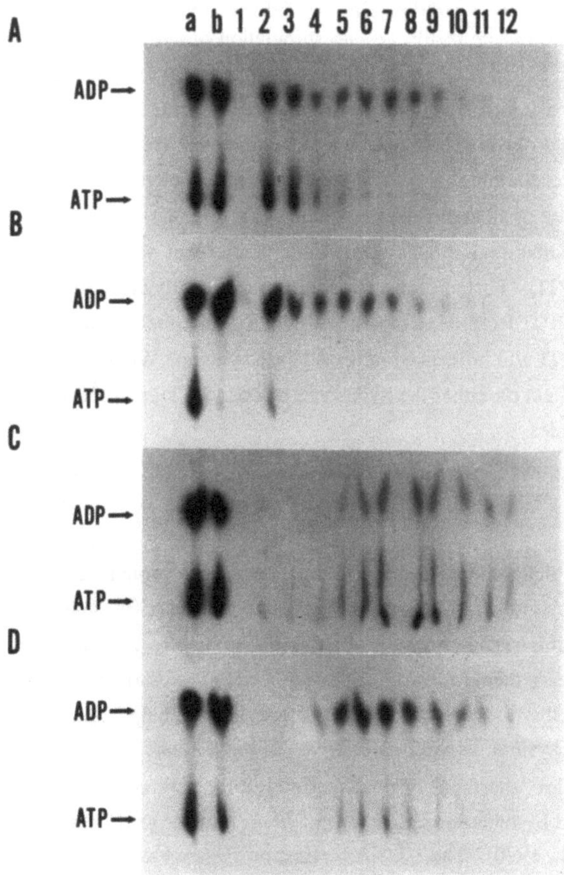

Fig. 4. Influence of dnaJ and grpE on the dnaK-nucleotide complex. As described by Liberek et al. (1991), dnaK bound to α-^{32}P-ATP was isolated by gel filtration. The dnaK-nucleotide complex (21.2 µM dnaK) was incubated for 1 min at 30°C as follows: **A**, alone; **B**, with dnaJ (3.1 µM); **C**, with grpE (2.4 µM); **D**, with grpE (2.4 µM) and dnaJ (3.1 µM). After incubation, mixtures were filtered on a P-60 (0.5 x 7 cm; BioRad) column. Fractiosn (1-12) were collected and analyzed by thin layer chromatography, followed by autoradiography. Controls: lane (1) ADP/ATP distribution in the dnaK-nucleotide complex before the 1 min incubation, and (b) distribution after incubation.

Acknowledgments

This work was supported by NIH grants to C.G., an NIH training grant to D.A., an NSF international exchange grant (to C.G. and M.Z.) and a Polish Ministry of Education grant to M.Z.

References

Alfano C, McMacken R (1989a) Heat shock protein-mediated disassembly of nucleoprotein structures is required for the initiation of bacteriophage λ DNA replication. J Biol Chem 264:10709-10718

Alfano C, McMacken R (1989b) Ordered assembly of nucleoprotein structures at the bacteriophage λ replication origin during the initiation of DNA replication. J Biol Chem 264:10699-10708

Ang D (1988) The role of the *Escherichia coli* heat shock protein, grpE, in *Escherichia coli* growth and λ DNA replication. Ph.D. Thesis University of Utah Salt Lake City

Ang D, Georgopoulos C (1989) The heat shock-regulated *grpE* gene of *Escherichia coli* is required for bacterial growth at all temperatures, but is dispensable in certain mutant backgrounds. J Bacteriol 171:2748-2755

Bramhill D, Kornberg A (1988) A model for initiation at origins of DNA replication. Cell 54:915-918

Dodson M, Echols H, Wickner S, Alfano C, Mensa-Wilmot K, Gomes B, LeBowitz JH, Roberts JD, McMacken R (1986) Specialized nucleoprotein structures at the origin of replication of bacteriophage λ: Localized unwinding of duplex DNA by six-protein reaction. Proc Natl Acad Sci USA 83:7638

Dodson M, Roberts J, McMacken R, Echols H (1985) Specialized nucleoprotein structures at the origin of replication of bacteriophage λ: complexes with λO, λP and *Escherichia coli* DnaB proteins. Proc Natl Acad Sci USA 82:4678-4682

Dodson M, McMacken R, Echols H (1989) Specialized nucleoprotein structures at the origin of replication of bacteriophage λ. J Biol Chem 264:10719-10725

Furth ME, Wickner SH (1983) Lambda DNA replication p 145-173 *In* Hendrix RW, Roberts JW, Stahl FW, Weisberg RA (ed) Lambda II Cold Spring Harbor Laboratory Cold Spring Harbor New York

Georgopoulos CP, Herskowitz I (1971) *Escherichia coli* mutants blocked in lambda DNA synthesis, p. 553-564. *In* Hershey AD (ed) The bacteriophage lambda. Cold Spring Harbor Laboratory Cold Spring Harbor New York

Georgopoulos C, Ang D, Liberek K, Zylicz M (1990) The role of the *E. coli* heat shock proteins in λ development. *In* Morimoto R, Tissieres A, Georgopoulos C (eds) Stress proteins in biology and medicine. pp 191-221 Cold Spring Harbor Laboratory Cold Spring Harbor New York

Gross CA, Straus DB, Erickson JW, Yura T (1990) *In* Morimoto R, Tissieres A, Georgopoulos C (eds) Stress proteins in biology and medicine. pp. 167-189 Cold Spring Harbor Laboratory Cold Spring Harbor New York

Keppel F, Fayet O, Georgopoulos C (1988) Strategies of bacteriophage DNA replication. *In* Calendar R (ed) The bacteriophages. 2:145-262 Plenum Press New York

LeBowitz JH, McMacken R (1986) The *Escherichia coli* DnaB replication protein is a DNA helicase. J Biol Chem 261:4738-4748

Liberek K, Georgopoulos C, Zylicz M (1988) The role of the *Escherichia coli* DnaK and DnaJ heat shock proteins in the initiation of bacteriophage λ DNA replication. Proc Natl Acad Sci USA 85:6632-6636

Liberek K, Osipiuk J, Zylicz M, Ang D, Skorko J, Georgopoulos C (1990) Physical interactions between bacteriophage and *Escherichia coli* proteins required for initiation of λ DNA replication. J Biol Chem 265:3022-3029

Liberek K, Marszalek J, Ang D, Georgopoulos C, Zylicz M (1991) The *Escherichia coli* DnaJ and GrpE heat shock proteins jointly stimulate DnaK's ATPase activity. Proc Natl Acad Sci USA in press

Mallory JB, Alfano C, McMacken R (1990) Host virus interactions in the initiation of bacteriophage λ DNA replication. J Biol Chem 265:13297-13307

McMacken R, Silver L, Georgopoulos C (1987) DNA replication. *In* Neidhardt FC et al. (eds), *Escherichia coli* and *Salmonella typhimurium*: Cellular and molecular biology. pp. 564-612 American Society for Microbiology Washington DC

Mensa-Wilmot K, Seaby R, Alfano C, Wold MS, Gomes B, McMacken R (1989) Reconstitution of a nine-protein system that initiates bacteriophage λ DNA replication. J Biol Chem 264:2853-2861

Morimoto RI, Milarski KL (1990) Expression and function of vertebrate hsp70 genes. *In* Morimoto R, Tissieres A, Georgopoulos C (eds) Stress proteins in biology and medicine. pp 323-359 Cold Spring Harbor Laboratory Cold Spring Harbor New York

Rothman JE (1989) Polypeptide chain binding proteins: Catalysts of protein folding and related processes in cells. Cell 59:591-601

Saito H, Uchida H (1977) Initiation of the DNA replication of bacteriophage lambda in *Escherichia coli* K12. J Mol Biol 113:1-25

Schnos M, Zahn K, Inman RB, Blattner FR (1988) Initiation portein induced helix destabilization at the λ origin: A prepriming step in DNA replication. Cell 52:385-395

Sunshine M, Feiss M, Stuart J, Jochem J (1977) A new host gene (*groPC*) necessary for lambda DNA replication. Mol Gen Genet 151:27-34

Wickner S (1979) DNA replication proteins of *Escherichia coli* and phage λ. Cold Spring Harbor Symp Quant Biol. 43:303-310

Wickner SH, Zahn K (1986) Characterization of the DNA binding domain of bacteriophage λ O protein. J Biol Chem 261:7537-7543

Zahn K, Blattner FR (1987) Direct evidence for DNA bending at the lambda replication origin. Science 236:416-422

Zylicz M, LeBowitz JH, McMacken R, Georgopoulos C (1983) The dnaK protein of *Escherichia coli* possesses an ATPase and autophosphorylating activity and is essential in an *in vitro* DNA replication system. Proc Natl Acad Sci USA 80:6431-6435

Zylicz M, Gorska I, Taylor K, Georgopoulos C (1984) Bacteriophage λ replication proteins: formation of a mixed oligomer and binding to the origin of λ DNA. Mol Gen Genet 196:401-406

Zylicz M, Ang D, Liberek K, Georgopoulos C (1989) Initiation of λ DNA replication with purified host- and bacteriophage-encoded proteins: The role of the dnaK, dnaJ and grpE heat shock proteins. EMBO J 8:1601-1608

SV40 DNA Replication with Purified Proteins: Functional Interactions Among the Initiation Proteins

Kathleen L. Collins, Lorne F. Erdile, Sandra K. Randall, Alicia A.R. Russo, Pamela R. Simancek, Christopher B. Umbricht, David M. Virshup, David H. Weinberg, Marc S. Wold, and Thomas J. Kelly
Department of Molecular Biology and Genetics, Johns Hopkins University School of Medicine, Baltimore, Maryland 21205, USA

Simian virus 40 (SV40), has proved to be a useful model system for the study of mammalian DNA replication (Challberg and Kelly 1989; Hurwitz *et al.* 1990; Kelly 1988; Stillman 1989). The SV40 genome is replicated by the host machinery in conjunction with a single viral protein, T antigen. Thus, the basic mechanisms involved in the replication of viral DNA are likely to be important for mammalian DNA replication. Synthesis of the viral genome begins at a well-defined origin and, like chromosomal DNA replication, proceeds in a bidirectional and semidiscontinuous manner. The utility of this relatively simple system is exemplified by the development of a cell-free replication system (Li and Kelly 1984; Li and Kelly 1985) which has allowed the identification of factors implicated in the replication of mammalian DNA. Replication *in vitro* was demonstrated to possess most of the characteristics of *in vivo* SV40 DNA replication (Li and Kelly 1984; Li and Kelly 1985; Li *et al.* 1986; Stillman *et al.* 1985; Stillman and Gluzman 1985; Wobbe *et al.* 1985). In particular, initiation depended on the viral protein T antigen, occured at the origin of DNA replication, and proceeded bidirectionally. Furthermore, replication intermediates formed *in vitro* were identical to those found *in vivo*. Subsequent fractionation of the crude extract revealed seven purified components necessary for DNA replication: DNA polymerase α-primase complex (Murakami *et al.* 1986); Topoisomerases I or II (Yang *et al.* 1987); replication protein A (RP-A) ; proliferating cell nuclear antigen (PCNA)(Prelich *et al.* 1987); protein phosphatase 2A ($PP2A_C$) (Virshup and Kelly 1989); DNA polymerase δ (Lee *et al.* 1989; Prelich *et al.* 1987; Weinberg *et al.* 1990; Weinberg and Kelly 1989); and RF-C (Tsurimoto and Stillman 1989b). It has been demonstrated that these seven cellular proteins plus T antigen are sufficient to reconstitute SV40 DNA replication *in vitro* (Tsurimoto *et al.* 1990; Weinberg *et al.* 1990).

Studies of the biochemical activities of purified replication proteins in isolation have shed light on the role each plays in DNA replication. T antigen specifically binds to the origin of DNA replication (Borowiec and Hurwitz 1988; Dean *et al.* 1987; Mastrangelo *et al.* 1989; Tegtmeyer *et al.* 1983) and, in the presence of a single strand binding protein (such as RP-A) unwinds the duplex DNA strands creating a replication bubble (Dean *et al.* 1987; Stahl *et al.* 1986; Wold *et al.* 1987). The unwinding reaction is further enhanced by the action of $PP2A_C$, which probably stimulates the unwinding activity of T antigen by the removal of an inhibitory phosphate (Virshup *et al.* 1989). Once single stranded DNA has been exposed, DNA polymerase α-primase can synthesize an oligoribonucleotide primer, which is extended with deoxynucleotides to form the first strand. DNA polymerase δ works together with PCNA and RF-C to elongate pre-formed DNA strands(Tsurimoto and Stillman 1989a).

The minimal proteins required for the initiation of DNA replication are T antigen, DNA polymerase α-primase and RP-A (Tsurimoto *et al.* 1990; Weinberg *et al.* 1990; Wobbe *et al.* 1987). Evidence that this process requires specific protein-protein interactions includes the following observations: 1) T antigen has been demonstrated to interact with DNA polymerase α–primase, and antibodies that disrupt this interaction are inhibitory to DNA replication (Gannon and Lane 1987; Gannon and Lane 1990; Smale and Tjian 1986). This physical interaction was recently demonstrated to be mediated directly through the catalytic subunit of DNA polymerase α-primase (Dornreiter *et al.* 1990). 2) Although both RP-A and *E. coli* SSB are single strand DNA binding proteins, the prokaryotic protein cannot initiate SV40 DNA synthesis (Matsumoto *et al.* 1990). This requirement for RP-A may involve an interaction with DNA polymerase α-primase, since RP-A, but not *E.coli* SSB, is stimulatory to DNA polymerase α-primase on pre-primed templates (Erdile *et al.* 1991; Kenny *et al.* 1989; Tsurimoto and Stillman 1989a). Although it is becoming more and more clear that these interactions are critical, the precise manner in which they promote efficient initiation of DNA replication is not yet understood.

In order to gain insight into how specific protein-protein interactions may influence the initiation of DNA synthesis, we have examined the effect of T antigen and RP-A on synthesis by DNA polymerase α. We discovered that T antigen stimulated the synthesis of oligoribonucleotide primers on single stranded M13 DNA. Studies with anti-T antigen monoclonal antibodies indicated that T antigen stimulated priming and DNA synthesis via direct interaction with DNA polymerase α-primase. In contrast, RP-A inhibited priming and DNA synthesis on unprimed single stranded DNA. T antigen was able to partially reverse the inhibitory effects of both RP-A and *E. coli* SSB on this template. Whole RP-A, *E. coli* SSB and the large subunit of RP-A also inhibited priming on an artificial replication fork. However, in this case T antigen reversed the inhibition due to whole RP-A but not that due to *E. coli* SSB or the large subunit of RP-A. Thus T antigen, DNA polymerase α-primase and RP-A appear to interact specifically on a forked substrate, and whole RP-A is required for this interaction. Based on these data we present a model for how protein-protein interactions might direct primer synthesis to promote coordinated DNA replication.

Materials and Methods

Protein purification. Purified replication proteins were isolated as described previously (Weinberg *et al.* 1990; Wold *et al.* 1989). The 70 kDa subunit of human RP-A was expressed in *E. coli* using a bacteriophage T7 expression construct(Erdile *et al.* 1991). The recombinant protein was purified by chromatography on hydroxylapatite (Calbiochem). The 70 mM KPO_4 eluate from hydroxylapaptite was loaded onto single-stranded DNA agarose (Bethesda Research Laboratory), and the column was washed with 2 M NaCl. The 70 kDa subunit was eluted with 2 M NaSCN. *E. coli* SSB was a generous gift from Roger McMacken (Johns Hopkins University). The hybridoma cell line for monoclonal antibody 419 was obtained from the ATCC. Purified monoclonal antibody 101 was the generous gift of Richard Piehl (Molecular Biology Resources). Hybridoma cell lines for monoclonal antibodies 205 and 414 were obtained from Keith Peden (N.I.H.). All monoclonals were purified using pro-A sepharose (Pharmacia), and were dialysed into a buffer containing 30 mM Hepes (pH 8.0), 15 mM KCl, 0.25 mM EDTA and 1.0 mM NaN_3.

DNA replication assays. Replication assays and product analyses were as described in Collins and Kelly, 1991, Weinberg *et al.*, 1990 and figure legends.

Construction of an artificial replication fork. Production of single stranded DNA from pBS+ and pBS- (Stratagene) was induced by G408 helper phage (Stratagene), (these plasmids are identical except that the F1 origin of replication is reversed in orientation). The circular single stranded DNA was linearized as follows: pBS+ and pBS- were annealed to 17mer oligonucleotides encompassing the Hinc II site in the multiple cloning region and were digested with Hinc II. Equimolar amounts of linearized pBS+ and pBS- were annealed to one another by incubation in 100 mM NaCl at 68°C for 10 min followed by a slow cool to room temperature. The annealed DNAs were purified away from residual single strands by electrophoresis on a 1.2 % low melting point agarose gel (Bethesda Research Labs). The gel slice containing the annealed DNA was digested with agarase (Calbiochem), phenol/chloroform extracted, and ethanol precipitated. The resulting 469 nucleotide single strand DNA bubble was asymmetrically located between two double stranded regions of 269 and 2200 nucleotides.

Results

Seven cellular proteins plus T antigen are sufficient to reconstitute SV40 DNA replication *in vitro*. Analysis of the products of this reaction by electron microscopy indicated the presence of replication intermediates that are indistinguishable from those formed either *in vivo* or *in vitro* using crude cellular extracts (Weinberg *et al.*, 1990). The size distribution of the DNA strands synthesized in the complete reconstituted reaction and in the absence of each of the protein components was analyzed by agarose gel electrophoresis (Figure 1). The products of the complete reaction were quite heterogeneous in length, ranging from a few hundred nucleotides to the length of the input DNA template (2886 nucleotides). Initiation of DNA synthesis was absolutely dependent upon the SV40 origin of DNA replication and three proteins; T antigen, RP-A and DNA polymerase α-primase complex. The reaction was stimulated significantly by topoisomerase I and $PP2A_c$, but neither protein significantly affected the size distribution of newly synthesized DNA strands. When RF-C, PCNA, or DNA polymerase δ were omitted from the reaction, the amount of DNA synthesis was reduced and the size distribution of newly synthesized strands was heavily weighted toward shorter species. Furthermore, in the absence of DNA polymerase δ, RF-C and PCNA, most of the products synthesized by DNA polymerase α-primase were derived form the lagging strand (Weinberg *et al.*, 1990). Thus it appears that in this partially reconstituted system, primer synthesis by DNA polymerase α-primase occured in a specific manner.

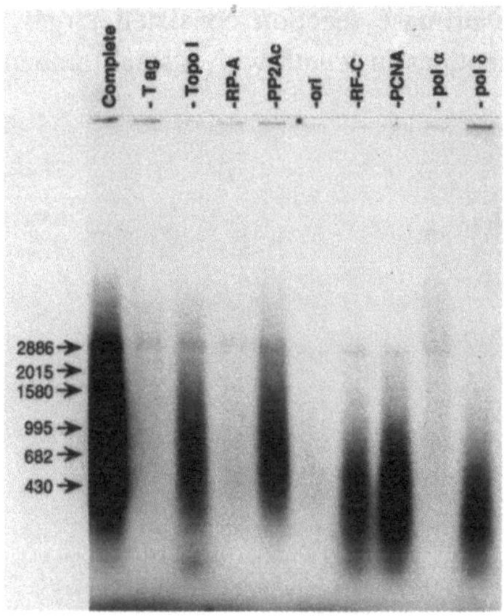

FIG. 1 Protein requirements for SV40 DNA replication. Individual proteins were omitted from the complete reaction as indicated. In the reaction designated ori-, the plasmid pUC8 was used as template. After incubation at 37°C for 90 minutes the reaction products were isolated, heat denatured, and subjected to agarose gel electrophoresis (modified from Weinberg *et al.*, 1990).

To better understand the functional interactions of the three proteins required for the initiation of DNA synthesis, we examined the effects of T antigen and RP-A on synthesis by DNA polymerase α-primase on a model single stranded DNA template. Under the standard conditions used for SV40 DNA replication *in vitro*, DNA polymerase α-primase was able to use M13 as a template to synthesize DNA products which ranged in size from about 400 to 3000 nucleotides. The addition of purified T antigen to this reaction stimulated DNA synthesis by DNA polymerase α-primase four fold. In contrast, the addition of RP-A or *E. coli* SSB strongly inhibited DNA synthesis. T antigen reversed the inhibitory effects of low levels of RP-A and *E. coli* SSB (Collins and Kelly 1991).

To determine whether the DNA primase activity of DNA polymerase α-primase complex was affected by T antigen and RP-A, we measured the synthesis of primers directly in reaction mixtures lacking deoxyribonucleoside triphosphates. With DNA polymerase α-primase alone, the products of the primase reaction consisted largely of ribonucleotide oligomers 6-12 nucleotides in length with a small amount of longer species (Figure 2).

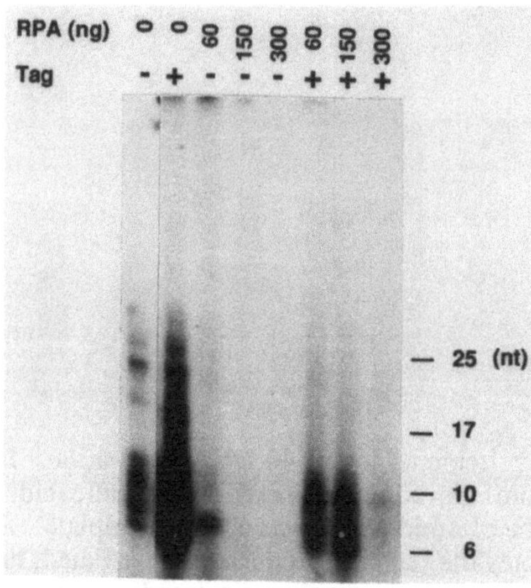

FIG. 2 Effects of T antigen and RP-A on the DNA primase activity of DNA polymerase α. Primase reactions (25 μl) containing 0.0025 units primase, 0.0125 units polymerase, 20 μCi [α-^{32}P]CTP (1 μM), 50 μM UTP, 50μM GTP, 4 mM ATP, 25 ng unprimed M13 DNA and 280 ng T antigen where indicated were assembled on ice. Reaction products were purified, electrophoresed on a denaturing 20 % acrylamide gel and visualized by autoradiography (from Collins and Kelly, 1991).

T antigen stimulated primer synthesis approximately 4-fold, while RP-A strongly inhibited the reaction. Neither protein significantly affected the length distribution of primers. T antigen partially reversed the inhibition of primer synthesis by RP-A. We have observed that the stimulatory effects of T antigen on priming were greatest at low polymerase and template concentrations (data not shown). These findings indicate that T antigen and

RP-A exert marked and antagonistic effects on the efficiency of primer synthesis by DNA polymerase α-primase complex.

In order to determine whether T antigen had an effect on the efficiency of primer elongation, we omitted ribonucleotides from the reaction and made use of a template containing a short oligodeoxynucletide primer. In the absence of T antigen, DNA polymerase α elongated the primer to a length of less than 100 nucleotides (Figure 3).

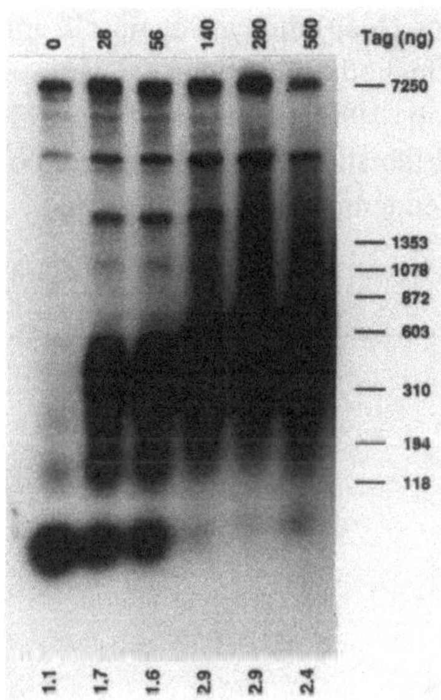

FIG. 3 Effect of T antigen on primer extension. Reaction mixtures (25 μl) lacking ribonucleoside triphosphates were assembled on ice with 0.012 units polymerase activity, 25 ng primed M13 mp18 template, and the indicated amount of T antigen. After incubation for 2 hours at 37°C, the reaction products were electrophoresed on a 2% agarose alkaline gel and visualized by autoradiography. Picomoles of DNA synthesized are indicated at the bottom of the figure (from Collins and Kelly, 1991).

In the presence of increasing concentrations of T antigen, the products became larger by relatively discrete increments. At the highest level of T antigen tested, almost all the products were greater than 300 nucleotides. Although the effect of T antigen on product length was most striking when a pre-primed template was used, we did observe a significant effect on product size with the unprimed template, particularly at low polymerase α concentration(Collins and Kelly 1991) .

One reasonable hypothesis consistent with our data is that T antigen stimulates priming and DNA synthesis via direct interaction with DNA polymerase α-primase. Indeed, several laboratories have demonstrated that the two proteins can associate specifically to form a complex (Dornreiter *et al.* 1990; Gannon and Lane 1987; Smale and Tjian 1986). The association of T antigen and DNA polymerase α-primase can be blocked by a specific subset of monoclonal antibodies (Mab) directed against T antigen (e.g. Mab 205 and Mab 414) that do not affect other functions of T antigen such as specific DNA binding (Gannon and Lane 1987; Smale and Tjian 1986). As shown in Figure 4 these Mabs abolished the stimulation of purified polymerase α-primase by T antigen on an unprimed template.

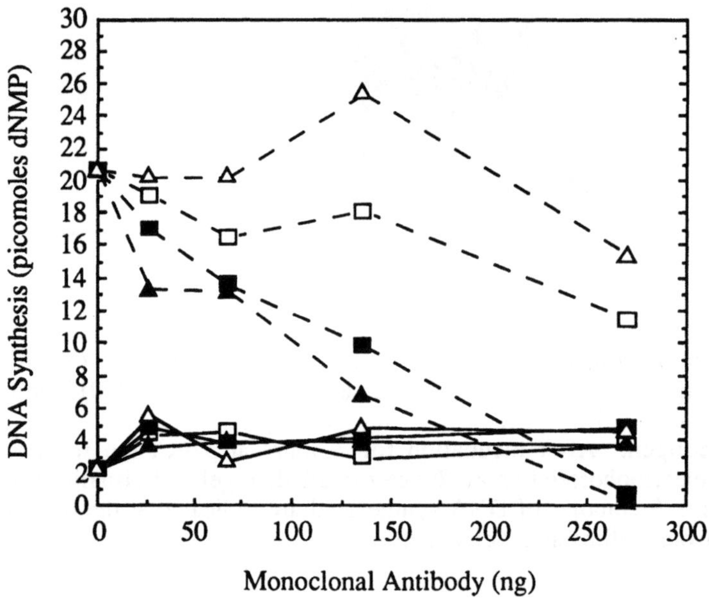

FIG. 4 Effects of monoclonal antibodies on the stimulation of DNA polymerase α activity by T antigen (unprimed template). Reaction mixtures (25 μl) contained 0.0125 units polymerase activity, 0.0025 units primase activity, 25 ng M13 DNA, and the indicated amount of anti-T antigen monoclonal antibody (Mab). Either no T antigen (—) or 350 ng T antigen (- -) was added to each reaction. The monoclonal antibodies tested were Mab

419 (*open triangles*), Mab 101 (*open squares*), Mab 205 (*solid triangle*) and Mab 414 (*solid square*). Mab 419 (which binds an N-terminal epitope (1-82), and Mab 101 (which binds a C-terminal epitope) do not inhibit SV40 DNA replication *in vitro* or formation of T antigen-DNA polymerase α complexes in crude extracts (Gannon and Lane 1987; Smale and Tjian 1986; Wiekowski *et al.* 1987). Mabs 205 and 414 bind the putative polymerase α interaction domain (363-708), inhibit SV40 DNA replication, and disrupt complex formation between T antigen and polymerase α in crude cellular extracts (Gannon and Lane 1987; Smale and Tjian 1986). The amount of DNA synthesized in 2 hours at 37°C was quantitated by excising the lanes from the dried gel and counting in liquid scintillation cocktail (from Collins and Kelly, 1991).

At antibody concentrations approximately equimolar with T antigen, DNA synthesis was actually depressed below the basal level observed in the absence of T antigen. One explanation for this observation is that the T antigen-antibody complexes bind to DNA and block DNA polymerase α-primase in a manner analogous to the SSB's. In contrast, anti-T antigen Mabs 419 and 101, which do not block the physical interaction between T antigen and DNA polymerase α-primase, had no significant effect on the stimulation of DNA synthesis (see Fig. 4 legend). These data indicate that the stimulation of priming that we have observed very likely requires formation of a complex between T antigen and DNA polymerase α.

In order to more closely approximate the conditions of DNA replication, we constructed an artificial replication bubble. This was accomplished by annealing two single stranded DNA molecules complementary over all but about 500 nucleotides. The region of non-homology between the strands created an open region of single stranded DNA which was used by DNA polymerase α-primase complex as a template for the synthesis of DNA strands *de novo*. As was seen for a completely single stranded DNA template, T antigen stimulated synthesis by DNA polymerase α-primase, whereas RP-A and *E. coli* SSB blocked it (Figure 5). T antigen was able to reverse the inhibitory effects of RP-A, but not *E. coli* SSB. In addition, the recombinant bacterially expressed large subunit of RP-A was tested. This subunit, which harbors the DNA binding activity of RP-A, behaved similarly to *E. coli* SSB on the artificial replication fork (Figure 5). Thus, as in DNA replication, there is a specific requirement for RP-A in this reaction which appears to involve at least one of the two low molecular weight RP-A subunits.

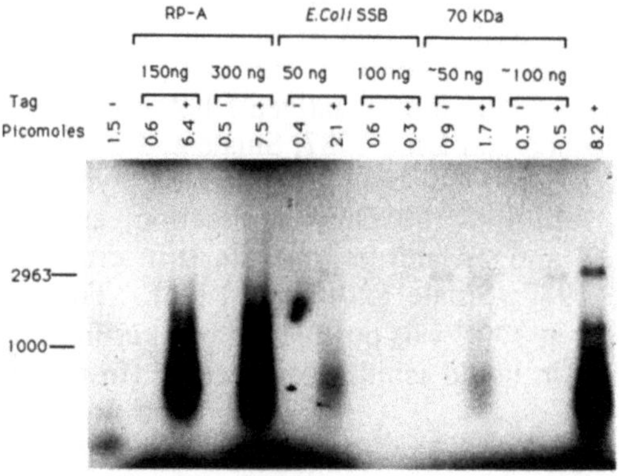

FIG. 5 DNA synthesis on an artificial replication fork. Reactions (25 µl) containing 0.015 units DNA polymerase, 0.0025 units primase and and 100 ng fork substrate were assembled on ice as described in Materials and Methods except that no creatine phosphate or creatine kinase was added. Where indicated reactions contained 300 ng T antigen. RP-A, *E. coli* SSB and the 70 kDa subunit of RP-A were added as indicated. After incubation for 2 hours at 37°C, samples were denatured by heating 5 minutes at 95°C, electrophoresed on a 1 % agarose gel, dried and visualized by autoradiography.

Discussion

It seems likely that the interaction between T antigen and DNA polymerase α-primase is important for determining the specificity of initiation events during SV40 DNA replication. In the initial phase of the replication reaction, T antigen binds specifically to the viral origin and then functions as a helicase to unwind the two strands of the duplex (Dean *et al.* 1987; Wold *et al.* 1987). Model studies indicate that a T antigen helicase translocates in the 3'-5' direction along each parental strand, thus generating two potential replication forks (Dean *et al.* 1987; Wold *et al.* 1987). As illustrated in Figure 6, we suggest that DNA polymerase α-primase forms a specific complex with T antigen at each fork and that the geometry of this complex facilitates primer synthesis on the lagging strand. This model, in which the T antigen-DNA polymerase α-primase complex functions as a mobile priming apparatus is analogous to many prokaryotic systems where primase activity and helicase activity reside in either a single polypepetide chain (phage T7) (reviewed in Richardson, 1983) or a complex of polypeptide chains (phages T4 and *E. coli*)(Alberts 1987; Kornberg 1980; LeBowitz and McMacken 1986). In our model, T antigen moves on the leading strand, but positions DNA polymerase α-primase so that primer synthesis takes place preferentially on the opposite (lagging) strand.

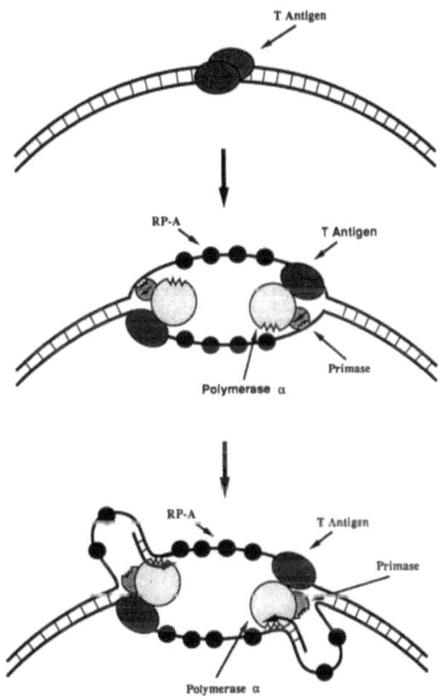

FIG. 6. Model of the initiation of DNA synthesis. See text for details (from Collins and Kelly, 1991)

Previous studies have demonstrated that the single-stranded DNA binding protein, RP-A has a stimulatory effect on the ability of DNA polymerase α to elongate a pre-formed primer (Kenny *et al.* 1989; Lee *et al.* 1989; Tsurimoto and Stillman 1989a). This observation, which we have confirmed, indicates that RP-A does not block the interaction of DNA polymerase α-primase with a primer terminus. Our finding that RP-A inhibits priming by DNA polymerase α-primase indicates that RP-A prevents the productive interaction of DNA primase with unprimed single stranded DNA, and defines an additional function for the protein in suppressing nonspecific priming events. We suggest that most of the single stranded DNA in replication intermediates will be complexed with RP-A and inaccessible to the primase activity of DNA polymerase α. Thus in a manner similar to gene 32 of bacteriophage T4 (Cha and Alberts 1990; Richardson and Nossal 1989) and the SSB of *E. coli* (Arai and Kornberg 1979), RP-A may prevent non-productive priming events by suppressing primase activity everywhere except at the replication fork (Figure 6).

This model in which DNA polymerase α–primase interacts specifically with T antigen on a fork structure is supported by experiments in which an artificial replication fork was used as a template. In the presence of T antigen, RP-A did not suppress synthesis by DNA polymerase α-primase on the artificial replication fork. One explanation for this observation is that T antigen positions DNA polymerase α-primase at the fork in a manner which allows it to compete more effectively with RP-A for binding to the newly unwound single strands. However, this simple explanation does not explain why whole RP-A, but not *E. coli* SSB allowed synthesis on this template. Moreover the 70 kDa DNA binding subunit of RP-A was unable to substitute for whole RP-A in this reaction. Thus it appears that at least one of the two smaller subunits of RP-A is essential. It is not yet clear whether this requirement for whole RP-A reflects a difference in the manner in which the RP-A protein complex interacts with the template, or whether there is a specific protein-protein interaction which is required for synthesis to occur.

It seems clear that the functional interactions of T antigen, DNA polymerase α-primase and RP-A play an important role in the initiation of DNA synthesis during SV40 DNA replication. It is plausible to speculate that similar interactions, involving cellular counterparts of T antigen may occur during initiation of cellular DNA synthesis. Moreover, modulation of such interactions, either by phosphorylation or by protein interactions could provide a mechanism for regulating DNA replication. In this regard, it is intriguing that the anti-oncogene p53, which is inhibitory to SV40 DNA replication (Braithwaite *et al.* 1987; Wang *et al.* 1989), competes with DNA polymerase α for binding to T antigen in crude cellular extracts (Gannon and Lane 1987; Gannon and Lane 1990).

References

Alberts, B. M. (1987). "Prokaryotic DNA replication mechanisms." Phil. Trans. R. Soc. Lond. **317**(supplement): 395-420.

Arai, K. and A. Kornberg. (1979). "A general priming system employing only dnaB protein and primase for DNA replication." Proc. Natl. Acad. Sci. USA. **76**(9): 4308-4312.

Borowiec, J. A. and J. Hurwitz. (1988). "Localized melting and structural changes in the SV40 origin of replication induced by T-antigen." EMBO J. **7**(10): 3149-3158.

Braithwaite, A. W., H. Sturzbecher, C. Addison, C. Palmer, K. Rudge and J. R. Jenkins. (1987). "Mouse p53 inhibits SV40 origin-dependent DNA replication." Nature. **329**: 458-460.

Cha, T. and B. M. Alberts. (1990). "Effects of the bacteriophage T4 gene 41 and gene 32 proteins on RNA primer synthesis: Coupling of leading- and lagging-strand DNA synthesis at a replication fork." Biochemistry. **29**: 1791-1798.

Challberg, M. D. and T. J. Kelly. (1989). "Animal virus DNA replication." Annu. Rev. Biochem. **58**: 671-717.

Collins, K. L. and T. J. Kelly. (1991). "Effects of T antigen and replication protein A on the initiation of DNA synthesis by DNA polymerase α-primase." Mol. Cell. Biol. **11**: 1542-1553.

Dean, F. B., J. A. Borowiec, Y. Ishimi, S. Deb, P. Tegtmeyer and J. Hurwitz. (1987). "Simian virus 40 large tumor antigen requires three core replication origin domains for DNA unwinding and replication *in vitro*." Proc. Natl. Acad. Sci. USA. **84**: 8267-8271.

Dean, F. B., M. Dodson, H. Echols and J. Hurwitz. (1987). "ATP-Dependent formation of a specialized nucleoprotein structure by simian virus 40 (SV40) large tumor antigen at the SV40 replication origin." Proc. Natl. Acad. Sci USA. **84**: 8981-8985.

Dornreiter, I., A. Hoss, A. K. Arthur and E. Fanning. (1990). "SV40 T antigen binds directly to the large subunit of purified DNA polymerase alpha." EMBO J. **9**(10): 3329-3336.

Erdile, L. F., W. Heyer, R. Kolodner and T. J. Kelly. (1991). "Characterization of a cDNA encoding the 70-kDa subunit of human replication protein A (RP-A), a single-stranded DNA-binding protein involved in DNA replication and recombination." J. Biol. Chem., *in press*.

Gannon, J. V. and D. P. Lane. (1987). "p53 and DNA polymerase alpha compete for binding to SV40 T antigen." Nature. **329**: 456-458.

Gannon, J. V. and D. P. Lane. (1990). "Interactions between SV40 T antigen and DNA polymerase α." New Biol. **2**(1): 84-92.

Hurwitz, J., F. B. Dean, A. D. Kwong and S. Lee. (1990). "The *in vitro* replication of DNA containing the SV40 origin." JBC. **265**(30): 18043-18046.

Kelly, T. J. (1988). "SV40 DNA replication." J. Biol. Chem. **263**: 17889-17892.

Kenny, M. K., S.-H. Lee, and J. Hurwitz. (1989). "Multiple functions of human single-stranded DNA binding protein in simian virus 40 DNA replication: single strand stabilization and stimulation of DNA polymerases alpha and delta." Proc. Natl. Acad. Sci. USA. **86**: 9757-9761.

Kornberg, A. (1980). DNA Replication. San Francisco,CA, Freeman Publications.

LeBowitz, J. H. and R. McMacken. (1986). "The *Escherichia coli* dnaB replication protein is a DNA helicase." J. Biol. Chem. **261**: 4738-4748.

Lee, S.-H., T. Eki and J. Hurwitz. (1989). "Synthesis of DNA containing the simian virus 40 origin of replication by the combined action of DNA polymerases alpha and delta." Proc. Natl. Acad. Sci. USA. **86**(19): 7361-7365.

Li, J. J. and T. J. Kelly. (1984). "Simian virus 40 DNA replication in vitro." Proc. Natl. Acad. Sci. USA. **81**: 6973-6977.

Li, J. J. and T. J. Kelly. (1985). "Simian virus 40 DNA replication *in vitro:* Specificity of initiation and evidence for bidirectional replication." Mol. Cell. Biol. **5**: 1238-1246.

Li, J. J., K. W. C. Peden, R. A. F. Dixon and T. J. Kelly. (1986). "Functional organization of the simian virus 40 origin of DNA replication." Mol. Cell. Biol. **6**: 1117-1128.

Mastrangelo, I. A., P. V. Hough, J. S. Wall, M. Dodson, F. B. Dean and J. Hurwitz. (1989). "ATP-dependent assembly of double hexamers of SV40 T antigen at the viral origin of DNA replication." Nature. **338**(6217): 658-662.

Matsumoto, T., T. Eki and J. Hurwitz. (1990). "Studies on the initiation and elongation reactions in the simian virus 40 DNA replication system." Proc. Natl. Acad. Sci. **87**: 9712-9716.

Murakami, Y., C. R. Wobbe, L. Weissbach, F. B. Dean and J. Hurwitz. (1986). "Role of DNA polymerase α and DNA primase in simian virus 40 DNA replication *in vitro.*" Proc. Natl. Acad. Sci. USA. **83**: 2869-2873.

Prelich, G., M. Kostura, D. R. Marshak, M. B. Mathews and B. Stillman. (1987). "The cell-cycle regulated proliferating cell nuclear antigen is required for SV40 DNA replication *in vitro.*" Nature. **326**(6112): 471-475.

Prelich, G., C. K. Tan, M. Kostura, M. B. Mathews, A. G. So, K. M. Downey and B. Stillman. (1987). "Functional identity of proliferating cell nuclear antigen and a DNA polymerase-delta auxiliary protein." Nature. **326**(6112): 517-520.

Richardson, C. C. (1983). "Bacteriophage T7: minimal requirements for the replication of a duplex DNA molecule." Cell. **33**: 315-317.

Richardson, R. W. and N. G. Nossal. (1989). "Trypsin cleavage in the COOH terminus of the bacteriophage T4 gene 41 DNA helicase alters the primase-helicase activities of the T4 replication complex *in vitro*." J. Biol. Chem. **264**(8): 4732-4739.

Smale, S. T. and R. Tjian. (1986). "T-antigen-DNA polymerase alpha complex implicated in simian virus 40 DNA replication." Mol. Cell. Biol. **6**(11): 4077-4087.

Stahl, H., P. Droge and R. Knippers. (1986). "DNA helicase activity of SV40 large tumor antigen." EMBO J. **5**: 1939-1944.

Stillman, B. (1989). "Initiation of eukaryotic DNA replication *in vitro*." Ann. Rev. Cell Biol. **5**: 197-245.

Stillman, B. W., R. D. Gerard, R. A. Guggenheimer and Y. Gluzman. (1985). "T antigen and template requirements for SV40 DNA replication *in vitro*." EMBO J. **4**: 2933-2939.

Stillman, B. W. and Y. Gluzman. (1985). "Replication and supercoiling of simian virus 40 DNA in cell extracts from human cells." Mol. Cell. Biol. **5**(8): 2051-2060.

Tegtmeyer, P., B. Lewton, A. Delucia, V. Wilson and K. Ryder. (1983). "Topography of simian virus 40 A protein-DNA complexes: arrangement of protein bound to the origin of replication." J. Virol. **46**: 151-161.

Tsurimoto, T., T. Melendy and B. Stillman. (1990). "Sequential initiation of lagging and leading strand synthesis by two different polymerase complexes at the SV40 DNA replication origin." Nature. **346**: 534-539.

Tsurimoto, T. and B. Stillman. (1989a). "Multiple replication factors augment DNA synthesis by the two eukaryotic DNA polymerases, alpha and delta." EMBO J. **8**(12): 3883-3889.

Tsurimoto, T. and B. Stillman. (1989b). "Purification of a cellular replication factor, RF-C, that is required for coordinated synthesis of leading and lagging strands during simian virus 40 DNA replication *in vitro*." Mol. Cell. Biol. **9**(2): 609-619.

Virshup, D. M., M. G. Kauffman and T. J. Kelly. (1989). "Activation of SV40 DNA replication *in vitro* by cellular protein phosphatase 2A." EMBO J. **8**(12): 3891-3898.

Virshup, D. M. and T. J. Kelly. (1989). "Purification of replication protein C, a cellular protein involved in the initial stages of SV40 DNA replication *in vitro*." Proc. Natl. Acad. Sci. USA. **86**(10): 3584-3588.

Wang, E. H., P. N. Friedman and C. Prives. (1989). "The murine p53 protein blocks replication of SV40 DNA *in vitro* by inhibiting the initiation functions of SV40 large T antigen." Cell. **57**: 379-392.

Weinberg, D. H., K. L. Collins, P. Simancek, A. Russo, M. S. Wold, D. M. Virshup and T. J. Kelly. (1990). "Reconstitution of SV40 DNA replication with purified proteins." PNAS. **87**: 8692-8696.

Weinberg, D. H. and T. J. Kelly. (1989). "Requirement for two DNA polymerases in the replication of simian virus 40 DNA *in vitro*." Proc. Natl. Acad. Sci. USA. **86**: 9742-9746.

Wiekowski, M., P. Droge and H. Stahl. (1987). "Monoclonal antibodies as probes for a function of large T antigen during the elongation process of simian virus 40 DNA replication." J. Virol. **61**: 411-418.

Wobbe, C. R., F. Dean, L. Weissbach and J. Hurwitz. (1985). "*In vitro* replication of duplex circular DNA containing the simian virus 40 DNA origin site." Proc. Natl. Acad. Sci. USA. **82**: 5710-5714.

Wobbe, C. R., L. Weissbach, J. A. Borowiec, F. B. Dean, Y. Murakami, P. Bullock and J. Hurwitz. (1987). "Replication of SV40-origin-containing DNA with purified proteins." Proc. Natl. Acad. Sci., USA. **84**: 1834-1838.

Wold, M. S., J. J. Li and T. J. Kelly. (1987). "Initiation of simian virus 40 DNA replication *in vitro*: Large-tumor-antigen- and origin-dependent unwinding of the template." Proc. Natl. Acad. Sci. USA. **84**: 3643-3647.

Wold, M. S., D. W. Weinberg, D. M. Virshup, J. J. Li and T. J. Kelly. (1989). "Identification of cellular proteins required for SV40 DNA replication." J. Biol. Chem. **264**: 2801-2809.

Yang, L., M. S. Wold, J. J. Li, T. J. Kelly and L. F. Liu. (1987). "Roles of DNA topoisomerases in SV40 DNA replication *in vitro*." Proc. Natl. Acad. Sci. USA. **84**: 950-954.

Bovine RP-A Functions in SV40 DNA Replication *in vitro*, but Bovine Polymerase *α*-Primase Inhibits Replication Competitively

Christine Schneider, Dorothea von Winkler, Irene Dornreiter, Heinz-Peter Nasheuer, Ilka Gilbert, Silke Dehde, Avril K. Arthur and Ellen Fanning
Institute for Biochemistry
Karlstr. 23
8000 Munich 2
Federal Republic of Germany

Introduction

The replication of SV40 DNA in a cell-free assay (Li and Kelly, 1984) containing primate cell extract has facilitated the identification and purification of a set of cellular replication proteins that, in concert with a single viral regulatory protein T antigen, are sufficient to replicate SV40 origin-containing DNA (Tsurimoto et al., 1990; Weinberg et al., 1990). Origin binding and unwinding to form a pre-initiation complex at the origin requires T antigen and the cellular single-stranded DNA binding protein RP-A (RF-A) (reviewed by Borowiec et al., 1990). The initial primer synthesis is thought to be catalyzed by DNA polymerase α-primase, but how the establishment of the pre-initiation complex is coupled to synthesis of RNA primers and DNA remains speculative. Protein-protein interactions among the initiation proteins, as observed in procaryotic model systems (reviewed by Bramhill and Kornberg, 1988), could serve to couple origin unwinding and DNA synthesis. Indeed, an association between T antigen and DNA polymerase α-primase in crude extracts of both primate and rodent cells has been reported (Smale and Tjian, 1986; Gough et al., 1988; Gannon and Lane, 1987, 1990). We have recently confirmed and extended these results, using highly purified T antigen and DNA polymerase α-primase from human and bovine cells to demonstrate a direct and specific interaction of T antigen with the large catalytic subunit of polymerase (Dornreiter et al., 1990).

SV40 DNA replication is supported by primate cell extracts, but not by extracts prepared from cells of other species (Li and Kelly, 1985), suggesting species-specific interactions between T antigen and one or more components of the replication machinery. This species specificity is thought to depend on DNA polymerase α-primase because the SV40 replication activity of human cell extracts depleted of polymerase α-primase could be restored by addition of the human, but not the bovine or murine polymerase-primase (Murakami et al., 1986). Moreover, human DNA polymerase α-primase was reported to render mouse cell extracts permissive for SV40 DNA replication, implying that the murine polymerase-primase did not compete with the human enzyme for T antigen or other replication factors (Murakami et al., 1986). Thus, the observed interaction of T antigen with polymerase-primase from both primate and non-primate cells presents a paradox, which we sought to resolve in the studies presented here.

Results

Interaction of DNA polymerase α-primase with SV40 T antigen and cellular replication protein A.

DNA polymerase α-primase was isolated from calf thymus by immunoaffinity chromatography (Dornreiter et al., 1990). The purified enzyme was made up of four different subunits: the large catalytic subunit p180, a p72 subunit of unknown function and the p58 and p48 subunits, which together constitute the primase activity (Fig.1, lane POL). Cellular replication protein A (RP-A) was isolated from calf thymus by conventional column chromatography (D. von Winkler, unpublished data). The purified RP-A was composed of three types of polypeptides resembling those described previously for the human and yeast homologs (Fairman and Stillman, 1988; Wold and Kelly, 1988; Brill and Stillman, 1989): p70, p32 and p14 (Fig. 1, lane RP-A). An additional band of about 52kDa, observed in some preparations, was shown by Southwestern blotting to share the DNA binding properties of the p70 subunit (D. von Winkler, in preparation) and is therefore very likely a degradation product of the p70 subunit, as shown previously for human RP-A (Wold and Kelly, 1988). SV40 large T antigen was purified from extracts of insect cells infected with a recombinant baculovirus (Höß et al., 1990).

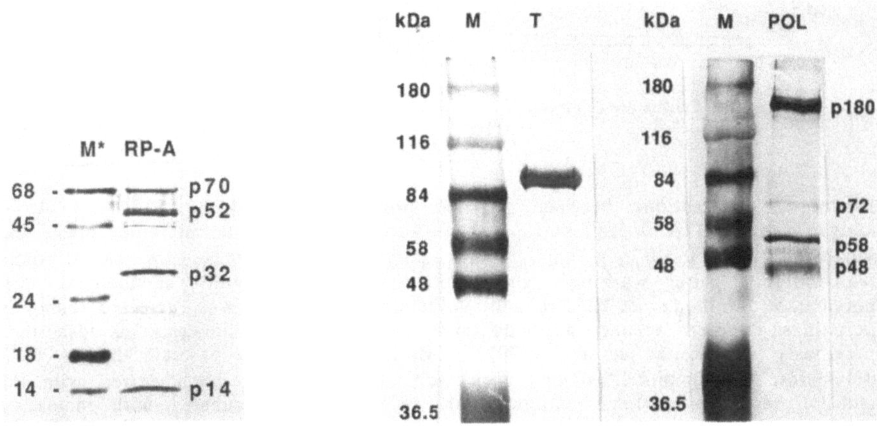

Figure 1. Silver-stained purified proteins separated by SDS polyacrylamide gel electrophoresis. **T;** 1μg immunopurified T antigen expressed from recombinant baculovirus in insect cells, 10% gel. **POL;** 2μg immunopurified DNA polymerase α-primase from calf thymus; 10% gel. **RP-A;** 2μg conventionally purified RP-A from calf thymus; 13.75% gel. **M*, M;** marker proteins of known molecular weights as indicated to the left of each lane in kD.

Protein-protein interactions between purified replication proteins were detected with two different assays: a modified immunoblotting method and an enzyme-linked immunosorbent assay (ELISA). Using the ELISA with native T antigen, RP-A or bovine serum albumin (BSA) immobilized in microtiter plates, binding of soluble DNA polymerase α-primase to T antigen and RP-A can be detected using a monoclonal antibody against polymerase-primase and a peroxidase-coupled second antibody (Fig. 2A). Polymerase-primase bound to both T antigen and RP-A in a concentration-dependent manner, while it failed to bind to the control protein BSA. Similar interactions were observed using purified human RP-A and calf thymus polymerase α-primase (I. Dornreiter, L. Erdile, T. Kelly and E. Fanning, in preparation). The association of DNA polymerase α-primase with RP-A and T antigen has also been confirmed by immunoblot analysis (Dornreiter et al., 1990), demonstrating that the interactions among these proteins are direct and specific.

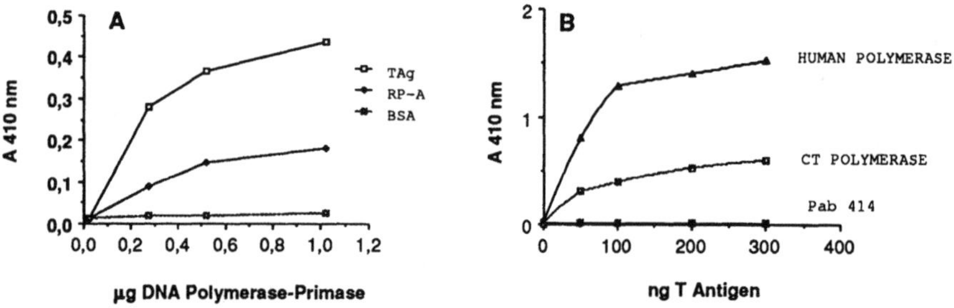

Figure 2. Interactions between purified proteins analysed by ELISA. **A;** T antigen (1µg), RP-A (0.5µg) or BSA (1µg) were immobilized in microtiter plates and incubated with 0-1.0 µg calf thymus polymerase α-primase for 1h at room temperature. After washing and incubation overnight with monoclonal anti-polymerase antibody 2CT25, bound polymerase-primase was detected using a peroxidase-coupled second antibody and a chromogenic substrate as described previously (Dornreiter et al., 1990). **B;** 1µg of human or calf thymus (CT) polymerase was immobilized in each well of a microtiter plate coated with the anti-polymerase monoclonal antibody SJK 132-20, and incubated with increasing amounts of T antigen for 1h at room temperature. T antigen was detected with peroxidase-coupled anti-T antigen monoclonal antibody Pab 419. Where indicated, the T antigen was preincubated with 10µg Pab 414 before addition to the microtiter wells.

DNA polymerase α-primase has been proposed to be the sole replication factor responsible for the ability of primate cells and primate cell extracts to support SV40 DNA replication (Murakami et al., 1986). In view of the proposed role of complex formation between T antigen and DNA polymerase α-primase in the initiation of replication at the SV40 origin, the ability of polymerase-primase from a non-primate source to form apparently specific complexes with T antigen was unexpected, raising the question whether the observed interactions are functionally relevant. To address this question, we purified DNA polymerase α-primase to near homogeneity from human tissue (Dornreiter et al., 1990). Complex formation of T antigen with the purified human and bovine enzymes was then compared directly in an ELISA (Fig. 2B). Equal amounts of human and bovine polymerase-primase were immobilized in microtiter plates coated with a monoclonal antibody against polymerase-primase. After addition of soluble T antigen in increasing amounts, bound T antigen was detected using a peroxidase-coupled monoclonal antibody against T antigen. T antigen bound to polymerase α-primase from both species in a concentration-dependent fashion, but its affinity for the human enzyme was nearly 10-fold greater (Fig. 2B). Preincubation of the soluble T antigen with saturating amounts of a monoclonal antibody, Pab 414, reduced complex formation with both the human and bovine enzymes to background levels (Fig. 2B). Pab 414 was shown previously to block T antigen-polymerase α-primase interaction in crude extracts, as well as SV40 DNA replication (Smale and Tjian, 1986; Gough et al., 1988; Gannon and Lane, 1987, 1990). Thus, the preferred association of T antigen with human DNA polymerase α-primase would appear to be correlated with the ability of human extracts to support SV40 DNA replication, and provides evidence that this interaction is of functional importance. However, since the bovine polymerase α-primase was also bound, albeit not so efficiently, and this interaction was also specifically blocked by the Pab 414, one might expect that non-primate cells or extracts could support SV40 DNA replication at a low level. Since replication of SV40 DNA was not detectable in non-primate extracts (Li and Kelly, 1985), the reduced affinity of non-primate polymerase-primase for T antigen is probably not sufficient to explain the species specificity of replication.

One possible explanation for the primate specificity of SV40 replication is that the T antigen-polymerase-primase complex must interact in a species specific manner with a third replication factor to support SV40 replication. Since we have shown that RP-A from both bovine and human sources associates with calf thymus DNA polymerase α–primase, RP-A would be a good candidate for such a factor. Moreover, yeast RP-A cannot replace human RP-A in the cell-free SV40 DNA replication assay (Brill and Stillman, 1989).

<u>Activity of bovine RP-A and bovine DNA polymerase α-primase in SV40 DNA replication.</u>

We set out to test the capacity of purified bovine RP-A to replace human RP-A in the SV40 DNA replication assay. A cell-free replication assay containing human cell extract replicated plasmid DNA carrying the

SV40 origin of replication in a T antigen-dependent fashion (Fig. 3A). Incorporation of radioactive nucleotides into DNA in the absence of T antigen was very low and no Dpn I-resistant plasmid DNA was detectable (Fig. 3A, lane 3, right panel), indicating that the assay is a reliable measure of SV40-specific DNA replication.

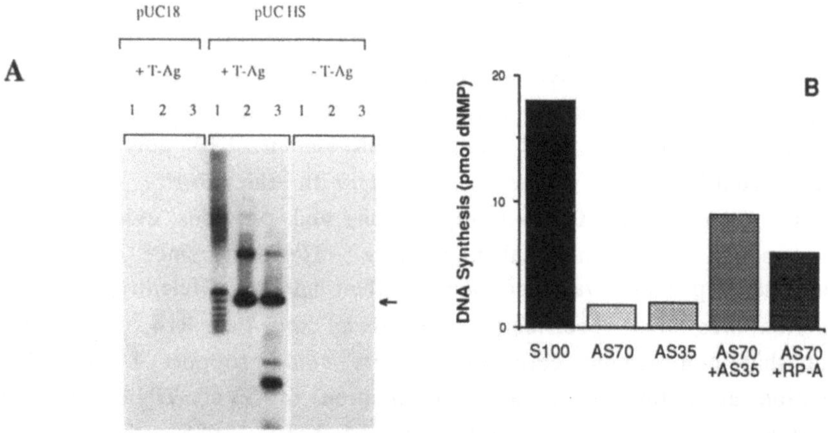

Figure 3. SV40 DNA replication in vitro. **A**; Cell free replication assays were carried out with plasmid carrying the SV40 origin of replication (pUC-HS) or lacking it (pUC18), and with or without T antigen, as indicated. The replication products were isolated, and analyzed either without restriction digestion (lane 1), or after digestion with Eco RI (lane 2) or with Eco RI and Dpn I (lane 3), by agarose gel electrophoresis and autoradiography. The arrow indicates the position of the Eco RI-linearized pUC-HS DNA. **B**; S100 extract (190μg), or AS35 (190μg) and AS70 (190μg) fractions prepared from 293S cell S100 extracts or a combinatión of AS35 (100μg) and AS70 (190μg) fractions were tested for SV40 replication activity in vitro. The AS35 fraction was substituted by 1μg of highly purified RP-A from calf thymus as indicated. Acid-precipitable radioactivity incorporated into DNA was taken as a measure of SV40 DNA replication.

To test the activity of bovine RP-A in SV40 replication, the human S100 extract was fractionated by ammonium sulfate precipitation into a 35% saturated fraction (AS35) containing RP-A and polymerase α-primase and a 70% saturated fraction (AS70) lacking RP-A, but containing polymerase α-primase and other replication factors (Wobbe et al., 1986). The AS35 and AS70 fractions were then assayed for SV40 DNA replication activity (Fig. 3B). Neither fraction alone was sufficient for replication activity, but AS35 and AS70 together reconstituted SV40

DNA replication activity, confirming the previous reports (Wobbe et al., 1986, 1987). When the human AS70 fraction was supplemented by purified bovine RP-A instead of the AS35 fraction, SV40 replication activity was comparable to that observed with the AS35 fraction (Fig. 3B), and in some experiments even greater (D. von Winkler, in preparation). Analysis of the replication products as in Fig. 3A confirmed the production of DpnI-resistant plasmid DNA in the reactions containing AS70 supplemented with AS35 or RP-A (not shown). These results demonstrate that bovine RP-A can functionally substitute for human RP-A in SV40 DNA replication in vitro, thus ruling out RP-A as a primate-specific factor involved in SV40 replication.

Thus, the strict requirement for primate DNA polymerase α-primase in the SV40 replication remains puzzling. If the lower affinity binding of T antigen to non-primate polymerase (Fig.2B) is solely responsible for its low SV40 replication activity, the bovine polymerase α-primase may be able to replace the human enzyme if added to the replication reaction in a very high concentration. Alternatively, the interaction of T antigen with non-primate polymerases may be non-productive, in that it fails to orient the polymerase α-primase correctly in the initiation complex and therefore does not permit initiation to occur. This premise would predict that a non-primate DNA polymerase α-primase might compete with the human enzyme in the initiation reaction, resulting in inhibition of replication.

To examine these alternatives, a human cell extract was depleted of polymerase α-primase by sequential incubations with anti-polymerase antibody-coupled Sepharose beads. The residual polymerase α activity of the depleted extract ranged from 10-20% of that in the original extract (Fig. 4A). The replication activity of SV40 in the depleted extract was correspondingly reduced to 20-50% of the original activity (Fig. 4B). The SV40 replication activity of the depleted extract was restored by the addition of either the AS35 or AS70 fraction from human extracts (Fig. 4A, B), although they themselves were inactive in SV40 replication (Fig. 3B). Both these fractions contain human polymerase α-primase activity (Fig.4A), as reported previously (Wobbe et al., 1987). When AS35 or AS70 fractions prepared from p53-negative murine Ehrlich ascites cells were used to supplement the depleted human extract, the residual SV40 replication activity dropped further to background level (Fig. 4B) and DpnI-resistant plasmid DNA could no longer be detected (not shown). These results suggest that addition of fractions containing human polymerase α-primase to the depleted extract is sufficient to reconstitute SV40 replication activity, but that the addition of fractions containing murine polymerase α-primase appears to inhibit the human enzyme's residual SV40 replication activity.

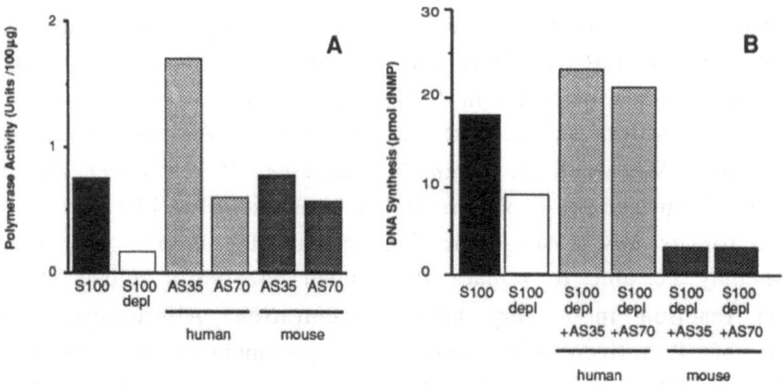

Figure 4. Restoration of SV40 DNA replication activity to polymerase α-primase-depleted human cell extracts by AS35 and AS70 fractions from human, but not murine cells. **A**; DNA polymerase α-primase activity of human 293S S100 extract, polymerase α-primase-depleted 293S S100 extract, AS35 and AS70 fractions prepared from 293S and Ehrlich ascites mouse cell extracts was measured using activated calf thymus DNA as template-primer as described (Dornreiter et al., 1990). **B**; In vitro replication assays were carried out using 190 µg of 293S S100 extract, 190µg polymerase α-primase-depleted S100 extract, or 190µg depleted S100 supplemented with 100µg of human or murine AS fractions as indicated. Incorporation of radioactive nucleotides into acid-precipitable material was taken as a measure of replication activity.

To test whether the inhibition was due directly to polymerase α-primase itself, and whether other non-primate polymerases have the same inhibitory effect, highly purified calf thymus DNA polymerase α-primase was added in increasing amounts to the depleted extract. Even low amounts of bovine polymerase α activity were sufficient to completely suppress the residual SV40 replication activity (Fig. 5). Additional calf thymus polymerase α-primase up to 50-fold higher activity failed to restore SV40 replication activity (Fig. 5). The slight increase in radioactive nucleotide incorporation observed in reactions containing 2 units or more calf thymus enzyme was not correlated with the appearance of DpnI-resistant plasmid DNA (not shown), indicating that little or no SV40 DNA replication had occurred. These results demonstrate that purified bovine polymerase α-primase not only is unable to substitute for the human enzyme in SV40 replication, but that it actually inhibits the SV40 replication activity of the human enzyme.

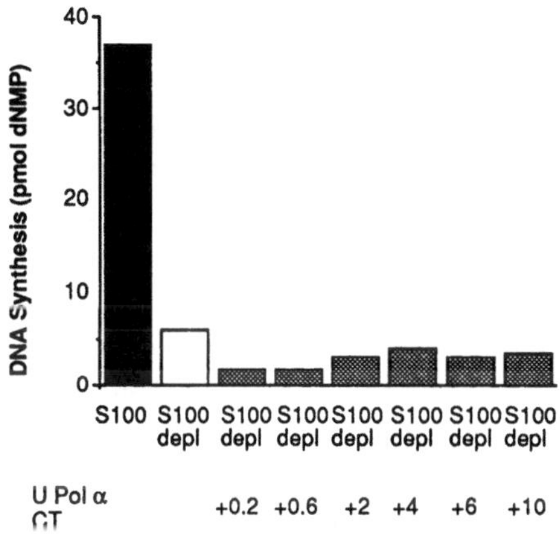

Figure 5. Purified calf thymus DNA polymerase α-primase inhibits the residual SV40 DNA replication activity of polymerase-primase-depleted human cell extracts. In vitro SV40 replication activity was assayed using 293S S100 extracts (190μg), polymerase α-primase-depleted 293S S100 (190μg) alone, and supplemented with purified bovine polymerase-primase (gray bars) in the amounts indicated in activity units below the graph. Incorporation of labelled nucleotides into acid-precipitable material is shown as a measure of SV40 DNA replication activity.

Discussion

We have presented evidence that purified DNA polymerase α-primase associates directly and specifically with SV40 T antigen and cellular replication protein A. T antigen binds to human polymerase α-primase with a nearly 10-fold higher affinity than to calf thymus polymerase-primase, in keeping with the ability of human but not non-primate cell extracts to support SV40 DNA replication. However, since the SV40 DNA replication activity of human cell extracts is at least two orders of magnitude greater than that of non-primate cell extracts (Li and Kelly, 1985), this difference in affinity cannot account for the difference in replication activity. The possibility that RP-A could be responsible in part for this differential activity was ruled out, since bovine RP-A substitutes quite efficiently for human RP-A in SV40 replication in vitro.

The observation that extracts containing human but not murine polymerase-primase could restore the SV40 replication activity of polymerase-primase-depleted human cell extracts (Murakami et al., 1986) was confirmed (Fig.4). Moreover, as reported previously, purified calf thymus polymerase-primase was unable to restore the replication activity of the depleted extract (Fig.5). However, in contrast with the conclusions drawn in the previous study, we demonstrate that extracts containing murine polymerase α-primase suppress the SV40 replication activity of human cell extracts (Fig.4B). The suppression is not caused by murine p53, which competes with polymerase α-primase binding to T-antigen (Gannon and Lane, 1987; Braithwaite et al., 1987), because the murine cells used do not express p53 (W.Deppert, personal communication). Moreover, highly purified calf thymus polymerase α-primase also inhibits the SV40 replication activity of human cell extracts (Fig.5). Our results suggest that non-primate DNA polymerase α-primase may act as a competitive inhibitor of SV40 replication, either by sequestering T antigen, RP-A and possibly other replication factors or by forming an inactive replication initiation complex at the SV40 origin, or both.

Significantly, this interpretation is also consistent with the experimental data reported by Murakami et al. (1986). Since the polymerase α-depleted extracts used in their experiments had no residual SV40 replication activity, a reduction caused by addition of non-primate polymerase-primase could not be detected. Furthermore, the addition of human polymerase-primase to murine extracts resulted in only a very low level of SV40 replication activity, equivalent to about 10-20% of the level resulting from the addition of human polymerase-primase to the depleted human extract (Murakami et al., 1986), implying that some component(s) in the murine extract inhibited SV40 replication.

In summary, the evidence presented here indicates that bovine DNA polymerase α-primase associates specifically with T antigen and RP-A, but it is not functional in SV40 DNA replication and instead, it acts as a competitive inhibitor of human polymerase α-primase in SV40 replication.

Methods:

In vitro SV40 DNA replication assay.

Cytoplasmic extracts were prepared from 293S monolayer cultures as described (Li and Kelly, 1985). The extracts were then adjusted to 100 mM NaCl, 1% Trasylol (Bayer, Leverkusen), 50 μM leupeptin, and ultracentrifuged (50 Ti rotor) for 1 h at 40 000 rpm and 4°C. The supernatant S100 protein concentration measured with a dye binding assay (Bio Rad, Munich) ranged from 8 to 14 mg/ml, using BSA to calibrate the assay. Supercoiled plasmid DNA pUC-HS, carrying the Hind III-Sph I origin fragment of wildtype SV40 DNA (strain SV-S) in pUC18 (Traut and Fanning, 1988) and the vector pUC18 (Yanisch-Perron et al., 1985) served as templates. The replication assays (60 μl) were assembled as described (Guo et al., 1989) with following modifications: 600 ng T antigen, 190 μg S100 extract, 75 ng DNA, 80 μM each CTP, UTP, GTP, 0.15 units pyruvate kinase (Boehringer, Mannheim) and without PEG. After 2h at 37°C, the reaction was adjusted to 1% SDS, 30mM EDTA and 0.5mg per ml proteinase K, and incubated for 1h at 37°C. Plasmid DNA isolated from each reaction was taken up in 20 μl of 10mM Tris-HCl pH 7.5, 1mM EDTA, and 5μl portions were analyzed by electrophoresis in 0.8% agarose gels, either directly or after restriction digestion (Peden et al., 1980). Alternatively, incorporation of labeled nucleotides into acid-insoluble material was determined by liquid scintillation counting, and the specific activity was used to calculate the total amount of newly synthesized DNA in pmol of dNMP as described (Li and Kelly, 1985).

Other experimental procedures

Enzymatic assays, purification of antibodies, T antigen and DNA polymerase α-primase, and ELISA were carried out exactly as described (Dornreiter et al., 1990; Höß et al., 1990). Ammonium sulfate fractionation of human 293S and murine Ehrlich ascites cell S100 extracts, immunodepletion of polymerase α-primase from 293S S100 extracts were performed essentially as described (Wobbe et al., 1986, 1987; Murakami et al., 1986). Purification of RP-A from calf thymus will be described elsewhere (D. von Winkler, in preparation).

Acknowledgements

We thank Lorne Erdile and Tom Kelly for samples of human RP-A and helpful criticism; Claudia Gruss, Frank Grosse, Robert Lanford, Wolfgang Deppert, Ed Harlow and Teresa Wang for valuable advice, baculovirus 941T, and cell lines; Karen van Zee for critical reading of the manuscript, and E. Rohrer and S. Kastenmüller for excellent secretarial assistance. The financial support of the DFG (grants to E. F. and H.-P. N.), the BMFT (Genzentrum grant to A. K. A. and E. F.), and Fonds der Chemischen Industrie (E. F.) is gratefully acknowledged.

References

Braithwhite AW, Sturzbecher H, Addison C, Palmer C, Rudge K, Jenkins JR (1987) Mouse p53 inhibits SV40 origin-dependent DNA replication. Nature 329: 458-460

Bramhill D, Kornberg A (1988) A model for initiation at origins of DNA replication. Cell 54: 915-918

Brill SJ, Stillman B (1989) Yeast replication factor-A functions in the unwinding of the SV40 origin of DNA replication. Nature 342: 92-95

Borowiec JA, Dean FB, Bullock PA, Hurwitz J (1990) Binding and unwinding-how T antigen engages the SV40 origin of DNA replication. Cell 60: 181-184

Dornreiter I, Höss A, Arthur AK, Fanning E (1990) SV40 T antigen binds directly to the catalytic subunit of DNA polymerase α. EMBO J 9: 3329-3336

Fairman MP, Stillman B (1988) Cellular factors required for multiple stages of SV40 replication in vitro. EMBO J 7: 1211-1218

Gannon JV, Lane DP (1987) p53 and DNA polymerase α compete for binding to SV40 T antigen. Nature 329: 456-458

Gannon JV, Lane DP (1990) Interactions between SV40 T antigen and DNA polymerase α. The New Biologist 2: 84-92

Gough G, Gannon JV, Lane DP (1988) Competition between DNA polymerase α and p53 for binding to SV40 T antigen. Cancer Cells 6:153-158

Guo ZS, Gutierrez C, Heine U, Sogo J, DePamphilis ML (1989) Origin auxiliary sequences can facilitate initiation of simian virus 40 DNA replication in vitro as they do in vivo. Mol Cell Biol 9: 3593-3602

Höß A, Moarefi I, Scheidtmann K-H, Cisek LJ, Corden JL, Dornreiter I, Arthur AK, Fanning E (1990) Altered phosphorylation pattern of SV40 T antigen expressed in insect cells by using a baculovirus vector. J Virol 64: 4799-4807

Li JJ, Kelly TJ (1984) Simian virus 40 DNA replication in vitro. Proc Natl Acad Sci USA 81: 6973-6977

Li JJ, Kelly TJ (1985) Simian virus 40 DNA replication in vitro: specificity of initiation and evidence for bidirectional replication. Mol Cell Biol 5: 1238-1246

Murakami Y, Wobbe CR, Weissbach L, Dean FB, Hurwitz J (1986) Role of DNA polymerase α and DNA primase in simian virus 40 DNA replication in vitro. Proc Natl Acad Sci USA 83: 2869-2873

Peden KWC, Pipas JM, Pearson-White S, Nathans D (1980) Isolation of mutants of an animal virus in bacteria. Science 209: 1392-1396

Smale ST, Tjian R (1986) T-antigen-DNA polymerase α complex implicated in simian virus 40 DNA replication. Mol Cell Biol 6: 4077-4087

Traut W, Fanning E (1988) Sequence-specific interactions between a cellular DNA-binding protein and the simian virus 40 origin of DNA replication. Mol Cell Biol 8: 903-911

Tsurimoto T, Melendy T, Stillman B (1990) Sequential initiation of lagging and leading strand synthesis by two different polymerase complexes at the SV40 DNA replication origin. Nature 346: 534-539

Weinberg DH, Collins KL, Simancek P, Russo A, Wold MS, Virshup DM, Kelly TJ (1990) Reconstitution of simian 40 DNA replication with purified proteins. Proc Natl Acad Sci USA 87: 8692-8696

Wobbe CR, Dean FB, Murakami Y, Weissenbach L, Hurwitz J (1986) Simian virus 40 DNA replicaton in vitro: Study of events preceding elongation of chains. Proc Natl Acad Sci USA 8: 4612-4616
Wobbe CR, Weissenbach L, Borowiec JA, Dean FB, Murakami Y, Bullock P, Hurwitz J (1987) Replication of simian virus 40 origin-containing DNA in vitro with purified proteins. Proc Natl Acad Sci USA 84: 1834-1838
Wold MS, Kelly TJ (1988) Purification and characterization of replication protein A, a cellular protein required for in vitro replication of simian virus 40 DNA. Proc Natl Acad Sci USA 85: 2523-2527
Yanisch-Perron C, Vieira J, Messing J (1985) Improved M13 phage cloning vectors and host strains: nuleotide sequences of the M13mp18 and pUC19 vectors. Gene 33: 103-119

Bovine Papillomavirus (BPV1) Replication Protein E1: Associated ATPase and Binding to the Viral E2 Transactivator

Sabine Santucci, Catherine Bonne-Andréa & Philippe Clertant
University of Nice / INSERM U.273
Centre de Biochimie
Parc Valrose
06034 NICE CEDEX
France

Summary : Recent genetic evidence has unambiguously established that the replication of the bovine papillomavirus type 1 (BPV1) plasmid DNA in virally-transformed cells requires the expression of the entire viral open reading frames E1 and E2 (Ustav & Stenlund, 91). We and others have recently identified in BPV1-transformed cells a 70 kDa nuclear phosphoprotein, encoded by the entire viral ORF E1 (Santucci *& al.*, 90, Sun *& al.*, 90). We present here evidence showing that an ATPase activity is associated with this protein in partially purified fractions and immunoprecipitates. Covalent affinity labeling made by the use of an ATP analogue of the E1 protein reinforces the idea of an intrinsic ATPase activity. Moreover, as recently shown by others (Mohr *& al.*, 90), we found *in vitro* indications of the existence of a molecular complex between the E1 protein and the viral E2 *trans*-activating transcription factor. ATPase activity was detected in association with this E1-E2 complex.

Rodent fibroblasts transformed in culture by bovine papillomavirus type 1 (BPV1) persistently harbor viral genomes as free nuclear plasmids (reviewed in Mecsas & Sugden, 87). The replication of BPV1 plasmids takes place during the S phase of the cell cycle (Gilbert & Cohen, 87), and is strictly limited to one duplication per cell generation, so that the DNA copy number per cell is maintained constant through generations (Roberts & Weintraub, 88). Genetic evidence has established that *cis* and *trans*-acting viral control mechanisms regulate BPV1 plasmid replication both positively and negatively (Roberts & Weintraub, 86, see also Sugden & Mecsas, 87, and Ustav & Stenlund, 91, for a most recent review). BPV1 plasmids thus offer a model system for studying the scheduled DNA replication in mammalian cells.

Genetic analysis helped to define the *cis* and *trans*-acting viral sequences, but until recently it was hampered by the low transfection efficiency of recipient rodent cells (Ustav & Stenlund, 91). Most of the genetic elements previously described have thus to be redefined. Long-term persistence of plasmidial replication requires at least one *cis*-acting element, named PMS (Lusky and Botchan, 84). On the BPV1 genomic map (Figure 1), electron microscopy helped to position the starting point of bidirectional replication within or very close to PMS (Waldeck *& al.*, 84). However, recent 2D-gel analyses of plasmid replicative forms present in transformed cells mapped the replication start point outside PMS (Yang & Botchan, 90). Recent results have also brought clarity with respect to the definition of the *trans*-acting positive factors, by establishing unambiguously that plasmid replication observed after efficient transfer of BPV1 DNA into mouse cells by electroporation requires the expression of two viral proteins, translated from the entire Open Reading Frames (ORFs) E1 and E2 (see Figure 1) (Ustav & Stenlund, 91). The E2 protein has been identified several years ago, as a 46 kDa DNA-binding activator of transcription (Androphy *& al.*, 87, Spalholz *& al.*, 85). Mutations which affect its expression abolish BPV1 long-term plasmid replication (DiMaio, 86, DiMaio & Settleman, 88, Rabson *& al.*, 86). From weak but significant sequence similarities with the SV40 large T antigen, it was suggested some time ago that the E1 ORF might encode a DNA replication initiator (Clertant & Seif, 84). This ORF is the longest and the most conserved among papillomaviruses and it has been suggested that it might act in *trans* on DNA replication in two opposite ways. The negative control, involved in limiting the DNA replication to a doubling per cell cycle, has been reported to require only the 5' third of the E1 ORF (*M* gene - Roberts & Weintraub, 86, Lusky & Botchan, 86b); a cDNA with

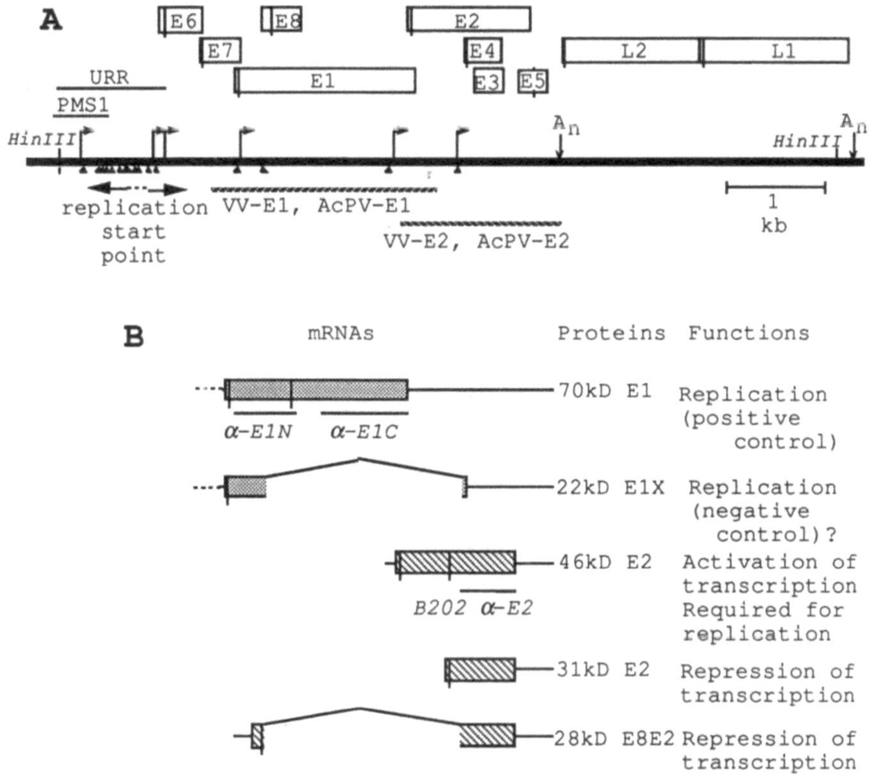

Figure 1 : ***A/ Map of the BPV1 genome*** (linearized at the *HindIII* site) (Lambert *& al.*, 88). Arrows mark the identified promoters, triangles the E2 binding sites (Androphy *& al.*, 87, Li *& al.*, 89), An the polyadenylation sites. The boxes aligned on the upper part correspond to the ORFs (E: early, L: late). URR: upstream regulatory region. PMS: plasmid maintenance sequence (Waldeck *& al.*, 83). Aligned under the genomic map are the fragments cloned for expression in baculo- (AcPV) or vaccinia- (VV) virus vectors. Indication of the actual start point of replication (Yang & Botchan, 90) is also indicated.
B/ Expression map of the E1-E2 region (Choe *& al.*, 89): messengers RNAs are aligned with indication of the translated ORF (shaded boxes). Methionine initiation codons are marked by vertical strokes. The encoded proteins and identified functions are indicated to the right. Under the E1 and E2 coding sequences are aligned the regions bearing epitopes for the antibodies used in this work : polyclonal rabbit antisera raised against two E1 bacterial fusion proteins (Santucci *& al.*, 90) and a mouse monoclonal antibody, B202, which recognizes the three E2 proteins (Mohr *& al.*, 90).

similar limited coding capacity has been identified, with expression of a 22 kDa nuclear phosphoprotein (Thorner *& al.*, 88). We and others have identified a second E1 product, which corresponds to the translation of the entire ORF, and consequently is the positive factor required for BPV1 plasmid replication (Santucci *& al.*, 90, Sun *& al.*, 91). It is a nuclear phosphoprotein, with expected apparent size of 70 kDa, which has the capacity to bind non-specifically to DNA. In mammalian cells this protein is subject to posttranslational modification, with appearance of larger forms, of unknown origin and significance.

It has been shown very recently that the E1 protein has the capacity to enter *in vitro* into molecular interactions with the E2 46 kDa protein (Mohr *& al.*,90, Blitz & Laimins, 91); the other E2-encoded DNA-binding proteins (Hubbert *& al.*, 88, Choe *& al.*, 89 - see Figure 1), which act as transcriptional repressors (Lambert *& al.*, 87), do not interact with E1. As do the three E2 proteins (Li *& al.*, 89), the E1-E2 complex has the capacity to specifically bind to E2-responsive elements within the URR; it was proposed that this mechanism helps to target the replication machinery to the viral origin of replication (Mohr *& al.*, 90).

We have expressed the BPV1 E1 and E2 proteins by using viral vectors, of vaccinia or baculovirus origin (Santucci *& al.*, 90, Blitz & Laimins, 91), and have achieved partial purification of both proteins. Here we report that both the E1 protein and the E1-E2 complex harbor associated ATPase activities, and we present evidence suggesting that this activity might be an intrinsic property of the E1 protein, alone or in complex.

The 70 kDa E1 protein binds nucleotides.

The comparison between the E1 proteins of various papillomaviruses and the large-T antigens of SV40 and polyoma viruses led us to the observation of polypeptide sequence similarities spanning the large-T ATPase domain (Clertant & Seif, 84), precisely centered on the invariant patterns that were reported to define a nucleotide-binding pocket (Bradley *& al.*, 87). We thus checked if the E1 protein expressed in

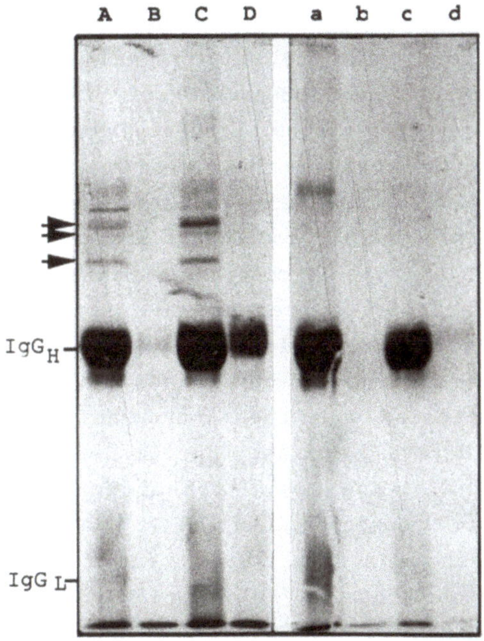

Figure 2 : *The E1 proteins expressed in mouse cells from a vaccinia vector are covalently labeled in vitro by the use of the ATP analogue [^{3}H]ATP-2',3'-dialdehyde.* After 14h infection at high multiplicity, cells were lysed (Tris 0.1M, LiCl 0.5M, NP40 1%, EDTA 1mM, DTT 1mM, PMSF 1mM, pH 8), and the clarified extracts submitted to immunoprecipitation with the aid of Protein A-Sepharose and anti-E1 antibodies. Beads were pelleted and washed, first with the lysis buffer, then with the appropriate buffer for covalent affinity labeling (Clertant *& al.*, 84). They were incubated overnight in the presence of 1µM ATP analogue and 10mM $NaCNBH_3$, then washed, solubilized in Laemmli buffer and analyzed by SDS-PAGE. Cells infected with an irrelevant vaccinia recombinant were used as negative control (small letters). Lanes A,a, and D, d, antibodies α-E1N and α-E1C (see Fig. 1), C,c, another anti-E1 antibody raised against the C-terminal decapeptide (Santucci *& al.*, 90), and B,b, non-immune serum. Arrows point to the E1 70, 85 and 88 kDa proteins. The position of the non specifically labeled IgG heavy and light chains is also marked.

mammalian cells by infection with the appropriate vaccinia recombinant (VV-E1) could bind nucleotides, by employing a covalent affinity labeling technique performed with an ATP analogue that we previously employed in the case of SV40 large T antigen (Clertant *& al.*, 84). Immunoprecipitates, prepared with the aid of various anti-E1 antibodies from extracts of infected cells, were incubated in the presence of periodate-oxidized [^{3}H]ATP, under conditions allowing optimal labeling of polyomavirus large-T antigen. As shown on Figure 2, by using two anti-E1 antibodies, directed against either an N-terminal part of the E1 polypeptide or against the C-terminal decapeptide, the immunoprecipitated 70 kDa (size of the unmodified translation product) E1 protein and its modified forms (85 and 88 kDa) were labeled (lanes A and C). In contrast, nothing specific could be observed in precipitates prepared by using non-immune serum (lane B) or from cells infected by an irrelevant vaccinia recombinant (lanes a-d). Significantly, affinity-labeled E1 could not be detected when the immunoprecipitate was prepared using antibodies raised against the 60% C-terminal part of the E1 polypeptide (lane D), although these antibodies quantitatively immunoprecipitated the E1 proteins present in the extracts (not shown). Thus, as expected from the observed similarities with other ATP-binding initiator proteins, and consistent with their position within the E1 polypeptide sequence, these results indicate that E1 binds nucleotides, presumably ATP or ADP, and that antibody binding to certain epitopes within the carboxy-terminal 60% of the E1 molecule neutralizes this activity.

An ATPase activity copurifies with E1.

To better characterize the molecular activities of the E1 protein, we have undertaken a classical purification of E1 from insect cells infected by an E1-expressing baculovirus vector (Santucci *& al.*, 90), since it provided production levels much higher than the E1-vaccinia recombinant (2-5 µg, instead of 20 ng E1 protein / 10^6 infected cells). Basically, cells were lysed, nuclei were extracted sequentially by varying pH and salt concentration (Santucci *& al.*, 90) and the resulting nuclear eluates were fractionated by FPLC chromatography. To date, we have achieved partial purification resulting in preparations in which the E1 protein represent more than 60% of the total protein content. Two preparations, obtained using different chromatographic techniques, were able to catalyze ATP hydrolysis, with comparable specific activities relative to

Table 1 : *ATPase activity associated to partially purified E1 protein.*

Material from Sf9 cells infected with :	protein, mg/ml*	E1, %*	ATPase°, u/µl	u/µg E1
E1 virus,				
Heparin-agarose eluate	0.20	10-15	0.44	18.7 ±5.7
MonoQ eluate	0.22	40-50	1.54	15.8 ±2.3
Immunoprecipitate made from monoQ eluate using α-E1C IgGs			1.10	11.2 ± 1.8
α-E1N "			1.30	13.4 ± 2.0
E1 + E2 viruses,				
MonoS eluate (see Fig. 4)	0.12	30-40	0.72	(18 ± 4)$^{+}$
Immunoprecipitate made from monoS eluate using α-E1C IgGs			0.44	10.8 ± 2.7
α-E1N "			0.57	13.8 ± 3.0
α-E2 "			0.63	15.5 ± 3.4

* Protein concentration in the assayed fractions was measured by the Biorad assay. E1 or E2 content was determined by Coomassie staining of the corresponding polypeptide on electrophoresis gels.
° ATPase was assayed as described in Fig. 5. 1 u ATPase = 1 pmol ATP hydrolyzed / hour at 30°C. Immunoprecipitates were prepared with the aid of Protein A-Sepharose and affinity-purified specific IgGs, and assayed for ATPase as described previously (Guizani *& al.*, 88); in this case, u/µl values refer to the ATPase activities found per µl of immunoprecipitated extract, after substraction of blank values measured on precipitates obtained in parallel by using non-immune IgGs. In all cases, the error on ATPase measurements was below 0.05 u/µl
$^{+}$ In addition to E1-associated ATPase, this fraction contains another ATPase, recovered in the supernatant of all the immunoprecipitates.

their E1 content (Table 1). Moreover, after having immunoprecipitated the E1 protein from such a partially purified fraction, ATPase activity was recovered in the pellet, with yields comparable to the input values (Table 1). We cannot exclude at this stage that the observed ATP hydrolysis would result from an E1-associated host protein. But, together with the affinity labeling data, these results suggest that the observed ATP hydrolysis might result from an intrinsic property of the E1 protein. The fact that anti-E1C antibodies, which have been found to block covalent affinity labeling by using periodate-oxidized ATP, efficiently precipitated active ATPase, does not disprove this idea, since it could be explained in several ways; trivial reasons could be invoked, like blocking of a lysine group, required to make the covalent bond with the analogue, but not involved in the functioning of the ATPase catalytic site; it merely reminded us a situation encountered in the case of polyomavirus large-T antigen, where two sets of polyclonal antibodies had been found to affect in inverse ways the ATPase and covalent labeling of the protein (Clertant *& al.*, 84), a phenomenon likely to be explained by immobilization of the antibody-bound large-T molecule into alternate conformations (Clertant & Bradley, manuscript in preparation).

E1 makes molecular complexes with E2 46 kDa protein.
Evidence has already been presented by others that the E1 protein interacts in vitro with the full-length E2 product, so as to form a complex with the ability to bind to at least some of the E2 binding sites within the URR on the viral genome (Mohr *& al.*, 90). We have been able to detect the presence of molecular complexes containing both E1 and E2 proteins by an indirect technique of enzyme-linked immunoassay (Figure 5). Extracts were prepared from cells singly or doubly infected with the E1 or E2-encoding baculoviruses, and were found to contain similar amounts of either protein. 96-wells plates were coated with a mouse monoclonal antibody, either anti-E2 or control anti-SV40 large-T IgG. They were incubated with increasing amounts of various extracts. After being washed, they were submitted to reaction with a specific rabbit polyclonal antibody, followed by reaction whith peroxydase-coupled anti-rabbit antibody and peroxydase assay. Background values were found in all negative controls, but anti-E1 antibodies consistently revealed the existence of a complex bound on the wells coated with anti-E2. Similar results were obtained in another experiment performed with mixed extracts of cells singly infected by the E1 and E2 vectors, indicating that the complex can be formed *in vitro* between preexisting E1 and E2 proteins. These results do not, however, preclude the presence within this complex of other proteins, of insect or baculovirus origin.

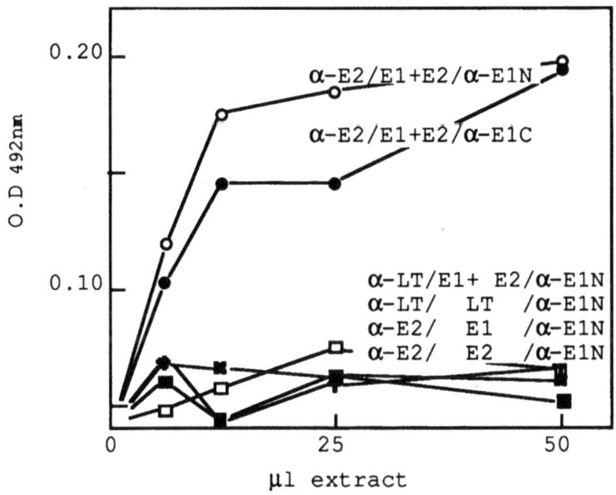

Figure 3 : *Detection of E1-E2 complexes by enzyme-linked sandwich immunoassay in extracts from insect cells coinfected with baculovirus recombinants.* E1- & E2-expressing viruses have been previously described (Santucci & *al.*, 1990, Blitz & Laimins, 91);we used a baculovirus expressing the SV40 large T antigen (Murphy & *al.*, 1988) as a control. After 40 h infection at high multiplicity, *Sf*9 cells were lysed with 0.3M NaCl, 50mM Hepes pH 7.5, 2mM DTT, 0.5% NP40 buffer containing a cocktail of protease inhibitors (1 ml/10^7 cells, 10 min on ice with with periodic vortexing); extracts were clarified by 30 min centrifugation and stored at -70°C. 96-wells plates were coated with mouse monoclonal B202 anti-E2 IgG (1 μg/ well), kindly provided by E. Androphy (Mohr & *al.*, 90); in parallel, anti-SV40 large T PAb419 mouse monoclonal IgG (Harlow & *al.*, 81) was used as a control. After washing and blocking the wells with serum albumin, they were supplemented in the cold with 50 μl serial dilutions of extracts from various infections and incubated for 2 h at 4°C. Extracts were removed, and the wells were washed three times with cold buffer and incubated overnight at 4°C with polyclonal rabbit anti-E1 sera (see Fig. 1); rabbit anti-large T serum was also used as a negative control (not shown). Sera were removed, plates washed again carefully, and revealed using an anti-rabbit IgG antibody coupled to peroxydase (Biorad). Peroxidase activity was measured using *o*-phenylenediamine by reading the optical density at 492 nm. Indications of the conditions are as follows : coating antibody / virus used to make the extract /2nd antibody.

In an attempt to purify the E1/E2 complex from insect cells coinfected with the two corresponding baculovirus vectors, we employed the same purification scheme as that used in parallel to partially purify the E2 protein from singly infected cells : selective extraction from infected nuclei, followed by monoS

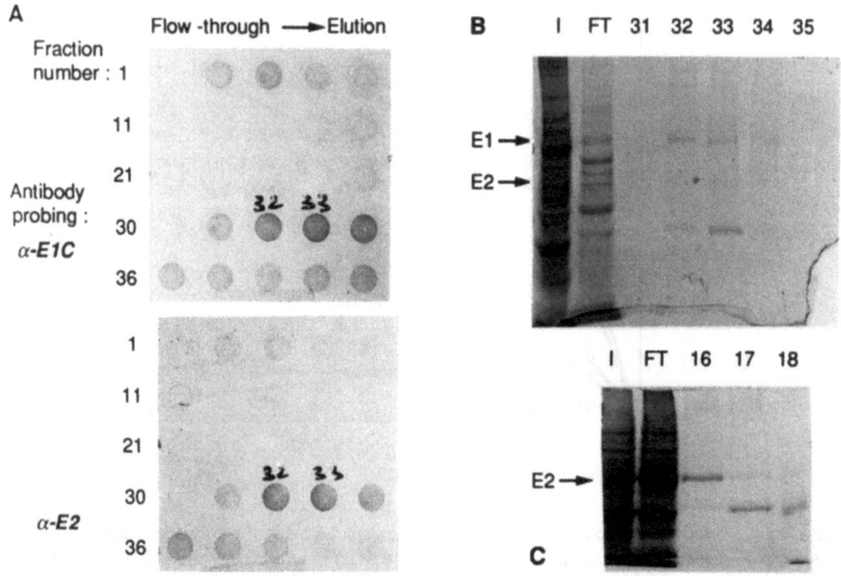

Figure 4 : *Copurification of E1 and E2 proteins from cells coinfected by E1 and E2 recombinant baculoviruses.* A soluble extract prepared by sequential extraction of nuclei isolated from coinfected cells was submitted to MonoS chromatography, and the column eluted with increasing salts. Aliquots of the fractions were examined for the presence of E1 or E2 proteins in a dot-blot immunoassay by using separately anti-E1 and anti-E2 antibodies (***panel A***: odd-numbered fractions were analyzed up to n° 29, and every fraction after n° 30; fraction 32 is 0.6 M NaCl), by Western blot with a mixture of both antibodies (not shown), and by Coomassie staining of a gel electrophoresis of the indicated fractions (***panel B*** : I, input; FT, flow-through, corresponding to fraction 3).
Panel C : gel electrophoresis (Coomassie staining) of peak fractions eluted from monoS column, in a similar experiment, performed on cells infected with the E2 virus only (fraction 16 is 0.6 M NaCl)

chromatography. As already documented (Santucci *& al.*, 90), the E1 protein, when expressed alone by baculovirus infection, is tightly bound to nuclear structure and requires high pH to be quantitatively extracted by salts from the nuclei. In contrast, when coexpressed with E2, most of the E1 protein was salt-extracted from the nuclei together with E2, under the same low pH conditions as those required for the extraction of the E2 protein from singly infected cells (not shown). Moreover, both E1 and E2 bound to monoS columns and were co-eluted as a single peak (Figure 4), by a salt concentration (0.6 M NaCl) identical to that found in parallel to achieve the elution of E2 alone. Low amounts of uncomplexed E1 could be detected, as being not retained by the monoS column (Figure 4, left panel). Coomassie staining of the peak fractions (right panel) revealed three main bands (E1 72kDa, E2 46 kDa, and a 32 kDa polypeptide); this pattern did not differ from that found in the case of E2 preparation from singly infected cells, except for the presence of the E1 polypeptide. The relative intensities of Coomassie-stained E1 and E2 bands, although imprecise, are consistent with molar ratios close to 1. Altogether, the changes in behaviour of E1 when purification is performed in the presence of E2 46kDa protein suggests that most of the E1 and E2 protein present in coinfected insect cellsis complexed together, perhaps with a 1/1 stoichiometry, as recently suggested by others (Mohr *& al.*, 90). However, the existence of this E1-E2 complex in BPV1-transformed cells, although suggested by the available genetic evidence (Ustav & Stenlund, 91), has still to be demonstrated.

E1-associated ATPase activity is retained in E1/E2 complex.

The partially purified E1/E2 complex was assayed for ATPase activity (Table 1). With respect to E1 amount, measured ATP hydrolysis values were higher than expected for E1-associated ATPase. By performing immunoprecipitations on this preparation with either anti-E1 or anti-E2 antibodies, we recovered in both cases a significant part of the input ATPase activity in the pellets, in addition to some activity kept by the supernatantb and probably due to contaminant host proteins. Similar values of specific activity, relative to the E1 content of these pellets, were measured, in the range of those already found for the E1 protein alone. This indicates that the E1-associated ATPase activity is also present in association with the E1/E2 complex, purified in a completely different way. This result reinforces the idea of an ATPase activity intrinsic to the E1 protein. Moreover, it suggests that the interaction with E2 does not quantitatively affect this activity.

To further document this point, we performed a mixing experiment between partially purified E1 and E2 proteins (Figure

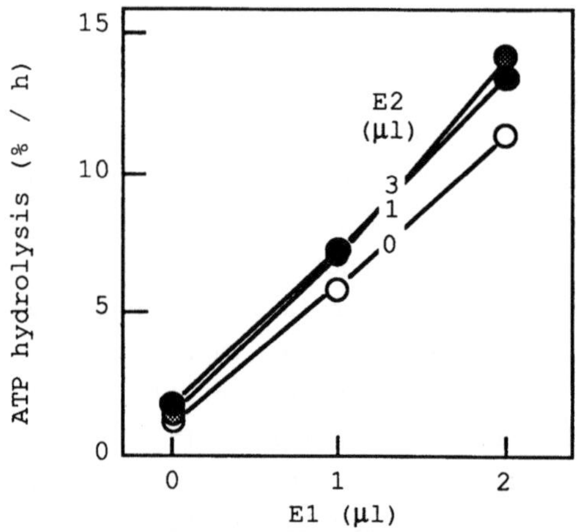

Figure 5 : *The ATPase activity of partially purified E1 is not enhanced by the E2 transactivator*. E1 and E2 proteins were purified partially from cells infected with the corresponding baculovirus vector, up to fractions where they represented ca. 50% of the total proteins (E1 fraction : 0.22 μg prot./ml, 40-50% E1, and E2 fraction : 50 μg prot./ml, 70% E2). Aliquots from these fractions were incubated, together or separately as indicated, in 20mM Hepes, 250mM K-Acetate, 5mM $MgSO_4$, 1mM DTT, 5% Glycerol, 0.4% DMSO, pH 8.0, firstly at 4°C for 1 h, then at 30°C in the presence of 1μM [γ-^{32}P] ATP. At various times, released inorganic phosphate was counted by the Charcoal assay (Clertant *& al.*, 84), to determine the extent of ATP hydrolysis, which was found to increase linearly over 2h of time.

5). No ATPase could be detected in the preparation containing E2 only. Preincubation of this preparation together with partially purified E1 did not significantly affect the ATPase activity found for E1 alone. Together with the results presented above, this suggests that the interaction between E1 and E2 proteins did not appear to modify the E1-associated ATPase activity.

Discussion

After having expressed the E1 protein by infection of insect cells with a baculovirus construct, we were able to detect an ATPase activity in partially purified E1 protein prepared by several different methods. This activity fulfills several criteria for being an intrinsic property of the E1 protein itself: *1/* the E1 molecule exhibits a nucleotide binding site, as shown by covalent affinity labeling. *2/* different schemes of purification allowed us to recover this activity in all partially purified E1 preparations assayed so far; *3/* the ATPase specific activity, relative to the E1 content of the various preparations, did not vary significantly from one preparation to another; *4/* immunoprecipitates, obtained from these preparations by using anti-E1 antibodies, retained ATPase activity, with levels similar to input. Significantly, this activity was also found in association with the E1-E2 complex; in this case, it could also be recovered by precipitation with the aid of anti-E2 monoclonal antibody. It seems to us unlikely that this activity could be the result of trivial contaminating proteins. Altogether, our results strongly suggest that ATPase is a property of the E1 protein molecule. Definite proof of this assumption could be obtained only by showing that this ATPase could be neutralized by monoclonal anti-E1 antibodies, or would be thermolabile in the case of thermosensitive mutants of E1, as this was shown for the SV40 large-T antigen (Clark & *al.*, 81)

The level of E1 ATPase activity is relatively low : at 30°C and 1μM ATP, 1μg E1 protein hydrolyzed 10-15 pmol ATP per hour; this corresponds to a turnover number of *ca.* 1 mol ATP/mol E1/h, a value lower by one order of magnitude than that measured under similar conditions for SV40 or polyomavirus large-T antigens (Clark & *al.*, 81, Clertant & *al.*, 84). A similar decrease was observed in the level of covalent labeling obtained by using periodate-oxidized ATP (Santucci & *al.*, 90). However at physiological concentrations of nucleotides, these differences could be counterbalanced by inverse variations in affinity. No variation in the E1 ATPase activity was observed to result from its interaction with the E2 protein. More work, especially to determine more precisely the kinetic parameters of E1 ATPase under various conditions, for instance in the presence of E2, in the presence of various DNA, including E2-responsive elements, etc..., is however required to substantiate this point.

Using three different approaches, indirect immunoassay, copurification, and immunoprecipitation with anti-E2 antibodies of E1-associated ATPase, we have found evidence of the existence of an E1-E2 molecular complex also observed by others (Mohr *& al.*, 90, Blitz & Laimins, 91). This complex might act *in vivo* as a trap for the E2 transcriptional activator, since viral transcription is enhanced in E1 mutants (Lambert& Howley, 88, Schiller *& al.*, 89). On the other hand, the signification of E1-E2 complex in the replication mechanism is open to speculation. The initiation of viral DNA replication obviously requires recognition of the starting point. The absence of specific binding to defined DNA sequences by the E1 protein (Santucci *& al.*, 90) led us to imagine that E2 helps to target E1 initiator protein to specific sites at the origin of replication. As a matter of fact, the E1-E2 complex has been shown to bind *in vitro* to the regulatory region of the BPV1 genome (Mohr *& al.*, 90), although it is not clear whether all E2 binding sites within this region are recognized equally well by the complex. Obviously, the discrepancy between the position of the PMS sequence, which is necessary and sufficient in *cis* for continuous replication of BPV1 plasmids (Waldeck *& al.*, 84, Lusky & Botchan, 84), and the location of the actual starting point of DNA replication, recently mapped outside PMS, close to a cluster of E2-responsive elements within the URR (Yang & Botchan, 90) has to be explained. Targeted mutagenesis within PMS has been used to tentatively define a critical region, postulated to be equivalent to SV40 core *ori* (Lusky & Botchan, 86a); this PMS critical domain is not limited to its unique E2-responsive element, but has been described to span over at least 140 bp, containing also a binding site for a cell protein and a promoter (Stenlund *& al.*, 87). The validity of the transfection technique used to defined this domain has been recently questioned however (Ustav & Stenlund, 91). Whatever the relevance of previous PMS analyses, it seems obvious that E2-responsive elements by themselves are not sufficient to define the BPV1 origin of replication. Most likely, either the relative topography of repeated E2 sites, like the four contained in the entire PMS segment, or the recognition of other defined features within the same region, are required to direct the start of the replication machinery to a certain distance in a bidirectional manner by a mechanism which necessarily involves the E1 and E2 proteins. Cell proteins might be involved also in this recognition process : it has to be stressed in this respect that the E2 46 kDa protein can also act as a general *trans*-activator, being able to activate heterologous promoters devoid of any E2-responsive element (Haugen *& al.*, 88). Identification of the proteins able to interact with the E1-E2 complex, as well as a detailed knowledge of the DNA recognition by this complex, might help to solve the question of how BPV1 replication is initiated.

The similarities of the E1 protein with SV40 large-T antigen (Clertant & Seif, 84) have suggested similar molecular functions for the E1-E2 complex in the initiation process. The E1-associated ATPase might reflect the existence of an E1 helicase activity, as it is the case for large-T (Stahl *& al.*, 86). It is not clear in the latter case whether a true helicase activity is needed to initiate DNA replication, or if it is merely involved in the later process of elongation. A mechanism which looks however to be absolutely required for the initiation of SV40 DNA replication, is the ATP-dependent opening of the DNA double helix by the large T antigen at its *ori* binding site II (Borowiec *& al.*, 90). It seems reasonable to speculate, as a working hypothesis, that the E1-E2 complex might act in the same way.

Acknowledgements : Special thanks are due to Elli Androphy, who kindly provided the antibodies used throughout this work, and to Lou Laimins, who made his E2 baculovirus vector available to us and for communicating unpublished results. We thank also Peggy Bradley for the gift of large-T baculovirus, as for many helpful suggestions, and François Cuzin for constant support. The skillful photographical work of Yann Fossat is acknowledged. This work was supported by grant n° 6112 of the french Association pour la Recherche sur le Cancer and by a NATO Collaborative grant together with M.K. Bradley. The critical reading of the manuscript by E. Fanning is also gratefully aknowledged.

References :

Androphy E.J., D.R. Lowy & J.T. Schiller (1987) Bovine papillomavirus E2 trans-activating gene product binds to specific sites in papilloma-virus DNA. Nature 325: 70-73

Blitz, I.L. & L.A. Laimins (1991) The 68 kDa E1 protein of bovine papillomavirus is a DNA-binding phosphoprotein which associates with the E2 transcriptional activator *in vitro*. J. Virol. 65: 649-656

Borowiec J.A., F.B. Dean, P.A. Bullock & J. Hurwitz (1990) Binding and unwinding - how T antigen engages the SV40 origin of DNA replication. Cell 60: 181-184.

Bradley M.K., T.F. Smith, R.H. Lathrop, D.M. Livingston & T.A. Webster (1987) Consensus topography in the ATP binding site of the simian virus 40 and polyomavirus large tumor antigens. Proc. Natl. Acad. Sci. USA 84: 4026-4030

Clark R., D.P. Lane & R. Tjian (1981) Use of monoclonal antibodies as probes for simina virus 40 T antigen ATPase activity. J. Biol. Chem. 254: 11854-11858

Clertant P., P. Gaudray, E. May & F. Cuzin (1984) The nucleotide binding site detected by affinity labeling of the large T proteins of polyoma and SV40 viruses. J. Biol. Chem. 259: 15196-15203.

Clertant P., & I. Seif (1984) A common function for polyomavirus large T and papillomavirus E1 proteins ? Nature 311: 276-279

Choe J., P. Vaillancourt, A. Stenlund & M. Botchan (1989) Bovine papillomavirus type 1 encodes two forms of a transcriptional repressor : structural and functional analysis of new viral cDNAs. J. Virol. 63: 1743-1755

DiMaio D. (1986) Non-sense mutation in open reading frame E2 of bovine papillomavirus DNA. J. Virol. 57: 475-480.

DiMaio D., & J. Settleman (1988) Bovine papillomavirus mutant temperature-sensitive for transformation, replication and trans-activation. EMBO J. 7: 1197-1204.

Gilbert D.M., & S.N. Cohen (1987) Bovine papillomavirus plasmids replicate randomly in mouse fibroblasts throughout S phase of the cell cycle. Cell 50: 59-63

Guizani I., M.P. Kieny, R. Lathe & P. Clertant (1988) Characterization of polyoma virus early proteins expressed from vaccinia virus recombinants. Gene 73: 163-173

Harlow E., L.V. Crawford, D.C. Pim & N.M. Williamson (1981) Monoclonal antibodies specific for simian virus 40 tumor antigen. J. Virol. 39: 861-869

Haugen T.H., L.P. Turek, F.M. Mercurio, T.P. Crispe, B.J. Olson, R.D. Anderson, D. Seidl, M. Karin & J.T. Schiller (1988) Sequence-specific and general transcriptional activation by the bovine papillomavirus E2 *trans*-activator require an N-terminal amphipathic helix-containing E2 domain. EMBO J. 7: 4245-4253

Hubbert N.L., J.T. Schiller, D.R. Lowy & E.J. Androphy (1988) Bovine papillomavirus-transformed cells contain multiple E2 proteins. Proc. Natl. Acad. Sci. USA 85: 5864-5868

Lambert P.F., C.C. Baker & P.M. Howley (1988) The genetics of bovine papillomavirus type 1. Annu. Rev. Genet. 22: 235-258

Lambert P.F., & P. M. Howley (1988) Bovine papillomavirus type 1 E1 replication-defective mutants are altered in their transcription regulation. J. Virol. 62: 4009-4015

Lambert P.F., B.A. Spalholz & P.M. Howley (1987) A transcriptional repressor encoded by BPV-1 shares a common carboxyterminal domain with the E2 transactivator. Cell 50: 69-78

Li R., J. Knight, G. Bream, A. Stenlund & M. Botchan (1989) Specific recognition nucleotides and their DNA context determines the affinity of E2 protein for 17 binding sites in the BPV1 genome. Genes Dev. 3: 510-526.

Lusky M. & M.R. Botchan (1984) Characterization of the bovine papilloma virus plasmid maintenance sequences. Cell 36: 391-401

Lusky M & M.R. Botchan (1986a) Transient replication of bovine papillomavirus type 1 plasmids : *cis* and *trans* requirements. Proc. Natl. Acad. Sci. USA 83: 3609-3613

Lusky M., & M. Botchan (1986b) A bovine papillomavirus type 1-encoded modulator function is dispensable for transient viral replication but is required for establishment of the stable plasmid state. J. Virol. 60: 729-742.

Mecsas J., & B. Sugden (1987) : Replication of plasmids derived from bovine papillomavirus type 1 and Epstein-Barr virus in cells in culture. Ann. Rev. Cell Biol. 3: 87-108.

Mohr I.J., R. Clark, S. Sun, E.J. Androphy, P. McPherson & M.R. Botchan (1990) Targeting the E1 replication protein to the papillomavirus origin of replication by complex formation with the E2 transactivator. Science 250: 1694-1699

Murphy C.I., B. Weiner, I. Bikel, H. Piwnica-Worms, M.K. Bradley, & D. M. Livingston (1988) Purification and functional properties of simian virus 40 large and small T antigens overproduced in insect cells. J. Virol. 62: 2951-2959

Rabson M.S., C. Yee, Y.C. Yang, & P.M. Howley (1986) Bovine papillomavirus type 1 3' early region transformation and plasmid maintenance functions. J. Virol. 60: 626-634.

Roberts J., & H. Weintraub (1986) Negative control of DNA replication in composite SV40-bovine papillomavirus plasmids. Cell 46: 741-752

Roberts J.M., & H. Weintraub (1988) *Cis*-acting negative control of DNA replication in eukaryotic cells. Cell 52: 397-404.

Santucci S., E.J. Androphy, C. Bonne-Andréa & P. Clertant (1990) Proteins encoded by the bovine papillomavirus E1 open reading frame : expression in heterologous systems and in virally transformed cells. J. Virol. 64: 6027 6039

Schiller J.T., E. Kleiner, E.J. Androphy, D.R. Lowy, and H. Pfister (1989) Identification of bovine papillomavirus E1 mutants with increased transforming and transcriptional activity. J. Virol. 63: 1775-1782

Spalholz B. A., Y.C. Yang, and P.M. Howley (1985) Transactivation of a bovine papillomavirus transcriptional regulatory element by the E2 gene product. Cell 42: 183-191

Stahl H., P. Dröge & H. Knippers (1986) DNA helicase activity of SV40 large tumor antigen. EMBO J. 5: 1939-1944

Stenlund A., G.L. Bream & M.R. Bochan (1987) A promoter with an internal regulatory domain is part of the origin of replication in BPV1. Science 236: 1666-1671

Sun S., L. Thorner, M. Lenz, P. McPherson & M. Botchan (1990) Identification of a 68 kilodalton nuclear ATP-binding phosphoprotein encoded by bovine papillomavirus type 1. J. Virol. 64: 5093-5105

Thorner L., N. Bucay, J. Choe & M. Botchan (1988) The product of the bovine papillomavirus type 1 modulator gene (*M*) is a phosphoprotein. J. Virol. 62: 2474-2482.

Ustav M., & A. Stenlund (1991) Transient replication of BPV1 requires two viral polypeptides encoded by the E1 and E2 open reading frames. The EMBO J. 10: 449-457

Waldeck W., F. Rösl & HW. Zentgraf (1984) Origin of replication in episomal bovine papillomavirus type 1 DNA isolated from transformed cells. The EMBO J. 3: 2173-2178
Yang L., & M. Botchan (1990) Replication of bovine papillomavirus type 1 DNA initiates within an E2-responsive enhancer element. J. Virol. 64: 5903-5911

Subject Index

MIX
Papier aus verantwortungsvollen Quellen
Paper from responsible sources
FSC® C105338

If you have any concerns about our products,
you can contact us on
ProductSafety@springernature.com

In case Publisher is established outside the EU,
the EU authorized representative is:
Springer Nature Customer Service Center GmbH
Europaplatz 3, 69115 Heidelberg, Germany

Printed by Libri Plureos GmbH
in Hamburg, Germany